Production and Operations Management Systems

Production and Operations Management Systems

Sushil Gupta and Martin Starr

CRC Press
Taylor & Francis Group
Boca Raton London New York

CRC Press is an imprint of the
Taylor & Francis Group, an **informa** business

CRC Press
Taylor & Francis Group
6000 Broken Sound Parkway NW, Suite 300
Boca Raton, FL 33487-2742

Version Date: 20131206

International Standard Book Number-13: 978-1-4665-0733-3 (Hardback)

Visit the Taylor & Francis Web site at
http://www.taylorandfrancis.com

and the CRC Press Web site at
http://www.crcpress.com

DEDICATION

We would be remiss if we did not thank our spouses, Lalita Gupta and Polly Starr, for putting up with intense working days and nights of writing and communicating by email and phone—all dedicated to bringing the book field of production and operations management (P/OM) into the twenty-first century.

Contents

Preface

THE PURPOSE OF THIS TEXTBOOK

Our goal is to make the subject of production and operations management (P/OM) interesting, even exciting, to those who are embarking on a career that involves *business* of any kind. This includes the business of making profit, as well as not-for-profit applications. Yes, P/OM applies directly to helping people who are under stress (as in humanitarian operations) as well as for everyone striving to have a better life. Since P/OM capabilities deal equally with goods and services, the fields of hospitality, travel, healthcare, education, entertainment, and agriculture are as vital a part of its purview as manufacturing.

This list of applications should be kept in mind when reading the section "To whom is this text directed?" Managers in all of the fields listed above will benefit from learning more about P/OM. We hope the reader will supply categories of personal interest that we skipped over such as sports and astronautics.

Since the beginning of mankind on earth, if the "busyness" process was successful, then some form of profit sustained it. There is nothing ugly about making money. Profitability is a requisite sign of smart management and productivity achievement. Return on investment is the essence of sustainability. Studies show that environmental concern enhances long-term profitability and generally is beneficial to short-term profits as well. P/OM (more than other functional areas of business, government, and not-for-profit organizations) impinges on the environment. We used "busyness" to reflect both personal task performance and work-related job performance. P/OM applies to both sides of the street. A disciplined manager is unlikely to be bogged down by inefficient processes at home.

P/OM's use of energy, water, and even air to make goods and provide services is functionally unique. The way that P/OM blends the talents of people with the power of technology resides in a special domain that no one else occupies. Cooperation with allies (e.g., specialists in energy, graphic arts, information systems, market research, and managers of other technologies) is increasingly essential.

The special phrases of P/OM lingo (special language) reveal the extent to which P/OM goals focus on critically important and totally unique endeavors. Let us look at a few of these "goal-oriented phrases," for example, "we want to—increase productivity, achieve zero defectives, reduce buffer stock, cut lead time, maintain just-in-time delivery, remove waste, eliminate idle time, surpass breakeven volume, protect environments." In a globally connected world, it is vital to recognize—P/OM controls the knobs that are instrumental in determining environmental sustainability.

Being good at business is a necessity because the business of life is the system of all transformation activities. It is the engagement of self and family, friends, and associates, in creating food, clothing, shelter, transportation, entertainment, as well as eating healthy foods, exercising, keeping stress under control, etc. P/OM is activity management. As a master of process control, P/OM is responsible for maintaining healthy and positive work environments and that includes work places that are *ergonomic.*

This particularly P/OM-oriented word is derived from two Greek words: *ergon* meaning *work*; *nomos* meaning *laws*. The *laws of work* relate to designing jobs for safety, comfort, and productivity. Wikipedia, calling this field HF&E, defines human factors and ergonomics as "a multidisciplinary field incorporating contributions from psychology, engineering, biomechanics, mechanobiology, industrial design, graphic design, statistics, operations research, and anthropometry." It is important to point out that only P/OM is comfortable in many of these domains, leading to eventual benefits and ensuing cooperation.

From product and process perspectives, the scientific study of measures and proportions of the human body (called anthropometry) is essential for designing products (like telephones and hammers) that fit well with all sizes of people including children. Market research combined with anthropometry provides major insights about likes and dislikes that people have for things that fit and those that do not. Comfort is high on the list of consumer preferences.

On the job, assembling tires to cars used to be done by workers bending down to put the tires on and drive home the bolts. Since P/OM began to catch the ear of top managers, assembly conveyors lift all cars so workers' positions are healthy (optimal) for assembly. There are fewer back problems and, therefore, less absenteeism. That is good HF&E (and competent P/OM). Let us not forget, we learned this and many other things about good production methods from Toyota. At the same time, Toyota's top management states unequivocally that Henry Ford laid the critical groundwork without which Toyota could not have become an archangel of automotive methodology.

Henry Ford established the value of sequenced assembly. There are no competitive claims for being first. Henry Ford gets that award. Automotive plants on a worldwide basis acknowledge Ford's role in revolutionizing the field of P/OM.

Managers lacking the knowledge of how P/OM creates excellence are fighting against the tide. P/OM is studied to allow for good management of every activity that is directly and indirectly responsible for making products (goods and services) that people want. The fundamentals are obvious: get the right inputs (materials, labor, money, and ideas); transform them into highly demanded, quality outputs; and make it available in time to the end consumer. This input-transformation-output model is fully developed in Chapter 1. It is a good idea for all students to read this chapter first.

The field of study concerned with input–output management—and doing it well—is P/OM. This is P/OM's responsibility and it is deep in content, broad in application, and powerful in determining organizational success.

TO WHOM IS THIS TEXT DIRECTED?

This text is an introduction to the *clout* of P/OM for those who have never studied P/OM. Just ask the workers who assemble tires to cars—hour after hour, day after day. This text is an educational reentry point for those whose backgrounds lack the systems approach to this subject. Those who studied P/OM using the systems approach are like Cordon Bleu chefs. Everyone is an authentic expert in distinguishing between Cordon Bleu quality and Greasy Spoon food. Those who have studied P/OM as an assortment of techniques applied to a variety of cases in a random-like way can profit greatly from the re-education that this text provides.

Our presentation is at a level that students for the first course in P/OM will welcome. Those who want to reach further can do so, but that is by choice. The main concepts and their applications are readily available without requisite training in mathematical manipulations. This textbook is particularly suitable for online courses where students require detailed explanations of the problems. The step by step solutions for all problems included in this book are available on the instructors' resource CD (IRCD). The students can review the material at their own pace.

The topics are modular, which means that instructors who like to start with supply chain management (Chapter 9) and then move on to inventory management (Chapter 5) and then quality management (Chapter 8) can do so in that order. Major changes have occurred in how quickly product line stability is disrupted. Competition reacts quickly and response must be rapid. Therefore, some instructors might prefer to begin with project management (Chapter 7) to reflect the continuous project mode required for fast redesign rapid response. Alternatively, some teachers may prefer to start with Chapter 11 on Innovation by P/OM for New Product Development (NPD) and Sustainability. The modular character of the text permits many different journeys through the materials.

This also applies to the "Problem" section and/or "Review Questions" that are presented at the end of the chapters. Instructors have access to Power Point presentations for each chapter, and these are equivalently modular. The key point to bear in mind is that while references may be made to materials in preceding or following chapters, understanding within all chapters is always independent of other chapters.

WHY THE SYSTEMS APPROACH IS ESSENTIAL

The label "systems approach" has been bandied about by a variety of fields for many years. Our use of the term is associated with direct problem-solving rather than the philosophical general systems theory (GST) that was developed by Karl Ludwig von Bertalanffy around 1950. GST had a biological foundation but was an interdisciplinary approach to the interacting components of complex systems. For P/OM, the systems approach is based on studying all elements related to creating better products and processes.

Our text is permeated by such systems thinking, for example, strong interactions abound between forecasting methods (Chapter 3), capacity management and aggregate production planning (Chapter 4), inventory management (Chapter 5), supply chain management (Chapter 9), and innovation by P/OM for new product development and sustainability (Chapter 11). Students will not have to search for them because they are delineated throughout the text.

This book does not hide the fact that its authors are subscribers to interdisciplinary and interfunctional solutions to problems. This proclivity follows from any honest application of the systems approach. For societal organizations (profit and nonprofit) to strive to solve problems, it is essential that all relevant disciplines team up, share data, and cooperate. There are no party lines here. Problem-solving needs to be based on optimizing organizational benefits.

Regarding cooperation, we need to clarify the field relationships of OM (operations management) to OR (operations research). In fact, traditional OR has now strongly associated itself with a new catchy title, viz., analytics. Classic OR dates back to World War I when the British scientific community led by physicist Patrick Blackett developed quantitative models for submarine search and military logistics (see Stephen Budiansky's *Blackett's War: The Men Who Defeated the Nazi U-Boats and Brought Science to the Art of Warfare*). Many developments of these 1940 scientists have been absorbed in the body of P/OM practice at the present time.

There have always been neophytes who cannot seem to differentiate between OR and OM. This Preface is an ideal place to set the record straight. Operations research is quantitative modeling of decision processes. When the decisions relate to planning for the production of goods and services, the modeling abilities of OR are used by OM. However, operations management is hardly limited to mathematical and statistical modeling. A great deal of P/OM problem-solving is done by experience with what works (know-how), common sense, logical analysis, and clever heuristics (rules of thumb).

Analytics is like a flavor of OR with its emphasis on the use of "big data" to resolve decision problems. The methods of "big data" include data-mining and predictive analytics. The responsibility for organizing massive amounts of data belongs to the information technology (IT) department. Apache's Hadoop (high-availability distributed object-oriented platform) is favored by many IT professionals. Although such matters are not within P/OM's sphere of expertise, it is in

P/OM's interest to understand how an open-source software framework supports data-intensive distributed applications.

This is because P/OM often embraces big data because solution of the problem requires using it. For example, when inventories constitute a significant percent of the operating budget of a company having thousands of SKUs (stock-keeping units), to be able to analyze huge amounts of data concerning demand levels, lead times by vendors, costs of materials, business-based carrying and order costs, etc., by computer provides a major payoff. Such databases are dynamic. They are changing all of the time, which requires big data methods to track the changes—and alter inventory policies accordingly on a regular basis.

Material requirements planning (MRP) is a manufacturing scheduling application which is extremely data-intensive. It is not included in this text because its proper use requires experienced MRP managers whose scheduling abilities are dependent on continuously updating their databases concerning the commonality of (modular) parts in parent products, modified demand levels, finished goods, and work in process (WIP) as well as materials inventories. Responsibilities for MRP are lodged with both P/OM and IT executives. The models for MRP cut across OM, IT, and OR analytics. As such, they rightfully belong to MRP—in its own domain.

Simulation is another systems-oriented approach of great power that gets used by OM and OR to solve extensive problems with complex linkages. Proper programs are written by computer experts from IT. Simulations are based on trying to connect all of the causal factors that relate to solutions such as how well a new airplane can fly in turbulence. Every person who has had a hand in any aspect of designing, making, or assembling parts is involved with the simulated performance of this airplane. This is a good illustration of interdependent responsibilities for problem-solving. The focus transcends departments.

Successful MRP or simulation leads to the question "whose success are we talking about?" The answer is straightforward from a systems perspective. The organization's success is paramount. Often, in seeking success, each department in an organization is faced with opportunities to look good (sometimes) at the expense of others. For instance, the sales department might find that it can sell more product at a lower price than production can deliver. The result is unhappy retail stores and customers. Top management receives reports that the sales campaign was successful but production's inability to deliver product hurt the company's bottom line. When departments cooperate and work together toward a common goal, such damaging tradeoffs are averted. Tradeoffs occur all the time that demand nonterritorial solutions.

Challenging problems cross department lines. Stage IV firms (Chapter 1) have shown that coordination of what used to be traditionally separate functions provides substantial benefits. Tradeoff situations (where as one department does better, another does worse) are commonplace in business and life. To do what is best for the organization, tradeoffs must be coordinated. P/OM is very conscious of tradeoff models because they occur in so many aspects of production and operations.

For example, in using quality control for acceptance *sampling*, it is always the case that stringent criteria for the protection of buyers from defective products will penalize suppliers by counting as defective a larger percentage of perfectly good products. Vice versa applies. By increasing supplier's protection, buyers will be forced to accept more defective products. Chapter 8 on quality management explains this effect.

Another important tradeoff example that persists for materials managers is the well-known relationship between the number of items purchased in an order—and the rate at which they are used up. With bigger orders, carrying costs go up and ordering costs go down. This could result in fewer purchasing agents but a larger warehouse is needed. Bigger orders are associated with the inventory philosophy of just-in-case when compared to smaller orders being just-in-time. Production and operations managers are generally very smart. They know it is better to overestimate order size (not only because of just-in-case risk management, but also because penalties for overestimating are lower than for underestimating).

INTERDISCIPLINARY COORDINATION

We continue to examine the implications of using the systems approach for P/OM problems. This text reveals many useful applications, but there are important loci (centers) of interdependencies that our Preface should address. Let us briefly consider, in turn, the important sets of relationships between production and operation managers and their colleagues in IT, market research, finance, accounting, and marketing. The need to cooperate with IT, OR, and analytics has been discussed above.

Market research provides the eyes and ears for awareness of the needs, wants, and complaints of customers from every demographic and market demand segmentation. Financial management is a very powerful, full-fledged associate in decisions about capitalization of technology. Process development is constrained by these decisions which are illuminated by the application of breakeven analysis (first explained in Chapter 10). See the Appendix for the fully developed breakeven analysis (as a straightforward quantitative model). After studying breakeven, students agree unanimously: accounting is an essential P/OM partner to properly develop the costs that breakeven analysis requires.

The interdisciplinary coordination is very important in designing supply chains particularly with a growing focus on e-business (see Chapter 9). e-Business is a multidimensional discipline involving the application of technology, the study of customers' attitudes, expectations, and satisfaction, the identification of internal organizational environment, the study of the relationships among partners in the supply chain, the development of collaborative strategies and coordination mechanisms, and the development of analytical models for operating (e.g., inventory and pricing) decisions. The e-business area has been influenced by the developments in many academic fields that include but are not limited to the following: behavioral sciences, computer science, economics, information systems, marketing, operations management, operations research/management science, and technology management.

Chapter 11 will help to explain the critical and special interrelationship of P/OM with marketing. There is an absolute necessity for strong cooperation between these two primary players in new product and process development. Simply stated, product life (in a growing number and variety of product classes) has gone from years to months. The mobile phone bought in May could be technologically out-of-date by September. Old dishwasher detergent (liquid or powder) has been replaced by capsules; high-efficiency (HE) laundry detergents are now used with the new washing machines; laptop computers are competing with a huge choice of digital gadgets. Constant communication and intense connections between family members, friends, and business associates are the essence of the information society which experiences unexpected repercussions and unintended disruptive consequences from the viral nature of sweeping attitudes.

New products are not sustainable if they are not conceived as a part of an evolutionary product line. Evolutionary product lines require evolutionary process designs. Chapter 11 describes these *platforms* for launching the next version of a line of successive replacement products. This dynamic situation applies immediately to goods—especially to electronics and telecommunications (E&T). As E&T become dominant marketing features of automobiles, entertainment centers, living rooms, kitchens, part of clothing (wearable technology), etc., the speeded-up disruptive effects of E&T will become increasingly pervasive.

The changes apply to consumable goods as well. The effects of social media lead to rapid alterations in a range of likes and dislikes. Whole Foods and The Fresh Market are not transients to the supermarket scene. Publix is morphing to match the organic and green appeal. Services may be slower to reflect the effects of significant variations of the product line, but they are occurring. Better and more personal services are the beneficiaries of the E&T revolution. This includes recommendations for restaurants and competitive pricing advice from personal assistant applications such as SIRI (on iOS devices), Google Now, Speaktoit, Robin, and Sherpa.

The demographic that relies on digital implementations is youthful and even younger. Not so gradually that demographic segment is replacing the part of the population that is less reliant on E&T. The effect of this is that cross-discipline communication and subsequent coordination between product designers, marketers, and P/OM are becoming essential. The huge change taking place requires reorientation of all organizations about how business will be conducted in the twenty-first century. P/OM will be at the core.

Thus, the project management component of P/OM is no longer an auxiliary aspect of the field. Instead, in the leading Stage IV companies, project management is the organizing activity for developing the platforms for evolving synchronized, balanced, and harmonious product and process designs. The entire system has to be viewed and understood. The whole system involves, for example, closed-loop supply chains that require detailed and specific plans for the return of parts after product life completion for remanufacturing, recycling, refurbishing, and retirement.

No wonder that the authors take a strong stand on the requirement for interdisciplinary organization of planning and problem-solving. They do not view it as a luxury, but a necessity.

Epilogue

The fundamentals of production and operations management (P/OM) have been undergoing substantial changes over the past two decades. P/OM books must reflect these changes. They cannot be made relevant by adding and deleting chapters. Repeated editions of existing books seem over-patched. We started this book with the new fundamentals of our field as our foundation. This book is current in topical coverage, succinct and affordable.

SKG
MKS

Acknowledgments

As the authors of this book, we are pleased to acknowledge the professional handling of this publication effort. Lara Zoble has been supportive of this project from day one. She has made the operations of her team efficient and transparent. In that regard, we want to include our appreciation for the coordination that Cynthia Klivecka and her cohorts including Syed Mohamad Shajahan have made possible. The results have been impressive. It is not surprising that we, as production and operations management (P/OM) academics and practitioners, are very sensitive to process prowess. We applaud the Taylor & Francis team.

Authors

Dr. Sushil K. Gupta is Professor and Knight Ridder Center Research Fellow in the College of Business, Florida International University (FIU), Miami, Florida, USA. He has taught at the Faculty of Management Studies (FMS), University of Delhi, Delhi, India (1969–1981), at the University of North Carolina, Greensboro, North Carolina, USA (1981–1983), and at FIU (since 1983). Dr. Gupta has also served FIU as Vice Provost, Academic Affairs, and associate dean, College of Engineering. Dr. Gupta is an Honorary Professor at Universidad Ricardo Palma, Lima, Peru.

Dr. Gupta is a highly esteemed scholar in the field of production and operations management. His research interests include production scheduling, mathematical modeling, project management, supply chain management, e-business, and disaster management. He has published in top-ranking professional journals, including *Production and Operations Management, Management Science, IIE Transactions, European Journal of Operational Research, International Journal of Production Research, Omega, Computers and Industrial Engineering,* and *Socio-Economic Planning Sciences.*

Dr. Gupta is one of the founding members of the Production and Operations Management Society (POMS) and has played a key role in its initiation, development, and growth. He was POMS Vice President for Member Activities (1994), POMS President (1996), and is currently serving as its Executive Director (since 1997). He also serves on the advisory board of *Production and Operations Management*, the flagship journal of POMS, the POMS Board, and the board of POMS College of Humanitarian Operations and Crisis Management.

Dr. Gupta was elected a POMS Fellow in 2004. POMS also bestowed an honor on Dr. Gupta by awarding him the first POMS Distinguished Service Award in 2012. This award, which began in 2013, is renamed as the Sushil K. Gupta POMS Distinguished Service Award to honor Dr. Gupta for his dedicated services to POMS.

Dr. Gupta is very active in community service. The Association of Indians in America (AIA), the oldest national association of Asian Indians in America, founded in 1967, gave Dr. Gupta a special recognition award for his academic achievements and for promoting a better understanding between the people of India and the United States in 2013.

Dr. Gupta has been the recipient of many other prestigious awards that include the FIU Foundation Excellence in Research/Scholarship award (1987), Distinguished Scholarly Research Award of the College of Business, FIU (1984), Fulbright Fellowship to do research work at the Sloan School of Management, Massachusetts Institute of Technology, USA (1977–1978), Commonwealth Scholarship for graduate work at the University of Toronto, Toronto, Canada (1972–1974), and the Dr. V.K.R.V. Rao Gold Medal for his first place ranking in the MBA graduating class (1969), Delhi, India.

Dr. Gupta is a coauthor of *POMS—Production and Operations Management Software*, a software package published by Allyn & Bacon (1986) for an operations management course. In addition, he has written two books, *Production Planning and Control, Text and Cases*, published by the University of Delhi, Delhi, India (1979), and *Production Scheduling Techniques*, published by K.P. Bagchi and Company, Delhi, India (1981).

Dr. Gupta has developed a website for online learning of production and operations management. He is very proficient in the use of Excel spreadsheets and has developed macro-based Excel spreadsheets to be used by instructors and students. He teaches fully online POM undergraduate and corporate MBA courses at FIU. Dr. Gupta has also received the teaching excellence award from the College of Business, FIU.

Dr. Gupta, in his capacity as the Director of Corporate and Global Programs in the College of Engineering, initiated the development and promotion of the College of Engineering's corporate and global programs in several countries that include China, Colombia, Dominican Republic, India, Jamaica, Mexico, Peru, Taiwan, Turkey, Venezuela, and Yugoslavia.

Dr. Gupta holds a Bachelor of Technology degree in mechanical engineering from Indian Institute of Technology Bombay, India, an MBA degree from FMS, University of Delhi, India, a Master of Applied Sciences degree in industrial engineering from the University of Toronto, Canada, and a PhD degree from FMS, University of Delhi, India.

Dr. Gupta migrated to the United States from India in 1981. He is a citizen of the United States.

Dr. Martin K. Starr is Distinguished Professor Emeritus of Production and Operations Management at the Crummer Graduate School of Business, Rollins College in Winter Park, Florida and Emeritus Senior Professor of Operations Research, Management Science, and Operations Management at the Graduate School of Business at Columbia University in New York City.

As the Director of the Center for Enterprise Management (CEM) at both schools, Dr. Starr spearheaded a series of comparative research studies of the performance of various manufacturing cultures over two decades starting in 1975. This resulted in an understanding of how Japanese manufacturing firms operating in the United States differed from their parent companies in Japan. Comparisons of Japanese, American, and European manufacturing firms operating in the United States revealed significant differences in their quality methods and process management. Later, the CEM at Rollins conducted studies on how to reduce waste and effectuate lean and agile approaches to healthcare services in line with Efficient Healthcare Consumer Response (EHCR) criteria. Significant progress was made along various paths of the healthcare supply chain.

Dr. Starr has been a visiting professor, including Hoover Fund Visiting Professor, at the University of New South Wales in Sydney, the University of Cape Town, the University of Southern California, and Ohio State University. Dr. Starr's activity with executive programs remains extensive. He has taught at Penn State at University Park, MIT—U.S. Naval War College, Newport, Rhode Island, Crummer GSB, Columbia University at Arden House, and many others. Additionally, he has made invited presentations at institutions such as the University of Western Ontario, Canada, Fundação Getúlio Vargas, São Paulo, Brazil, and Centro Studi D'Impresa, Valmadrera, Lecco, Italy.

Dr. Starr was elected a Fellow of the Production and Operations Management Society (POMS) in 2004. He is also a Fellow of INFORMS and Fellow of the American Association for the Advancement of Science (AAAS) since 1969. He is presently Chair of the Council of POMS Presidents and was President of POMS (1995–1996). He was President of TIMS (The Institute of Management Sciences; now INFORMS) (1974–1975), Editor-in-Chief of *Management Science* (January 1967–June 1982), and Vice Dean, Graduate School of Business, Columbia University (1974–1975). He is now on the board of various organizations, including The Cloisters, the College of Humanitarian Operations and Crisis Management, and POMS. In 1983, he received the George Kimball Medal.

Dr. Starr has authored 26 books on business topics and over 100 papers. His research concerns crisis management (e.g., *EurOMA Amsterdam Proceedings*, 2012) and modular production (Modular production—A 45-year-old concept; Emerald; *International Journal of Operations & Production Management*, 2010, Vol. 30, Issue 1, pp. 7–19). Dr. Starr has received teaching excellence awards from various schools, including the Crummer GSB at Rollins and the GBS at Columbia University. *Production and Operations Management* (POM), the flagship journal of POMS, published an article (2007) about his activities over the years—*Martin K. Starr: A Visionary Proponent for Systems Integration, Modular Production, and Catastrophe Avoidance* (by A. Roth and S. Gupta). The *Martin K. Starr Excellence in Production and Operations Management Practice Award* was instituted by POMS in 2006. It has been awarded to eight global leaders of operations management practice. He is the coeditor of the *Special Issue on Humanitarian Operations and Crisis Management*, *POM*, 2014.

Dr. Starr's consultancies have included Alabama Aircraft, American Express, AT&T, ABB, Boston Consulting Group, Bristol-Myers Squibb, Citibank, CNL, Chrysler, DuPont, Eastman Kodak, ExxonMobil, Fiat, GE Capital, Hendry Corporation, HP, IBM, Lever Brothers, McGraw-Hill Book Company, Merrill Lynch, Philips Electronics, Philips Healthcare, Unilever, United Nations, YPF de Argentina, and Young & Rubicam. Dr. Starr is presently director of MKS Associates, which offers POM, supply chain management, and analytics consulting services.

Dr. Starr holds a Bachelor of Science degree from Massachusetts Institute of Technology, Cambridge, Massachusetts, and MS and PhD degrees from Columbia University in New York City.

Chapter 1

Introduction to Production and Operations Management

Readers' Choice: P/OM History Archive

This archive (available at http://www.nytimes.com/ads/peoplesoft/) presents a series of New York Times archival articles outlining the evolution of manufacturing. These articles were specifically chosen to chronologically outline the manufacturing innovations man has made through time. Readers will be taken back to the time of Henry Ford and the Industrial Revolution to the leaders of today's Fortune 500. The importance of history for the study of P/OM is that it is the only way to explain the present state of the system and to predict changes that are likely. Managers who are ignorant of the history of P/OM are operating at a great disadvantage. Historical literacy pays off.

ARCHIVAL ARTICLES

Jan. 3, 1909; System: The Secret of Ford's Success; By Henry Ford. http://www.nytimes.com/ads/peoplesoft/article1.html

Jan. 11, 1914; Henry Ford Explains Why He Gives Away $10,000,000; By Thos. A. Edison. http://www.nytimes.com/ads/peoplesoft/article2.html

Dec. 18, 1927; What Lies Ahead For America? We Dream of Utopias and Dread Domination by the Machine; By Evans Clark. http://www.nytimes.com/ads/peoplesoft/article3.html

Sep. 9, 1945; Detroit: Reconversion Laboratory: In the Motor City are Tested the Social and Industrial Problems of a Nation Returning to the Ways of Peace; By Russell Porter. http://www.nytimes.com/ads/peoplesoft/article4.html

Sep. 24, 1950; Inexpensive TV Sets; Mass Production Techniques Account For A Wider Range of Receivers; By Alfred Zipser Jr. http://www.nytimes.com/ads/peoplesoft/article5.html

Feb. 16, 1964; 50 Years Ago; By Sherwin D. Smith. http://www.nytimes.com/ads/peoplesoft/article6.html

Jan. 9, 1967; Computers: New Values For Society; By Melvin Kranzberg. http://www.nytimes.com/ads/peoplesoft/article7.html

Oct. 25, 1979; Technology: Light Industry Adding Robots; By Peter J. Schuyten. http://www.nytimes.com/ads/peoplesoft/article8.html

Sep. 28, 1980; Technology—Elixir for U.S. Industry; By Anthony J. Parisi. http://www.nytimes.com/ads/peoplesoft/article9.html

Aug. 25, 1985; Break the Habits of Mass Production Putting America's Factories Back On Top; By Wickham Skinner. http://www.nytimes.com/ads/peoplesoft/article10.html

Dec. 7, 1998; Six Sigma Enlightenment: Managers Seek Corporate Nirvana Through Quality Control; By Claudia H. Deutsch. http://www.nytimes.com/ads/peoplesoft/article11.html

Aug. 21, 2000; New Economy: An Old-style Deal Maker Takes Up The Cause of High Technology With Manufacturing Over the Internet; By Barnaby J. Feder. http://www.nytimes.com/ads/peoplesoft/article12.html

---

This chapter defines production and operations management (P/OM) and explains how this management field is applicable to both manufacturing and services, as well as to both profit-making and not-for-profit organizations. This chapter also elaborates the advantages of using the systems perspective, which links P/OM to all other managerial functions in the organization.

After reading this chapter, you should be able to:

- Define and explain operations management and contrast it with production management.

- Explain categories of the systems approach and why they are important to P/OM.
- Detail the systems approach that is used by P/OM.
- Understand how P/OM—using the systems approach—increases the competitive effectiveness of the organization.
- Understand why this book is titled *Production and Operations Management Systems*.
- Distinguish between the application of P/OM to manufacturing and services.
- Explain how special P/OM capabilities provide competitive advantages.
- Relate information systems to the distinction between production and operations.
- Explain how the input–output (I/O) model defines production and operations.
- Describe the stages of development of companies with respect to OM and P/OM.
- Discuss positions in P/OM that exist in the organization and career success in P/OM as a function of process types.
- Explain the effects of globalization on P/OM careers.

1.1 The Systems Viewpoint

Organizations are created to provide goods and services to the public. Goods refer to manufactured, assembled, and processed items. Goods are tangibles that can be produced before their actual use and they can be inventoried. Services, on the other hand, are intangibles that cannot be inventoried. Services are provided at the time when the customers need them. The study of P/OM is the study of operations and processes leading to the creation of goods and services.

The terms "manufacturing," "production," and "operations" are used interchangeably in the literature and also in this book to represent the P/OM field. Similarities and differences between these terms do exist and have been explained in this chapter. However, if one term is more representative than the other to stand for both P/OM, that term is operations management. Most production managers will accept the appellation of operations manager but not vice versa. Therefore, operations management can be used instead of P/OM. Also, OM can be used in place of P/OM; however, PM is not a good substitute for P/OM unless only manufacturing is clearly involved.

In the current information age, the importance of scientific decision-making has increased hugely. Organizations have grown in size; they continue to become bigger and more global. With greater size, the complexity of operations has also increased and the number of decisions to be made has expanded. Accurate and timely decisions are expected from every business executive to achieve organizational goals and to compete effectively in the marketplace. For making decisions in this complex and dynamic environment, we propose and highlight the use of "systems approach" in this book for studying, analyzing, and applying P/OM functions. The systems point of view requires consideration of P/OM dealing with all business functions, such as marketing and finance.

If the part that operations managers play is to be effective, it should be systems-based. Compare the operations manager to the coach of a sports team. What is the coach's job for a baseball, football, basketball, or soccer team? It is to guide the team to achieve competitive excellence. The coach knows that making the team win requires coordinating the contributions of the individual players. Winning takes teamwork, and the coach tries to develop that cooperative ability. Teamwork skills require a systems viewpoint. The systems viewpoint means that everything that is important to goal achievement is included in the analysis. If the goals cannot be achieved, then the strategies must be changed. Instead of the sports analogy, product line development can be used. The same teamwork requirement applies.

1.2 Strategic Thinking

The systems viewpoint requires strategic planning. Goals and strategies must be congruent and realistic. Assume that the game being coached is called business and that players' positions are known as marketing, finance, operations, etc. The successful coach emphasizes coordination of these functions to pursue a strategy aimed at achieving the objectives. Apply this same statement to football, baseball, basketball, soccer, etc., and it works as well. Managers of all functional areas need to understand P/OM, and P/OM managers need to understand areas that interact with their own. Understanding global competitors requires understanding their strategies within the context of the international character of their operations management systems. That is why this text directs P/OM managers to focus on the use of the systems approach for strategic planning and tactical actions. The need begins with the development of strategies for product line planning. P/OM strategy is discussed in more detail in Chapter 2; innovation and new product development are discussed in Chapter 11.

1.3 Explaining P/OM

P/OM is the work function that oversees making goods and providing services. Because it provides what others sell, finance, and account for, it is an indisputable partner in any business. Product line planning is the starting point for strategic planning.

This book will familiarize students with the language and abbreviations used by production and operations managers—such as writing P/OM (or even faster, OM) for operations management. Important P/OM terms are explained in the text along with their definitions.

The OM language describes methods, tools, procedures, goals, and concepts that relate to the management of people, materials, facilities, energy, information, and technology. Operations managers learn how to study a process by observing it and mapping its flow; from that platform, its performance can be improved. OM allows for the state of a production process to be assessed. P/OM often starts from scratch with a new product line. In that case, what is known from prior experiences must be brought to bear.

1.4 Use of Models by P/OM

The Greyhound bus driver is an operations manager assessing highway-driving conditions. The driver knows how rain slows velocity (v), which cuts down miles that can be driven per day (m). The manager in charge of operating the fleet of buses could describe this relationship as follows: $m = vt$, where m is the driver's output in miles driven per 8-hour day, v is the velocity, measured in miles per hour, and $t = 8$ hours. If v is 50 mph in clear conditions and 30 mph when it rains, then the driver can achieve only 240 miles when it rains. The strategy of making up the 160-mile difference must be clear to the driver in planning stops and achieving the bus' final destination.

This method of quantitative description is often used by P/OM to build a model—a representation of the real situation. The model permits P/OM to test the effect of different t's and v's. A general quantitative model that describes output is $O = pt$, where O is the output per day. O changes as a function of the production rate per hour (p) and the hours worked (t). P/OM develops models to describe productivity (p) as a function of scheduling, training, technology, and capacity.

There are various P/OM models used to make equipment selection, workforce and production scheduling, quality control, inventory, distribution, plant location, output capacity, maintenance, and transportation decisions, among others. Decision models organize the elements of a problem into actions that can be taken, forecasts of things that can happen that will affect the results, and thereby, the relative likelihood of the various outcomes occurring. Thus, decision models organize all of the vital elements in a systematic way.

1.5 The Systems Approach

There are only two approaches that P/OM can use:

1. The functional field approach, and
2. The systems approach.

With the functional field approach, operations management is expected to perform its P/OM function with minimum reference to other parts of the business, such as marketing and finance. The functional field approach concentrates on the specific tasks that must be done to make the product or deliver the service. This approach is tactical, not strategic. Many marginal firms use the functional field approach (by default) because it is human nature to be territorial. Teamwork requires caring and takes effort.

A typical organization chart—but with hierarchical details mainly for P/OM—is shown in Figure 1.1. The P/OM department is headed by a senior vice president of operations, who has the general manager as well as staff heads (for quality, materials, and engineering) reporting to him or her. The chart also shows, without details, marketing, finance, accounting, R&D, and other functions.

There are no lines connecting people in the other functional areas to people in P/OM. The only connection is at the president's level. Within the P/OM area, there are a limited number of connections and these are hierarchically structured. The traditional organization chart does not reflect the systems approach wherein anyone can talk to anyone else if they are part of the problem or part of the solution. Teamwork is difficult to achieve with self-contained functions.

This is called a "stovepipe" organization because each function operates as if it has its own separate compartment with its own chimney.

The systems approach integrates P/OM decisions with those of all other business functions. This is an integrated and coordinated team-playing model of the

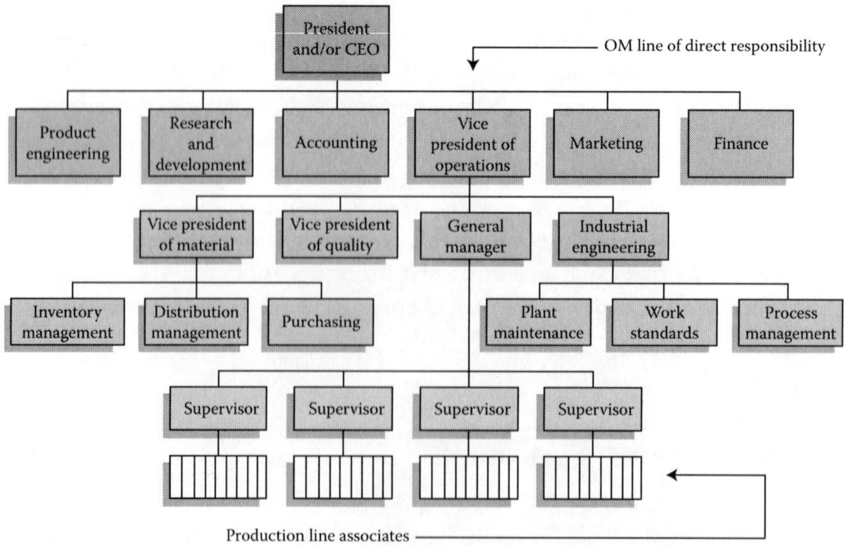

Figure 1.1 Traditional organization chart showing self-contained functional areas.

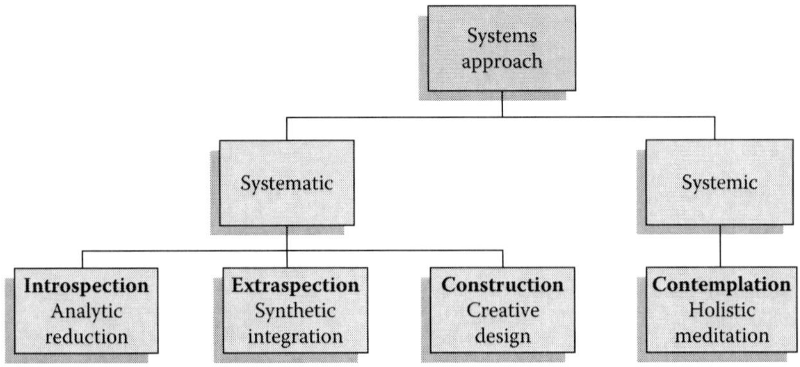

Figure 1.2 Taxonomy (hierarchical classification) of the systems approach.

organization. The challenge is to make the firm perform as a team. The systems approach entails having all participants cooperate in solving problems that require mutual involvement. It begins with strategic planning and moves to tactical accomplishments.

Muller-Merbach (1994) has provided a useful way of describing the systems approach as a combination of concepts. Figure 1.2 depicts the various meanings that he has associated with the systems approach.

> "The systems approach focuses the consideration of wholes and of their relations to their parts. The systems approach is necessarily comprehensive, holistic and interdisciplinary. However, there are several types of systems approaches around, quite different from each other and competing with one another," Muller-Merbach (1994) notes.

1.5.1 Using the Systematic–Constructive Approach

The systematic systems approach is considered the Western tradition, whereas the systemic systems approach is characteristic of Eastern philosophies. The systems approach this book uses is systematic: analytic, synthetic (i.e., synthesis), and constructive.

The systems approach called introspection is based on the analytic reduction of systems into their parts, which is characteristic of the sciences.

The systems approach called extraspection is characteristic of philosophy and the humanities. It strives to integrate objects and ideas into higher-order systems using synthesis. The field of general systems is closely associated with this effort to develop meta-systems of knowledge.

Say that a computer stops working. Following introspection, it is opened up and taken apart. By means of analysis, components are tested to find the cause of the trouble.

Synthesis is required to reassemble the computer. Using extraspection, perhaps a better overall configuration can be found. In fact, a faster and better method for maintaining all computers in the office may be invented.

Combine analysis and synthesis to obtain the third systems approach, called construction. It is "characteristic of the engineering sciences and their creative design of systems for practical purposes," Muller-Merbach (1994).

Teamwork increases the effectiveness of introspection, extraspection, and construction. Creative design that uses analysis, synthesis, and construction is the systems approach described in this book.

1.5.2 Why Is the Systems Approach Required?

The systems approach is needed because it produces better solutions than any other approach, especially the functional field approach. It leads to better decisions and provides better problem-solving for complex situations, enabling those that use it to be more successful.

Think of the systems approach in terms of the sports team. If the players are coordinated by communication and training, they play better, and win more games. Similarly, in business, those using the systems approach are the leading competitors in every industry.

1.5.3 Defining the System

Elements that qualify to be part of a system are those that have a direct *or indirect* impact on the problem, or its solution—on the plan or the decision. Thus, a P/OM system is everything that affects product line formulation, process planning, capacity decisions, quality standards, inventory levels, and production schedules. A P/OM system incorporates all relevant factors, that is, those related to P/OM with an effect on the purposes and goals of the organization.

Figure 1.3 is a symbolic picture of a system. The shape (or core) encloses all factors that have a strong effect on the purposes and goals of the system. Weak forces outside the core may also have to be considered.

Figure 1.4 shows an organization chart with the system's shape mapped as a circle across certain functions. This is meant to reflect the fact that problems and opportunities include various people, departments, etc. The problem map cuts across P/OM as well as certain specific parts of marketing and accounting.

This is meant as a symbolic representation of the fact that people in departments falling inside the shaded area are involved with the problem being considered. Major responsibilities appear to fall within the domain of the general manager, a particular supervisor, and plant maintenance. All other departments have only partial involvement.

The key to understanding the relevant system is to identify all of the main players and elements that interact to create the system in which the real problem resides.

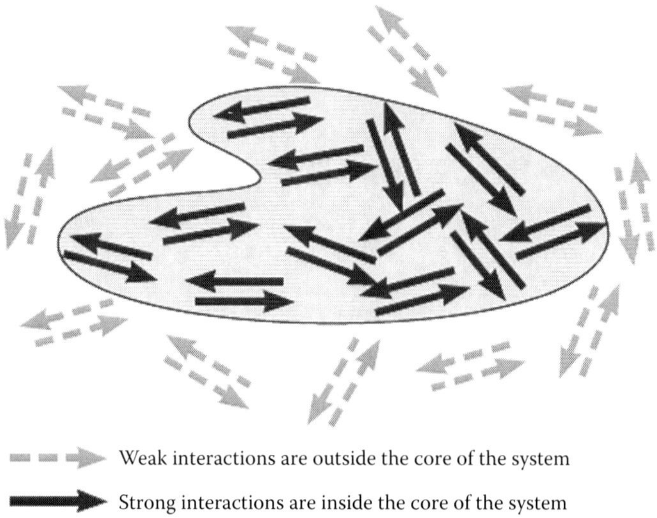

Figure 1.3 Representation of a system. Weak interactions are outside the core of the system; strong interactions are inside the core of the system.

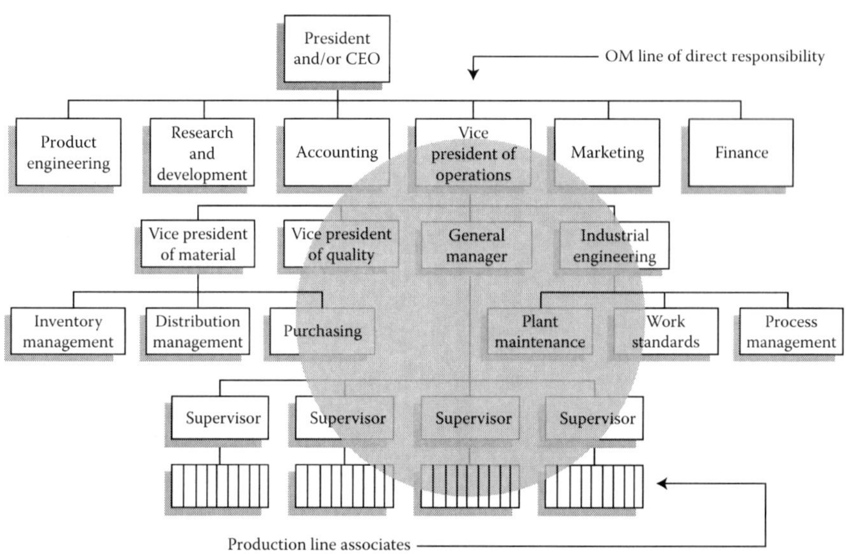

Figure 1.4 Systems problems are mapped over a traditional organization chart.

Even though the problem solution may be assigned to the operations management team, the resolution requires cooperation of all the organizational participants in the problem.

1.5.4 Structure of the Systems Approach

The systems approach requires identification of all the elements related to purposes and goals. The question to be answered: What accounts for the attainment of the goals?

1. The visual concept depicted in Figure 1.3 is one way of answering the question.
2. Another way is to use a mathematical model that shows what accounts for the performance of the system and the attainment of its goals. The equation shown here can be read as "The goals y_i are a function of all relevant factors x_j and t_j." That is,

$$\{y_i\} = f\{x_1,x_2,\cdots,x_j;t_1,t_2,\cdots,t_j\}. \quad (1.1)$$

3. The systems approach requires control of timing. It is to be noted that Equation 1.1 includes measures of time (t_j) as an important systems parameter. By recording conditions over time, it is possible to measure rates of change which are also important systems descriptors. Timing is critical to the performance of orchestras, sports teams, and business organizations. To achieve synchronization of functions, and harmony, which are time-dependent processes, the systems approach is required.
4. The systems approach demands teamwork. Coordination of all participants is essential. Designing a productive process for making sandwiches or toothpaste, delivering packages, or servicing cars requires cooperation among all members of the system. To operate a process properly, it is necessary for all participants to have the systems perspective. In particular, the attainment of quality calls for a dedicated team effort. A mistake by anyone involved in the attainment of quality is like the weak link of a chain, which causes the whole system to fail. For example, "the surgery was successful in every regard except for the missing scalpel which may still be in the patient."

1.5.5 Examples of the Systems Approach

As previously pointed out, managing a sports team is an excellent example of a purposeful effort that is enhanced by using the systems approach. Another fine example is the symphony orchestra whose conductor makes certain that all participants are synchronized (on the same timeline). If the violins, woodwinds, and brass treat their participation as if they were separate functional fields, bedlam would result. Everyone looks to the conductor to keep the components of the system related and balanced. A well-run restaurant, hospital, school, or theater exemplifies the critical importance of competent synchronization.

For current purposes, the main example to explore is a generic business model whereby operations managers make products and/or deliver services. Using the systems approach to coordinate the business-unit team is essential to balance supply and demand, meet schedules, minimize costs, guarantee quality-standards fulfillment, maximize productivity, and optimize the use of critical resources.

For fun, a more abstract example that is widely known to both children and adults might be useful at this point. Jigsaw puzzles strike a chord because being able to assemble them smartly requires a systems perspective. Vision is needed to relate the interdependent elements of pieces using clues of various kinds. Puzzle difficulty increases when pieces are cut to look alike and little color differentiation is provided. Edges are important when internal spatial characteristics send no real hints concerning congruent outlines.

Puzzles become geometrically more difficult as the number of pieces increases. Similarly, as a system becomes larger with more complex interactions, it becomes more difficult to fathom its structure and to understand how it functions. Operations management problems are composed of complex subsystems, which require interfunctional communications to uncover patterns that relate the subsystems to the whole system.

1.5.6 Designing the Product Line Using the Systems Approach

The product line (goods and/or services) is the starting point of strategic thinking for the firm as discussed in Chapter 11. Every factor relevant to the success of the product line must be included in deliberations among all functional managers of the business. The product line is tested against marketing assumptions.

Market research starts with the concept and later, after prototypes are made, tests them in the marketplace. If services are the products, the same considerations apply. Price points are conceived that should generate an expected volume of demand for the chosen qualities of the products.

If the products test well, then P/OM designs processes for making and delivering them. Most of the time, process improvements can be suggested based on changes in the design of the products. The costs of making and delivering the products, and the qualities of the products, are a function of materials and processes.

The discussion between marketing and P/OM involves finance as well. The kind of processes used will determine investments required by P/OM to be underwritten by the financial managers. All business functions are involved in strategic planning, which means that the systems approach is essential.

1.6 Information Systems for Manufacturing and Services

The growing recognition of the importance of the service function in manufacturing has broadened the situations to which the term operations is applied. Manufacturers have become more comfortable with the notion that they must cater to the customer's service requirements. Information systems provide the necessary data about customer needs so that operations management can supply the required services.

Both services and manufacturing are increasingly responsive to—and controlled by—information systems. Therefore, knowledge of computers, computer programming, networking, and telecommunications is essential in both the manufacturing and service environment. The field of analytics, which combines computer power with huge amounts of data, has been growing exponentially as the comfort level of managers with computers analyzing big data has improved. Massive amounts of information can be stored and analyzed for the first time in the history of management. P/OM is at the forefront of this systems-oriented evolutionary capability.

Schools of business include both goods and services under the term operations, whereas industrial engineering departments are still inclined to teach "production" courses. Nevertheless, there is inevitable convergence of both to an information-dominated workplace.

Operations is the familiar management term for an information systems environment, so the word "operations" fits nicely.

Programming and maintenance (both service functions) have become increasingly important to manufacturing. Further, the relevance of service to customers increasingly is viewed as a part of the total package that the manufacturer must deliver. Manufacturing joins such distinguished service industries as transportation, banking, entertainment, education, and healthcare. In this regard, note the following trends for manufacturing:

1. The labor component (the input of blue-collar workers) has been decreasing as a percent of the cost of goods at an accelerating rate for over 50 years. In part this explains the persistence of unemployment in the developed economies of the world.
2. The technological (and capital assets) component as a percent of the cost of goods has been increasing for many years. In the past 20 years, this effect has become multiplicative, with computers controlling sophisticated and costly equipment across vast distances via satellites and networks.
3. As information systems play a larger part in manufacturing, highly trained computer programmers (sometimes called gold-collar workers) and white-collar supervisors add to growing sales and administrative (overhead) costs, which have to be partitioned into the cost of goods. These costs are an increasing percent of the cost of goods. Traditional methods for assigning these costs

can lead to detrimental P/OM decisions. New accounting methods, called activity-based costing (ABC), should be used to improve overhead accounting. A good introduction to ABC can be found in Kaplan and Cooper (1988) and Kaplan and Anderson (2007). Operations managers need to discuss these issues with their colleagues from accounting.

4. The systems approach requires communication between functions and the sharing of what used to be (and still are, in many traditional firms) mutually exclusive databases. The databases of marketing and sales, P/OM, R&D, engineering, and finance are cross-linked when advantageous. This sharing is crucial to enabling the systems approach to work. There are many examples of both manufacturing and service industries where shared databases have been installed and utilized successfully.
5. The technology of the twentieth century is moving rapidly into retirement along with a lot of executives who grew up with its characteristics. The 21st century is a different ball game with new players who feel free to deal with the distinction between services and manufacturing as well as between operations and production in their own way.

Practitioners now have stepped into the twenty-first century, but they have yet to get accustomed to it. It is a good bet that the taxonomy of the twenty-first century will categorize production as a subheading under operations, and services will be an integral part of manufacturing. Inter-functional planning with shared information will be the norm and not the exception.

When a discussion applies equally well to both manufacturing and services, it is often referred to as P/OM. As explained earlier, it is increasingly common to call it OM. However, in this book, we will use the terms P/OM and OM interchangeably. Because we are straddling the 20th and 21st centuries, we tend to use P/OM most often. We do so to be as inclusive as possible.

1.7 Defining Operations

Operations are purposeful actions (or activities) methodically done as part of a plan of work by a process that is designed to achieve practical ends and concrete objectives. This definition is applicable to both manufacturing and services organizations. This interpretation further justifies the use of the term *operations* for manufacturing. An operations manager is responsible for planning, organizing, coordinating, and controlling organizational resources to produce desired goods and services, which is the subject matter of this textbook.

1.7.1 Manufacturing Operations

Manufacturing operations transform materials into desired goods and products. Operations can be described using different verbs and object phrases such as pressing and turning metal (on a lathe), cutting paper, sewing clothes, sawing and drilling wood, sandblasting glass, forming plastics, shaping clay, heat-treating materials, soldering contacts, weaving fabric, blending fuels, filling cans, and extruding wires. Similarly, there are a variety of assembly phrases, such as snapping together parts, gluing sheets, fitting components, joining pieces together, preparing (assembling) a burger. Products such as automobiles, airplanes, televisions, furniture suites, computers, refrigerators, and light bulbs are made in factories. On the other hand, fast-food chains such as McDonald's and Burger King view the assembly of sandwiches from meat, buns, and condiments as a manufacturing application. Goods also include processed items such as paint, milk, cheese, chemicals, etc. While there is a notable distinction between fresh foods and factory foods, much of agriculture is a production process.

1.7.2 Service Operations

Service operations in the office environment are quite familiar, that is, filing documents, typing input for the word processor, and answering the phone. There are similar lists of verbs and objects that apply to jobs done in banks, hospitals, and schools: granting loans, taking X-rays, and teaching classes are a few examples. Movies are one of the biggest export products of the United States. Operations management applies directly to entertainment, film-making, and sports. Administration of the law is a major service industry that requires operations management. Law firms are well aware of the importance of productivity management, information systems, and quality improvement. Jobs available for operations managers of law firms are one indication of the extent to which OM savvy lawyers are highly valued.

Those who have worked for UPS, Federal Express, or the post office are able to list the various service operations related to delivering mail and packages. Those who have worked for the IRS will have another set of job descriptions to define specific operations that characterize tax collection activities of the federal government. Also, if one has worked for The Gap, Banana Republic, Eddie Bauer, The Limited, Wal-Mart, The Sharper Image, Kmart, Sears, or other retail operators, he or she will be able to define processes that are pertinent to merchandise selection and pricing, outsourcing, distribution logistics, display, and store retailing. The experience will have similarities and differences with supermarkets that must also cope with dated products such as milk and greens that speak for themselves (besides being labeled) regarding freshness.

Successful mail-order (and Internet) companies like Lands' End, Amazon, L.L.Bean, Victoria's Secret, Norm Thompson, and Barnes & Noble are good examples of entrepreneurial firms that have struggled to master changes in technologies to gain the operational advantages of smart logistics. Distribution, in retail, mail order, and Internet B2C (business-to-consumer web customers), is a production process that lends itself to all of the benefits that excellent information systems and new technology can bestow.

The credit card business is another splendid example of a situation that combines many aspects of service functions. MasterCard, VISA, and American Express are totally dependent on smart operations management to provide profit margin excellence. The IT component of the credit card business model is almost transparent, but it is still very difficult to do well with consistent performance. These production processes are good examples of information systems operating under high-volume flow shop conditions.

Disaster (crisis) management, another service area, is an emerging field within the realm of P/OM. A disaster can be man-made through error and/or terror; or nature-driven such as flooding, fires, earthquakes, hurricanes, and volcanic eruptions. With an increase in the frequency of disasters—both natural and man-made—the importance of crisis management has increased. First responders to crises must cope with needs of people buried, injured, starving, thirsty, and requiring shelter. They need supplies and equipment. In this instance, crisis management anticipates needs before they arise based on understanding the type of disaster that is predicted. Dealing with the aftermaths of tragedies requires a priori analysis of supply chains to deliver healthcare and subsistence. This point of view is understandable since alleviating human suffering deserves top priority.

From a systems perspective, the anticipation of catastrophes can sometimes result in mitigation of damage. Under some circumstances, it can even prevent the catastrophe from occurring. P/OM concepts can be used during various phases of catastrophic events. Various phases include anticipating and stopping catastrophes, mitigating catastrophes, and preparing for catastrophes.

P/OM procedures can stop catastrophes. There is no record of how many disasters did not occur because good operational procedures detected problems and corrected them before calamities resulted. There are, however, numerous instances of where official reports confirmed that correctable conditions were detected but the corrections were not made in time and disaster resulted.

As is later discussed in Chapter 11 (where sustainability is the P/OM topic under consideration), operational procedures did not require a delay in the Challenger space shot (STS-51-L) on January 28, 1986. Postponement *should have been required* because O-ring conditions had not been tested nor certified in very low temperatures which prevailed on that day.

Many other examples can be cited including Space Shuttle Columbia (STS-107) which experienced vehicle disintegration on February 1, 2003. Learning from this event, design changes were instituted which would prevent similar tragedies from occurring in future space launchings. The list of preventable accidents is extremely large and further examples will be presented in Chapter 11 and other suitable parts of this textbook.

P/OM can control impact severity. Bringing to bear mobile firefighters and equipment of sufficient capacity to manage forest wildfires has helped reduce the size of the burning areas. Similarly, taking steps to build up walls of sandbags has helped limit damage in many flood areas. P/OM is well known for bringing adequate supplies to staging areas *before* catastrophic events occur. Thus, Home Depot had repair supplies on the move from dozens of states in the USA before hurricane

Katrina struck New Orleans in August of 2005. Supermarket Publix moved food and water to selected storage sites before hurricane Andrew decimated parts of South Florida in August of 1992.

How much time is needed after a prediction of a stage 4 hurricane to strengthen levees and dikes? There was more than enough time for the Army Corp of Engineers to reinforce the levees around New Orleans. It was not done because the people in control were politicians, not managers. Preparation for events that are highly likely to occur can be short term (for necessities such as food and drugs) and also long term for engineering construction projects.

P/OM dreams up various process crisis scenarios (like Toyota Production System—TPS) and inculcates "best practice" response drills. P/OM knows how to monitor the system's dashboard which shows the degree to which various strategies would have succeeded in saving the Titanic. Toyota designed the Andon (a signal board or dashboard that flashes lights) to indicate root causes of problems. This helps trace problems to their origins.

The methodology of P/OM was first developed by and for manufacturing, but it has now been extended to services with great success. Service industries involve an increasing percent of the workforce. Thus, more attention needs to be directed toward achieving coherent and efficient operations for services. The service industries include: hospitals, banks, restaurants, airlines, hotels, tourism, cruises, educational institutions, departmental stores, government agencies, knowledge management, and so on.

In the current era, the distinction between managing goods and service organizations is diminishing, and there is a common body of knowledge that can be used to manage both types of organizations effectively and efficiently.

1.7.3 Similarities and Differences between Services and Manufacturing

There is less difference than similarity between P/OM in manufacturing and service organizations. Manufacturing is the fabrication and assembly of goods, whereas services generate revenues either independent of goods or to help the user of those goods. Banking, transportation, healthcare, and entertainment are all services. They change the customer's location, financial condition, and sense of well-being. Increasingly, manufacturers recognize the importance of servicing customers, and service systems recognize the value of using manufacturing capabilities.

Similarities between services and manufacturing can be noted when service operations are based on repetitive steps in information processing. Almost identical methods apply with respect to production scheduling, job design and design of the workplace, process configurations, and quality achievement. High-volume repetitive operations on physical items (i.e., for fast foods or blood testing) constitute production irrespective of whether they are categorized as manufacturing or services. Similar analogies can be made for lower volumes of production and services delivered.

The similarities stop and significant differences occur when the operations involve contact between people. Person-to-person activities that require transfer of information and/or treatments offered by one to another are difficult to schedule; activity times vary more than with machines. Human-to-human interactions involve many more intangibles than interactions between people and machines. The contact aspect of services requires different methods for analysis and synthesis than are needed for manufacturing systems.

At the same time, care should be taken to avoid stereotyping services as being all too human and, therefore, difficult to control for quality and productivity. It does a disservice to services to consider them quixotic or flawed by humanism, while manufacturing is admired for its elegant, efficient technological component. A most respected thinker, Levitt (1972) has written, "Until we think of service in more positive and encompassing terms, until it is enthusiastically viewed as manufacturing in the field, receptive to the same kind of technological approaches that are used in the factory, the results are likely to be … costly and idiosyncratic."

The point to make is that services that are currently rendered in an inherently inefficient way often can be transformed into rational repeatable activities that emulate the best of manufacturing environments. However, *often* is not *always*. Some services are not amenable to the concept of manufacturing in the field. One might be fearful if the services rendered by a doctor were based on a repetitive manufacturing model. At the same time, many aspects of open-heart surgery are the better for such systematization. The same can be said for blood testing, taking X-rays, and other repetitious aspects of the healthcare business. Juxtaposed to this, some products, such as artwork, are epitomized by being custom-made. They lose most of their value if they are manufactured.

Another significant difference between the provision of services and manufacturing occurs because of inventory. It is not considered possible, usually, to stock services. For example, when the machine-repairing person is idle, there is no way to build up an inventory of repair hours that can be used when two machines go down at the same time. In most service businesses, supply being greater than demand is one of the great waste factors.

On the other hand, many companies use automated systems to provide custom-tailored information such as stock, bond, and mutual fund quotes for anyone knowing the symbols. Phone-call requests for product information are answered by a digitized voice that instructs the caller to use a Touch-Tone phone to input the product of interest and his or her fax number. The appropriate fax is automatically transmitted within a minute. This entire service transaction, without human intervention, is becoming increasingly common and epitomizes automated manufacturing and/or service processes. In this case, the supply of service hours is limited only by the technology and there is no cost for supply waiting for demand.

Voice recognition technology is getting so good that a new era is about to occur that will likely revolutionize the service function. Contacts by computer have become kindly and comfortable to customers. Machine's ability to understand customers' responses has also altered the contact relationship. Logical reasoning by computers for service requests is likely to be far better than what can be supplied by outsourced, call-center employees whose native language and cultural milieu are different from the caller's. This voice recognition advantage of computers using the Internet will alter outsourcing of call-center functions for banks, e-stores, etc. Voice-directed picking technology (as well as goods receiving, put-away, replenishment, and dispatch) is finding its way in streamlining warehousing operations; source: http://www.manufacturingdigital.com/technology/talking-sense-voice-recognition-technology. SIRI, the voice of iPhone's activated personal assistant, is a noteworthy example of mobile voice interface.

1.8 Working Definitions of Production and Operations

The generic or collective definition of operations emphasizes rational design, careful control, and the systematic approach that characterizes the methodology of P/OM. Production/operations is a big umbrella that always includes services and often includes manufacturing.

P/OM is the systematic planning, execution, and control of operations. This definition implies that management is needed to ensure that actions are purposeful—designed to achieve practical goals and targets. P/OM makes sure that the work is done methodically, that is, characterized by method and order. The fact that a process is used suggests the presence of management to install a procedure for working systematically.

Operations management is responsible for a plan of work—a thoughtful progression from one step to another. Plans require details for accomplishing work. These details are often called the tactics of the plan. Practical ends are not realized without operations management that is able to provide strategies and tactics for public service objectives, which can include the ability to gain market share on a bus route or participation in a recycling plan. Everyone wants to be able to gain market share. Strangely, the same does not apply to profit. Non-profit organizations pay salaries and provide services which are profits transformed—always labeled as expenditures. There is a need to review why some organizations consider it embarrassing to make a profit.

Operations management uses methodology that consists of procedures, rules of thumb, and algorithms for analyzing situations and setting policies. They apply to many different kinds of service and manufacturing processes. In brief, operations management consists of tactics such as scheduling work, assigning resources including people and equipment, managing inventories, assuring quality standards, process-type decisions that include capacity decisions, maintenance policies, equipment selection, worker-training options, and the sequence for making individual items in a product-mix set.

1.9 Contrasting Production Management and Operations Management

What is the difference between production management and operations management? Production is an old and venerable term used by engineers, economists, entrepreneurs, and managers to describe physical work both in homes and in factories to produce a material product.

Operations management is a more recent term associated with services performed by organizations such as banks, insurance companies, fast-food servers, and airlines. Government jobs are also in the services. Healthcare providers, including hospitals and schools, belong to the services category. Running the Olympic Games is an operations management job. It is not surprising that there has been a rapid growth of service jobs in the US economy. The current ratio of service jobs to manufacturing jobs is nearly 4:1, compared to an approximate 1:1 ratio in the 1950's. As a result, the number of people that are now engaged in operations is far larger (and still growing) than the number of people that work in manufacturing.

Manufacturers have come to view service to the customer as part of the quality of the product line. This includes repairing defective products as well as providing regularly scheduled maintenance. Auto manufacturers learned a great deal from Acura when Honda launched that division with a service mission that eclipsed anything in auto service that preceded it. One of the best automotive service operations before Acura was Honda itself, so there was precedent to follow. Xerox, after losing significant market share, began to establish strict guidelines for the maximum allowable downtimes that would be tolerated for its copying machines. Until the 1990's, IBM provided almost no service to its personal computer customers. After a serious fall from grace, IBM changed its policy and became a full-service company to all of its customers.

Having developed a successful consulting business model, which accounted for most of its profits, IBM sold its personal computer product line to Lenovo, China's leading PC-maker, for $1.75 billion dollars in December 2004. Lenovo completed the purchase of IBM's PC division in May 2005, after receiving clearance from the US Committee on Foreign Investment in March 2005. Lenovo became "the world's third-largest PC maker," according to The New York Times, Business Day, p. C5, May 2, 2005. After a rocky transition, earnings swung back to black in the fourth quarter of fiscal 2007.

Lenovo inherited from IBM the right to sponsor the 2008 Beijing Olympics. This included designing the 2008 Olympic Torch and supplying the digital foundation of the Olympics administration. A year-long test period was designed by Lenovo to check on all aspects of operations. Using 14,000 pieces of computing equipment, Lenovo gathered and stored participants' data, display scores, and organized all activities of the BOCOG (Beijing Organizing Committee for the Games of the XXIX Olympiad). This major undertaking dwarfed all operational control systems employed by OM at prior Olympics.

In 2013, Lenovo became the "biggest PC supplier in the world." This is according to IDC and Gartner who track PC sales. In the second quarter of 2013, Lenovo had 16.7%, HP had 16.3%, and Dell was third with 11.89%. This is an interesting business case in which Lenovo exhibits the excellent characteristics of a Stage IV company (see Section 1.15). Source: Mashable.com/2013/07/11/lenova-emerges-as-worlds-biggest-pc-maker/

1.10 P/OM—The Hub of the Business Model

The product line of goods and services determines the operations needed to match supply and demand. The business model combines marketing forces (including competition), financial investments, and operating costs. This business model had to be thought out in detail during strategic planning.

A product that cannot be made or delivered on time, with quality, and at an acceptable cost must be referred back to marketing and general management. If financial support is insufficient to develop a satisfactory process, that fact must be referred back to finance and marketing. If employee resources are inadequate to operate the processes, that fact must be referred back to HRM, marketing, finance, and general management. These and other issues place P/OM at the hub (core and center) of the business model.

The planning details (of the model), once accepted, have to be adhered to by all of the business functions. When results do not jibe with plans, it is essential that all parties reexamine original assumptions and make adjustments as soon as practicable. As will be recognized from the following discussion, the functional field approach cannot be accepted. Although this is true in general, it is particularly so in a global business environment. The systems approach is essential.

1.11 Transformation Process

All operations management and production systems involve transformation. The goal of the production/operations department is to transform the inputs (using labor, machines, and materials) into desired qualities of goods and services at the minimum cost. Alteration of materials and components adds value and changes them into goods and services that customers want to own. The raw materials and components before transformation could not be used—and therefore had no utility—for the customer. Service conversions have customer utility even if no transfer of goods takes place. The conversion may be a change of location or related to the customer's state of well-being (e.g., visiting the doctor, the repairperson fixes the air conditioner on a hot summer day, or a rescue from a disaster area during a hurricane).

The manufacturing transformation of raw materials into finished goods is successful if customers are willing to pay more for the goods than it costs to make them including selling expenses and other general and administrative costs. Consider what has to be done to make a product. The raw materials for glass, steel, food, and paper have no utility without technological transformations. New processes are constantly being invented for improving the transformations and the products that can be obtained from them.

The same transformation rules apply to services. The conversion is successful if customers are willing to pay more for the services than it costs to provide them, including selling expenses and other general and administrative costs. To illustrate a service transformation, consider an information system in a bank. Depositing a check in a bank results in the electronic transfer of funds (ETF) from the paying account to the paid account, which is clearly an input–output transformation.

Another information transformation is to take raw data and turn it into averages and standard deviations. The latter is characteristic of the operations aspect of market research. As another service example, consider the transformation that is at the heart of the airline business—moving people from one place (input) to another (output) for profit. Other airline inputs are fuel, food, and the attention of the flight attendants.

Figure 1.5 represents a generalized input–output transformation model. Generalized means that it is a standard form that could be applied to any system where conversions are taking place. Inputs are fed into the transformation box—representing the process. The "process" often includes many subprocesses. If the process is to make a burger sandwich, then important subprocesses include cooking the hamburger and toasting the buns.

The inputs are combined by the process, resulting in the production of units of goods or the creation of types of services. The transformed units emerge from the facility (factory, office, etc.) at a given rate. Time needed to carry out the transformation determines the production rate. The transformed inputs emerge as outputs to be sold or used beneficially. The transformation model depicts work being done. This work involves the use of resources made up of people, materials, energy, and machines to achieve transformations.

Figure 1.6 illustrates an expanded version of the input–output transformation model. There are many boxes now within the transformation grid. Each box represents operations that generate the product line, which can be goods or services. Productive P/OM systems have well-designed transformations.

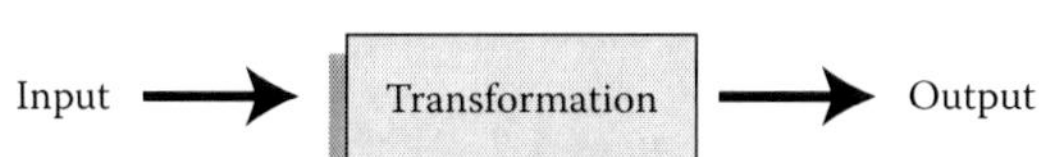

Input–output is the basic P/OM model

Figure 1.5 Input–output transformation model.

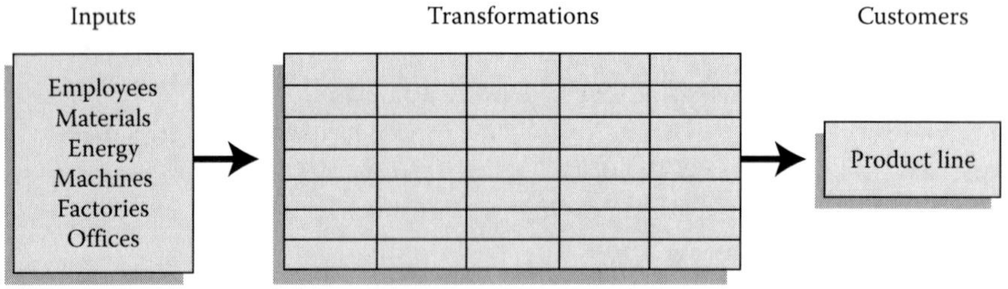

Figure 1.6 Expanded input–output P/OM model.

Transformations are being accomplished when people are served chili at Wendy's or when they give blood to the Red Cross, have their teeth cleaned by the dentist, or visit Disney World to be entertained. A travel agency will have secured the necessary reservations and tickets for the customer's flight to Orlando, Florida, and for the hotel. The travel agent designs the trip and fits it to the customer's specifications concerning dates and costs using reservation and other information systems to complete all necessary transactions. It is a sign of the times that an increasing number of customers are content to act as their own travel agent doing all the transactions on the Internet. The explanations for their willingness to do it yourself (DIY) are many, and travel agencies are adapting to the practices of Priceline, Expedia, etc. Whoever makes the arrangements, when the desired outputs are fully specified, the transformations can be planned, along with the inputs, and the plan can be carried through to completion. The culmination of the transformation process constitutes the desired output—a visit with Mickey Mouse.

1.12 Costs and Revenues Associated with Input–Output (I/O) Models

Cost management is a key function associated with all aspects of P/OM. A major portion of the cost of goods or services originates with operations. Figure 1.7 is meant to illustrate how costs are related to input–output models. Controlling costs is of prime concern to all managers.

For the most part, costs are readily categorized into variable costs and fixed costs. Generally, costs are considered to be easily measured, although the treatment of overhead costs is subject to debate. Also, a variety of accounting methodologies exist. The differences between them are not trivial because they can impact P/OM decisions in significantly disparate ways. P/OM and accounting coexist in the same system, and they are interdependent when the measurement of costs interacts with P/OM decision-making. Quality, another key criteria associated with all aspects of P/OM, interacts with costs in a variety of ways, as do productivity, timeliness of delivery, and styles and sizes of products and services.

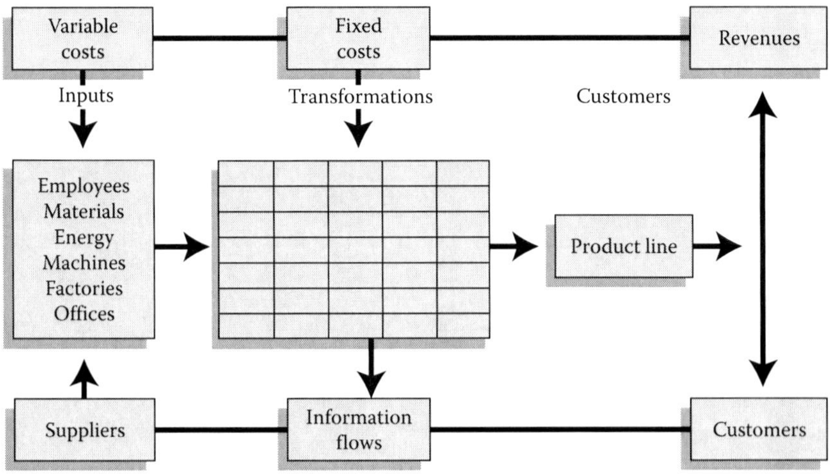

Figure 1.7 Input–output P/OM model cost and revenue structure.

1.12.1 Inputs Associated with Variable (or Direct) Costs

The input components of the transformation model that apply to an airline transportation process include fuel, food, crew pay, maintenance, and other costs. Variable operating costs increase as there are more flights flown and more people flying. Variable costs are also called direct costs because they can be applied directly, without ambiguity, to each unit that is processed.

The same reasoning applies to a manufacturing example. The variable costs for the inputs include labor, energy, and all of the materials purchased from suppliers and used to make the product. Materials include raw materials, subassemblies, semi-finished materials, and components. The more finished the purchased materials, the less work that has to be done by the purchaser (i.e., less value can be added by the purchaser). Less value-adding generally translates into less profit.

1.12.2 Transformations Associated with Fixed (or Indirect) Costs

When Delta or American Airlines buys aircraft from Boeing or Airbus Industries, the airline increases its already substantial fixed cost investment in planes. Fixed costs are also called indirect costs because they are part of overhead and must be allocated to units of output by some formula. Often the charge per year—called depreciation—is calculated by dividing the cost of the investment by the number of years in the estimated lifetime of the investment. For example, a $30 million aircraft with a 15-year lifetime would generate $2 million of depreciation per year. This is called straight-line depreciation because the amount per year does not change.

DEPRECIATION

Dividing the total cost of an asset to reflect how much of that cost applies to each period of its useful life.

Finance with P/OM's assistance must control how much overhead has to be paid even when no positive cash flow (revenue) is coming into the company.

There still remains the question of how to allocate a portion of the $2 million as an applicable charge for a particular passenger flying from Milan to New York on that aircraft. Determining the appropriate fixed costs to be charged to each job, unit made, or passenger mile flown is a joint responsibility of P/OM and accounting.

Delta and American airlines also have investments in maintenance facilities, airport terminals, and training and education systems, as well as in their workers and management. The payments that airlines make to support the operations of airports generally are fixed and not variable costs. Airports—like factories—are major fixed-cost facilities; treat them as fixed costs because the same expenses must be met no matter how many flights depart or arrive there. However, if part of the airport charges is based on the number of flights an airline makes, then both fixed and variable (input) costs must be considered.

1.12.3 Outputs Associated with Revenues and Profits

Passengers pay the airline for transportation. The number of passengers (units) that are transported (processed) by the airline is the output (sometimes called throughput to emphasize the output rate) of the system. It usually is measured by passenger miles flown system wide (or between two points) in a period of time. Throughput is managed to balance supply (seat capacity) and demand (for seats). The demand level for transportation between any two points is related to marketing factors, not the least of which is the price for a round-trip ticket.

All airlines do not charge the same amount for a round-trip ticket. By adjusting price, airlines often can affect the percent occupancy of their flights. Such marketing decisions are part of the total system that affects operations. These marketing decisions typify the need for systems coordination to relate P/OM with the other functional areas within the framework of the transformation model. Southwest Airlines (SWA) has used efficient operations to maintain low costs. This allows them to charge low prices. This business model has made SWA uniquely profitable even though others have been trying to emulate SWA including Spirit which may have even fewer frills. JetBlue, Virgin America, and EasyJet (in Europe) are among other "low-cost airlines."

The manufacturer can measure output in terms of the number of units of each kind of product it produces. Because there may be many varieties such as sizes and flavors, it is usual P/OM practice to aggregate the output into some common unit, such as standard units of toothpaste produced. Both forms of information would reflect the variability of demand, but the aggregate measure much less so than the detailed product reports. Depending on the demand levels, the marketing department could stimulate sales by dropping prices or raising prices to slow demand that is exceeding supply capabilities.

If lower prices are effective in generating new business, then the demand is said to be elastic to price. When there is price elasticity, operations must keep costs down so that the advantage of low prices can be obtained. Marketing ultimately controls the volume of business that production must process. Financial planning has determined the operating capacity for peak demand. This, in turn, translates back to the dollar amount of inputs that need to be purchased to meet demand.

The systems perspective is required to make sure that all participants are connected directly to the revenue-generating capacity of the I/O system. The information system helps to foster the process of keeping connected. Many kinds of data are regularly transmitted between participants. For example, information about what is selling—and what is in stock—leads to production scheduling decisions. It also leads to initiatives by the sales department. The levels of inventory are perpetually examined to make sure that no stock outages occur, and care is taken to keep track of what is in the finished goods inventory.

1.13 P/OM Input–Output Profit Model

The purpose of this section is to link the equation for profit—which is critical for people in business—to the input–output transformation model. The I/O profit model is derived from the costs and revenues shown in Figure 1.7. The model assigns the costs and revenues of the traditional equation of profit to the inputs, the outputs, and the transformation process—all based on a specific period of time (t).

In Equation 1.2, total costs, TC, are subtracted from revenue, R, yielding profit, P:

$$P = R - TC. \tag{1.2}$$

In Equation 1.3, total costs, TC, equal the sum of total fixed costs, FC, and total variable costs, vQ. The latter is the variable costs per unit (v) multiplied by (Q), which is quantity (volume) in production units for the time period (t):

$$TC = FC + vQ. \tag{1.3}$$

Equation 1.4 represents the calculation of revenue, R, where price per unit, p, is multiplied by Q, sold in (t):

$$R = pQ. \tag{1.4}$$

Equation 1.5 is the statement of how the factors of profit (P) come together. It is useful in other ways as well. Setting profit equal to zero and solving for demand volume provides the breakeven volume. Breakeven analysis is discussed in detail in Appendix A. Questions surface about how demand volume changes according to the prices charged. Another interesting issue is the relationship of FC with v as discussed below:

$$P = R - TC = pQ - FC - vQ = (p - v)Q - FC. \tag{1.5}$$

Nothing in the structure of the profit model has changed over time; the equations are the same. What has changed is the technology that the fixed costs can buy, and, in turn, that investment affects the variable costs. In addition, there have been major changes in the knowledge base about the productivity of processes and the quality they can deliver. Knowledge has been developed that affects the volume of output that the fixed costs can deliver. It is the production system (or the system of operations) embedded in the profit model that has changed a great deal. The performance of the input–output transformation system has been altered, resulting in increased productivity of the system. The architecture of the operating system has been changing for many years, but recently at an accelerated rate. The system is more productive and less labor-intensive.

1.14 Productivity—A Major P/OM Issue

Productivity is a critical business variable that directly impacts the "bottom line"; improved productivity raises net profits. P/OM is responsible for the productivity of the process. This is such a critical factor in a company's overall success that excellence in productivity achievement is a major P/OM issue. Productivity measures the performance of the organization's processes for doing work. Productivity is defined as the ratio measure of output (O) divided by input (I). Operations management views the measurement of productivity as essential for assessing the performance of an organization's productive capacity over a specific time period and in comparison to the competition. When outputs are high and inputs are low, the system is said to be efficient and productive, but everything is relative or should we say competitive?

Productivity is relatively easy to measure for physical goods. It is more difficult to find appropriate measures for some services outputs such as units of education or healthcare. Creative knowledge workers provide other instances of intangible outputs that are highly valued, but elusive to calculate. The effort has to be made to appraise the value of these outputs in a standardized way to provide a benchmark (or standard) for measurement. The detailed discussion of the concept of productivity and its measurement are discussed in Chapter 2.

1.15 The Stages of P/OM Development

The profit model operates differently according to the stage of development of the company's input–output operating system. This strategic issue must be understood by the functional area managers and *addressed in a coordinated fashion*. The stage reflects the degree to which a company's activities have been coordinated and carried out. Thus, stage determines company effectiveness (ability to do the right thing) and efficiency (ability to do the thing right). As a company improves its stage of operations, it is expected that its profitability will increase. However, it is necessary to relate the company's stage of development to that of its competitors.

Each company's input–output profit model indirectly and directly reflects the impact of the competitors' input–output models. The cost structures, prices, volumes, and profit margins reflect the influence and extent of the competition. If the competitors all are at the same stage and one of them starts to move to a higher stage, the expected result is that the advancing company will gain market share, while the remaining companies experience a decrease in share and volume. This then gets translated into higher variable costs and lower profit margins. P/OMs must be involved in competitive analysis, which means that the company's planning participants fully understand the stages of development of all competing companies.

How a company manages its profit model provides insights into the role that P/OM can play in a company. Decisions concerning capacity and the resulting economies of scale capture only a portion of the story; the extent to which new technology is utilized to offset high variable labor costs also plays a part. The development stage relates as well to the management of the throughput rate, quality attainment, and variety achievement.

Stage I companies operate on the premise that there is no competitive advantage to be gained by changing the production process. Therefore, the process usually is relatively unrefined and often rudimentary and out-of-date. Management, seeing no leverage in processes, pays minimal attention to production and operations. Stage I firms squeak by on quality. Because the competitors are not much better, everyone appears indifferent. In such firms, management and its resources are insufficient to do more than keep up with demand. Survival occurs only when and if the competition is in the same boat. The indifference to quality also happens when the company believes that repeat business is not going to occur no matter how good the product they deliver. For example, how likely is it that a tourist to Antartica will return in the near future?

	Internally	Externally
Neutral	Stage I	Stage II
Supportive	Stage III	Stage IV

Figure 1.8 Stages of P/OM development.

Internet-based communication, which is fast and available to everyone, also influences repeat business and evaluations. So the family that will not return to Antartica may write: although we will not return soon, we want others to know how terrible X's service was and how indifferent all of X's employees were to our needs.

Marginal firms do not survive long in a market that is dominated by higher-stage companies. Figure 1.8 provides a simple matrix for categorizing stages of a company's P/OM development. This is applicable to manufacturing and service product lines.

At the other end of the spectrum, there are companies that use operations to gain unique basic advantages through the development of special capabilities. The concept of competing on capabilities is clearly formulated by Stalk et al. (1992). The advantages gained can vary from speeding up distribution to tactical superiority in managing inventories, or product preeminence obtained from effective product development and total quality management (TQM).

Stage IV companies practice continuous improvement (CI), which means that they persistently remove waste. They aggressively seek to innovate in their unswerving pursuit of quality for competitive dominance. Stage IV companies have a high level of basic advantages that are unique to them, whereas Stage I companies have virtually none of these advantages. Stage II and III companies fall in-between on various scales of performance and degrees of advantage.

This approach is based on work done by Wheelwright and Hayes (1985), cited in their article in the *Harvard Business Review*. They have created a framework applicable to manufacturing firms. This approach also utilizes work by Chase and Hayes (1991) detailed in their article in the *Sloan Management Review*. Their framework is applicable to service firms. The discussion that follows puts together the concept of stages of development for both manufacturing and service firms.

1. A *Stage I* company is centered on meeting shipment quotas and providing service when requested. C&H call it "available for service" and cite, as an example, a government agency. A Stage I company has no planning horizon and is predisposed to be indifferent to P/OM goals. It is reactive to orders and has no quality agenda. Worker control is stressed. The company is not conscious of special capabilities for itself or for its competitors. W&H describe

such firms as being internally neutral, which connotes that top management does not consider P/OM as being able to promote competitive advantage and, therefore, P/OM is kept in neutral gear.

2. A *Stage II* company manages traditional P/OM processes and has a relatively short-term planning horizon. It makes efforts to secure orders and to meet customers' service desires. The primary goal of Stage II companies is to control costs. Quality tends to be defined as products or services that are not worse than some standard. These companies consider the most important advantages to be derived from economies of scale, which means that as output volume increases, costs go down. W&H describe such firms as being externally neutral; they strive to have parity in P/OM matters with the competition.
3. A *Stage III* company installs and manages manufacturing and service processes that are equivalent to those used by the leading companies. C&H describe this as "distinctive service competence." A Stage III company makes efforts to emulate the special capabilities of the best companies. Quality and productivity improvement programs are utilized in an effort to be as good as the best. Stage III firms have a relatively long-term planning horizon supported by a detailed P/OM strategy. W&H describe such firms as being internally supportive, meaning that P/OM activities support the Stage III company's competitive position.
4. A *Stage IV* company is a P/OM innovator. It has short- and long-term planning horizons that are integrated. Long-term P/OM planning requires excellence in project management to bring about changes needed to adapt to new circumstances, conditions, and environments. Short-term P/OM involves meeting standards by controlling the production process. Both short- and long-term considerations are crucial elements in the success of the firm. Both require that P/OM be a part of the top management strategy team because the production processes are held to be a source of unique advantage gained through special capabilities, as are product and service design.

Stage IV companies aggressively seek to innovate product design and development while assisting environmental improvements in pursuit of sustainability goals. For example, Wal-Mart, teamed with GE, announced (August 2006) that in the next 12 months the plan was to sell 100 million compact fluorescent lamps (CFL). The idea is to replace incandescent bulbs on a broad basis. In October 2013, Wal-Mart launched a new line of LEDs on a nationwide basis at competitive pricing. There is no doubt that LED lamps will replace CFLs in the next wave of innovation. Market replacements, on this scale, are a great deal of hard work for P/OM and others. Thomas Edison, inventor of the incandescent bulb, said that "Genius is 1 percent inspiration, 99 percent perspiration." Stage IV companies do not shirk the hard work.

Project management is a P/OM responsibility that offers significant advantage to those in operations management who know how-to-use project management methods to innovate quickly and successfully for competitive advantage. Wheelwright and Hayes (1985) describe such firms as externally supportive, which means that competitive strategy "rests to a significant degree on the firm's manufacturing capability." Chase and Hayes (1991) conclude that Stage IV firms offer services that "raise customer expectations." Stage IV firms use the systems approach to integrate service and manufacturing activities.

It takes a lot of work to progress through successive stages. It is unlikely that an existing P/OM organization can skip a stage. Only with total reorganization is it possible for a *Stage I* company to become a *Stage III* or *IV* company. Reengineering (REE), which is defined as starting from scratch to redesign a system, is an appealing way to circumvent bureaucratic arthritis and jump stages. However, it is costly and, if not done right, has a high risk of failure. Mastery of this P/OM text will lower that risk.

1.16 Organizational Positions and Career Opportunities in P/OM

To qualify for an operations management job, there are several reasons it helps to have an undergraduate degree in business or an MBA. First, P/OM is at the hub of the business model. This requires an understanding of the various functional business partners to achieve successful strategic planning. Second, there are many concepts to learn and a special P/OM language to master. A single, introductory course in P/OM will not suffice. The system's perspective is instrumental for success. This requires knowledge about the various business functions including marketing, finance, accounting, and human resources management. Vice versa, a marketing career is enhanced by an understanding of P/OM. The same applies to careers in all other functional areas.

1.16.1 Career Success and Types of Processes

It is essential when talking about careers in P/OM to recognize that one of the major differences in P/OM jobs relates to the kinds of processes that are involved in the transformation of inputs into outputs. This means such things as the continuity of processing, the number of units processed at one time, the volume of throughput between setups, and the degree of repetition of the operations.

It can be seen from the historical development of P/OM that manufacturing started with custom work, which in many ways resembles an artist at work. For example, the shoemaker who fits and makes the entire shoe for each customer is an artist in leather. Most often, the left and right shoes differ. However, for store-bought shoes, customized attention to fitting the customer is not possible. The science of sizing clothing, shoes, etc., is not well developed, so everything that is not customized fits almost everyone poorly. Different systems of sizing are used throughout the world which increases customer dissatisfaction. The situation provides entrepreneurial opportunities for innovators in the P/OM field.

Services often are of the custom variety. The medical doctor sees one patient at a time and treats that patient as warranted. Service processes can prosper by making them more like manufacturing. In time, manufacturing learned how to process small batches efficiently. Some service systems, like elevators, lend themselves to batch processes. Various medical experiments have been conducted for treating a batch of patients at one time with a team of healthcare providers. Shouldice Hospital in Toronto repairs hernias on a flow shop-type production line.

Continuous flow processes were developed by a variety of industries, including chemical processors, refineries, and auto assembly manufacturers. Fast-food chains try to emulate this kind of process to handle a continuous flow of information and to assemble sandwiches. Until the late 1970's, there were basically three different ways to get work done. A fourth (flexible processes) was added when computers began to change the way processes were designed. The four categories are

1. *Project.* Each project is a unique process, done once, like launching a new product, building a plant, or writing a book. Both service providers and manufacturers need to know how to plan and complete projects that are associated with the evolving goals of "temporary" organizations. Projects appeal to people who prefer non-repetitive, constantly evolving, creative challenges. Projects do not attract people who opt for a stable environment and the security of fixed goals—associated with the flow shop. There is a unique profile of people who prefer the project environment to other process types and who excel in that milieu.
2. *Batch processing.* Facilities are set up, and n units are made or processed at a time. Then the facility is reset for another job. When $n = 1$, or a very few, it is called custom work, and it is done in a custom shop. When n is more than a few, and the work is done in batches, it is called a job shop. The average batch size in job shops is 50. The work arrangement ceases to be a job shop when the work is done in serial flow shop fashion.

   With the job shop, many different kinds of goods and/or services can be processed. As the batch size gets larger for manufacturing or services, more effort is warranted to make the process efficient and to convert it to a serialized production system. Job shops, with their batch production systems, appeal to people who prefer repetitive assignments within a relatively hectic environment. The job shop generally involves a lot of people interactions and negotiations. The tempo of batch production is related to the variety of the product line and thereby the number of setups, cleanups, and changeovers.
3. *Flow shop processing.* As the batch size increases so that production can be serialized, either continuously or intermittently, it is rational for both

manufacturing and services to pre-engineer the system. This means that balanced flow is designed for the process before it is ever run. It is expected that variable costs will decrease as the fixed cost investments in equipment increase. Continuous process systems require a great deal of planning and investment. Flow shops run the gamut from crude setups arranged to run for short periods of times (such as days or weeks) to continuous process systems that have been carefully designed and preengineered for automation. The more automated processes appeal to people who like a controlled, stable, and well-planned system. The lower costs of flow shop production are related to economies of scale, for example, the investment in consistent process quality can be justified.

4. *Flexible (programmable) processing systems.* As far back as the 1980's, a new process category began to emerge that continues to grow faster than any other P/OM segment. Flexibility is derived from the combination of computers controlling machines, making this option the high-tech career choice. People who enjoy working with computers prefer these technologically based environments. There are two aspects to this attraction. First is the application of the technology to do the work, and second is the programming of the computers to instruct and control the equipment that does the work. Associated with the adoption of the new technology is much experimentation. Openness to learning is essential because the systems are continuously changing and need high levels of adaptability. People who like to work with high technology are attracted to this process configuration which is related to the achievement of mass customization (discussed in Chapter 11).

   An assessment has found that this category continues to grow, but the extent of its application has narrowed. Flexibility has been stymied by design constraints and higher costs than had been expected. Each product design decision removes degrees of freedom for further design opportunities. The second crankshaft is easier to make than the seventh one. While progress has been slowed, there is belief that mastery of flexible technology will continue to be improved. Investments in flexible manufacturing systems are conditioned by the payoffs resulting from being able to increase variety without incurring large setup costs for each new product design made on the same production line (called economies of scope). As noted above, this is also interrelated with the achievement of mass customization. There is increasing evidence that the tipping point for mass customization is near.

Many people prefer working with a specific type of process. There are also people who prefer to work in either manufacturing or services; these issues usually are more important than type of industry preferences. For an example of the first kind, autos, airplanes, and computers are associated with assembly-oriented industries. Real advantages often accrue to companies that hire employees from similar but not the same industries. Alan R. Mulally exemplifies this versatility. As President of Boeing Commercial Airlines he was a force behind the 767 and 777. William

Clay Ford (Bill) hired Mulally as President and CEO of the Ford Motor Company. When asked how he could make that switch, he is said to have replied: "An automobile has about 10,000 moving parts, right? An airplane has two million, and it has to stay up in the air." Since becoming CEO of Ford Motor Co. in September of 2006, Ford has had a resurgence of success.

Hiring across service industry types is also popular. A person having expertise in the hotel business is likely to be courted for employment by resorts, theme parks, and restaurants. The Ritz Carlton Corporation has made some remarkable competitive strides with respect to the quality of hotel service that can be applied broadly to the entire hospitality class of service. Club Med, which represents one of the best of the resort industry, has a very strong—transferable—P/OM orientation. Club Med, Cirque du Soleil, and Four Seasons Hotels and Resorts are operations management cases in the Harvard Business School series (see Horovitz, 1990). Media and entertainment are two other service areas with a strong draw on career selection.

Certain industries and services have intense regionality, that is, Florida, Hawaii, Mexico, and the Caribbean represent a cross-section for the resort business. Thailand was building a reputation for an exotic holiday destination until the tsunami destruction (December 26–27, 2004). It has since rebuilt its beaches, hotels, restaurants, and reputation. Brazil, Greece, Tahiti, and Bali are some other resort destinations to analyze.

Michigan, OH, and within the last 30 years, many Southeastern states of the United States are beehives of automotive activities. At one time, only Detroit was known as the center of the carmaker's world. Then Toyota, Honda, Subaru, Hyundai, Mercedes, and BMW found new locations far away from Detroit.

New York City, a leader in financial markets, is also a prominent location for product lines in publishing and entertainment; Amsterdam and New York City are preeminent sources for diamond cutting and sales, respectively. Starbucks is growing in Japan, and tourists walk past Starbuck's Cafe in Beijing's Forbidden City in China. Global locations compound the complexity of career decisions in operations.

1.16.2 Operations Management Career Paths

There are many different kinds of P/OM careers. No one can describe all of them because the scenarios of opportunity are always changing and expanding. Consider the actual pattern of developments that started with Y2K (the year 2000). The Y2K fear was that computers would be unreliable when the calendar shifted from 1999 to 2000. Computers, with fixes for changing 2-digit dates into 4-digit dates, worked well. However, as the year 2000 loomed, people became irrational. Technology hardly failed at all. Billions were spent in anticipation and preparation for a crisis that did not occur.

Events have helped to shape the character of P/OM jobs. The Enron failure led to Sarbanes–Oxley revitalizing the importance of operations-oriented accounting to identify real costs and revenues. The World Trade Center disaster of

September 11, 2001, completely altered the management of security operations at airports in scope and importance. New P/OM jobs are being added continuously as crises and disasters receive increasing attention from the media.

Career paths differ according to whether line or staff positions are chosen. Line positions mean responsibility for producing products or services. The term comes from working on the production line. Staff positions by definition are not on the production line, but supportive of the line. Staff positions provide information, guidance, and advice on topics such as cost, quality, suppliers, inventories, and work schedules. Titles for both line and staff positions vary in different companies, and specific details of responsibilities would differ for each position. However, the generalities regarding accountability remain the same.

Knowing about career paths provides a useful perspective for a person starting the study of P/OM. It should be noted, however, that the field is dynamic and changing. It is involved in organizational experimentation with teamwork and the systems approach. The use of multifunctional teams is increasing and likely to spawn new kinds of positions and career opportunities. Also affecting the P/OM role is evolving technology. The rate of change is accelerating.

1.16.3 Global Aspects of Career Paths

Among the vast possibilities of exciting new careers are managers of global P/OM support networks, which need to connect and synchronize factories and service systems located all over the world. P/OM is an international endeavor. With the North American Free Trade Agreement (NAFTA), and the General Agreement on Tariffs and Trade (GATT), the European Union (EU) has become a giant market for goods and services as well as a new environment for manufacturing and service operations (see "View Partnerships Through Trade" at http://wwwustr.gov).

The Pacific Rim has come alive with manufacturing, and great new markets are opening up in Southeast Asia. The off-and-on again agreement to create a free trade zone for the Americas stretching from Alaska to Argentina (34 countries including the United States) is yet another potential indicator of the internationalization of operations management. Suppliers from everywhere will be competing in the global market. A career in P/OM will involve much travel and an equally large amount of global communication. P/OM careers will require ability to coordinate and synchronize systems on a global scale. The global P/OM managers will need to be familiar with local culture and customs for effective management of their operations, and being multilingual will definitely be an asset.

In the following list, 12 traditional career paths are provided that highlight differences between line and staff positions. Manufacturing jobs are followed by a list of service professions. Both include line and staff positions. Titles only approximate the progression from top through middle management to first-level management. For services, there is an even greater problem to get titles that fairly represent progression through the management hierarchy.

P/OM Careers in Manufacturing—Line

1. Corporate Vice President of Manufacturing
2. Divisional Manager of Production
3. Plant Manager
4. Vice President of Materials Management (and Supply Chain Management)
5. Project Manager of Transitions
6. Department Foreman or Forelady, Department Supervisor

P/OM Careers in Manufacturing—Staff

7. Director of Quality
8. Inventory Manager, Materials Manager, or Purchasing Agent (and Supply Chain Manager)
9. Production Schedule Controller
10. Project Manager/Consultant (Internal or External)
11. Performance Improvement Manager
12. Methods Analyst

P/OM Careers in Services—Line

1. Corporate Vice President of Operations
2. Divisional Manager of Operations
3. Administrative Head
4. Department Manager or Supervisor
5. Facilities Manager
6. Branch Manager or Store Manager

P/OM Careers in Services—Staff

7. Quality Supervisor
8. Materials Manager or Purchasing Agent or Supply Chain Manager
9. Project Manager/Consultant (Internal or External)
10. Staff Schedule Controller
11. Performance Improvement Manager
12. Systems or Methods Analyst

By explaining a few of these titles in more detail, it is possible to provide quick insights into various aspects of the P/OM field. Also, it further clarifies differences between manufacturing and service systems.

1.16.4 Manager of Production or Operations: Manufacturing or Services

The manager of operations in services and the production manager in a manufacturing plant are in line positions, meaning that they are responsible for the inputs, the outputs, and the transformation process. These managers oversee the people and technology doing the job, which could be preparing and serving food; taking blood and giving shots; or making DVDs and programming robots on the line. As a rule, they report to a corporate vice president, who has multifunctional responsibilities, or they report directly to the president of the company. Middle managers and selected staff functions report to the manager of production or operations.

For example, department supervisors report to the manager of production and operations. The supervisor's position is a line job. In manufacturing, the supervisor, who is often called foreman or forelady, oversees some part of the production process. In service operations, the supervisor title describes the person responsible for some specific function such as reservations or insurance premium collection. The supervisor of a call center can have hundreds of people working for him or her.

This is a high-level position, but the character of the job depends upon the stage of the company. In *Stage IV* firms, this person will regularly be invited to lunch in the board room because of top management's keen interest in operations.

1.16.5 Inventory Manager, Materials Manager, or Purchasing Agent (and Supply Chain Manager)

The inventory or materials manager holds a staff position that is accountable for controlling the flow of input materials to the line. The function of this job is to determine when and how much to order, and how much stock to keep on hand. There are myriad titles associated with these jobs. Many manufacturing and some service firms have vice presidents of materials management because the cost of materials as a percent of the total cost of goods sold is high. In service firms such as Starbucks and JetBlue Airways, materials are similarly important. Service firms tend to have a larger labor cost component, and the airlines have high fixed costs. Nevertheless, the cost of coffee has been rising, and fuel costs have been volatile and very high. The inventory managers might consider hedging when purchasing coffee and kerosene.

1.16.6 Director of Quality

There are a great number of jobs in the quality assurance area and even more titles. Most of these jobs are staff positions that range from auditing quality levels to doing statistical analyses for control charts. The director of quality or quality manager (who could be a vice president) is in charge of the various quality activities that are going on in the firm.

In some companies, line workers have been given quality responsibilities, so it is possible to find supervisors with quality team assignments and organizing quality circles. Quality adjustments to inputs, including vendors, and to the transformation process are common. Quality positions usually focus on improving the quality of outputs by inspecting for defects, preventing them from happening, and correcting their causes. Quality management is just as important in service firms as it is in firms manufacturing goods, but it is more elusive to measure what customers consider quality and, therefore, more challenging to deal with service quality.

1.16.7 Project Manager/Consultant (Internal or External)

There are important P/OM jobs relating to projects. These can include the development of new products and services as well as the processes to make and deliver them. Constructing a refinery, putting the space station into orbit, writing this text online, and producing a paper textbook are all projects. Consultants, both internal and external, are usually engaged in project management. External consultants are employees of a consulting firm. P/OM is an excellent background for providing entree for a consulting career. Internal consulting applies only to the company for which the employee works (e.g., GE, UPS, FedEx, W. L. Gore & Associates, and Amgen are companies that have very successful records using their own internal employees as project consultants).

During volatile business times, transition management is used with various scenarios such as downsizing, turnarounds, and business process redesign. Companies have created positions that indicate responsibility for some form of transition in the job title. For example, a project manager of transitions is in charge of downsizing or rightsizing the company, turnarounds (restoring a company that is in trouble to good financial health), and reengineering (starting from scratch to redesign the firm). Various new job titles have appeared such as outsourcing manager and the transformational CEO, which indicate that the management of change is taking place. For a variety of good references, see Change Management in Wikipedia.

Summary

Chapter 1 begins with an explanation of P/OM. It presents the role of P/OM in strategic planning for the product line and subsequently for the processes that are required to make, fashion, and deliver the products. It explains why the systems approach is essential for successful strategic planning of the business model using coordination of all functions of the business.

Next, the systems approach is defined as systematic and constructive methodology using systems taxonomy. The implementation of the systems approach employed by P/OM is detailed and explored with examples applied to manufacturing of goods and to services. In this context, production management is compared to operations management. Chapter 1 explains how information systems play a vital role in both manufacturing and services. Also, the basic model of P/OM—the input–output transformation model—is developed, along with the costs and revenue that are associated with its use. This leads to a discussion of the input–output profit model and how it has changed over time.

Stages of P/OM development are introduced with attention paid to special capabilities (strategic and tactical) derived from P/OM that allow a company to reach highly competitive stages of development. This leads to deliberation about the kind of P/OM positions that might be encountered by other functional managers in the organization, as well as the type of careers that P/OM offers.

Review Questions

1. What is operations management? Define OM. Compare it to P/OM. Distinguish between P/OM and POM.
2. What are the differentiating characteristics of services when compared to those of manufacturing? Illustrate the distinctive aspects of each by naming industries and citing specific companies that represent each.
3. To what category of P/OM does hotel management belong? What fixed and variable costs are appropriate for this industry?
4. To what category of P/OM does agricultural management belong? What fixed and variable costs are appropriate for this industry?
5. To what category of P/OM does healthcare management belong? What fixed and variable costs are appropriate for this industry?
6. How do information systems relate to operations management?
7. Why is the systems approach essential to assure real participation of P/OM in a firm? What does the systems approach have to do with company strategy?
8. What types of costs are usually associated with inputs? Give specific examples from the education field.
9. What costs are identified with the equipment that provides transformations? Give specific examples from the fields of manufacturing and transportation.
10. How do outputs convert into dollars of revenue? How are these dollars related to fixed and variable costs as described in Questions 3, 4, and 5?
11. Explain the stages of P/OM development and try to identify some companies that might be representative of each stage.
12. What changes have occurred over time in the following profit model?

$$P = (p - v)Q - FC.$$

13. Describe career paths of P/OM for both manufacturing and service systems.
14. Would a human factor analysis (statistical) of the length of feet correlated with the width of feet and the height and weight of a person (for each gender) result in shoe sizes that better accommodated customers?
15. Would an ergonomic analysis (statistical) of the length of legs for slacks correlated with the waist of those slacks and the height and weight of a person (for each gender) result in slack and trouser sizes that fit their customers better?
16. What differences exist between human factor studies and ergonomic studies?

Problems

1. Develop the transformation process to (a) make wine, (b) bake bread, and (c) produce wooden lead pencils (consult http://science.howstuffworks.com/question465.htm).
2. Draw the input–output model that could be used to run a successful restaurant. Label the variable costs for labor and materials, and detail those costs. What fixed costs apply to this model?
3. Draw an input–output model for a Starbucks store. Label as many of the specific and detailed inputs and outputs as you can. (Figure 1.7 is a model of categories to include.) What transformations link inputs with outputs? (It is helpful to visit a Starbucks store. Observe operations. If possible, talk to a manager about how drinks are assembled.)
4. Repeat what you have done in Problem 3 for a Dunkin' Donuts store. Make a detailed comparison between Starbucks and Dunkin' Donuts.
5. A manufacturing plant and equipment cost $150 million and are estimated to have a lifetime of 25 years. Straight-line depreciation is to be used. Additional fixed costs per year are $4 million. Variable costs are $1.25 and price is set at $3.25. State annual profit when annual volume, in million units, is (a) 10, (b) 2.5, (c) 5 and (d) 8. What is the breakeven volume in millions of units for each level of demand?
6. A call center and its equipment cost $120 million and are estimated to have a lifetime of 30 years. Straight-line depreciation is to be used. Additional fixed costs per year are $6 million. Variable costs are $2.50; price is set at $3.75. State annual profit (or loss) if annual volume in millions of units is (a) 10 and (b) 8.
7. A manufacturing plant and equipment cost $200 million and are estimated to have a lifetime of 20 years. Straight-line depreciation is to be used. Additional fixed costs per year are $5.5 million. Variable costs are $1.50 and price is set at $2.50. What will annual profit be if the annual volume is 10 million units?
8. A service center has installed a new computer system with local area networking at a cost of $1.6 million. The system is expected to serve for 8 years, and straight-line depreciation is acceptable. There are additional fixed costs of $300,000 per year. This service repair center charges each customer a flat fee of $30. The variable costs are $20. What will profit be if the annual volume is (a) 50,000 units? (b) 25,000 units? and (c) 75,000 units? What is the breakeven point?
9. Four companies are described as follows. Characterize each in terms of the stage of P/OM development that seems appropriate. Classify each situation and explain the reason for the classification used.
   a. This manufacturer pays little attention to the quality of the product. The owner is convinced that customers are plentiful but not loyal. Hard sell is stressed.

b. This service organization tries to keep up with its competitors by copying everything they do as soon as possible. The president believes that development costs are saved and that being a fast imitator results in a competitive advantage.
c. This manufacturer is constantly working at being as good as the best of the competitors. They have improved quality repeatedly while holding costs constant. The production manager participates in strategy formulation.
d. This service organization aims at global leadership based on the finest operations in the world. The service is constantly being improved, which gives the organization a proactive leadership role in its industry.

P/OM History Archive

This archive presents a series of New York Times archival articles outlining the evolution of manufacturing. These articles were specifically chosen to chronologically outline the manufacturing innovations man has made through time. Readers will be taken back to the time of Henry Ford and the Industrial Revolution and up to the leaders of today's Fortune 500.

Archival Articles

Jan. 3, 1909; System: The Secret of Ford's Success; By Henry Ford. http://www.nytimes.com/ads/peoplesoft/article1.html

Jan. 11, 1914; Henry Ford Explains Why He Gives Away $10,000,000; By Thos. A. Edison. http://www.nytimes.com/ads/peoplesoft/article2.html

Dec. 18, 1927; What Lies Ahead For America? We Dream of Utopias and Dread Domination by the Machine; By Evans Clark. http://www.nytimes.com/ads/peoplesoft/article3.html

Sep. 9, 1945; Detroit: Reconversion Laboratory: In the Motor City are Tested the Social and Industrial Problems of a Nation Returning to the Ways of Peace; By Russell Porter. http://www.nytimes.com/ads/peoplesoft/article4.html

Sep. 24, 1950; Inexpensive TV Sets; Mass Production Techniques Account For A Wider Range of Receivers; By Alfred Zipser Jr. http://www.nytimes.com/ads/peoplesoft/article5.html

Feb. 16, 1964; 50 Years Ago; By Sherwin D. Smith. http://www.nytimes.com/ads/peoplesoft/article6.html

Jan. 9, 1967; Computers: New Values For Society; By Melvin Kranzberg. http://www.nytimes.com/ads/peoplesoft/article7.html

Oct. 25, 1979; Technology; Light Industry Adding Robots; By Peter J. Schuyten. http://www.nytimes.com/ads/peoplesoft/article8.html

Sep. 28, 1980; Technology—Elixir for U.S. Industry; By Anthony J. Parisi. http://www.nytimes.com/ads/peoplesoft/article9.html

Aug. 25, 1985; Break the Habits of Mass Production Putting America's Factories Back On Top; By Wickham Skinner. http://www.nytimes.com/ads/peoplesoft/article10.html

Dec. 7, 1998; Six Sigma Enlightenment: Managers Seek Corporate Nirvana Through Quality Control; By Claudia H. Deutsch. http://www.nytimes.com/ads/peoplesoft/article11.html

Aug. 21, 2000; New Economy: An old-style deal maker takes up the cause of high technology with manufacturing over the Internet; By Barnaby J. Feder. http://www.nytimes.com/ads/peoplesoft/article12.html

Readings and References

Beinhocker, E., I. Davis, and L. Mendonca, The 10 Trends You Have to Watch, *Harvard Business Review*, July–August 2009, pp. 2–6.

Chase, R.B., and R.H. Hayes, Beefing-Up Operations in Service Firms, *Sloan Management Review*, Fall 1991, pp. 17–28.

Cole, R.E., What Really Happened to Toyota? *MIT Sloan Management Review*, 52, 4, Summer 2011.

Drucker, P.F., The Emerging Theory of Manufacturing, *Harvard Business Review*, May–June 1990, pp. 94–102.

Freeman, K.W., The Right Way to Close an Operation, *Harvard Business Review*, May 2009, pp. 2–8.

Hayes, R.H., and G.P. Pisano, Beyond World Class: The New Manufacturing Strategy, *Harvard Business Review*, January–February 1994, pp. 77–86.

Horovitz, J. *Winning Ways: Achieving Zero-Defect Service*. Productivity Press, Boca Raton, FL, 1990.

Immelt, J.R., The CEO of General Electric on Sparkling: An American Manufacturing Renewal, *Harvard Business Review*, March 2012, pp. 43–46.

Kaplan, R.S., and S. Anderson, *Time-Driven Activity Based Costing*, Harvard Business School Press, 2007.

Kaplan, R.S., and R. Cooper, Make Cost Right: Make the Right Decisions, *Harvard Business Review*, Sept.–Oct. 1988.

Kumar, N., Strategies to Fight Low-Cost Rivals, *Harvard Business Review*, December 2006, pp. 1–9.

Levitt, T., Production-Line Approach to Service, *Harvard Business Review*, September–October 1972, pp. 2–11.

Merrifield, R., J. Calhoun, and D. Stevens, The Next Revolution in Productivity, *Harvard Business Review*, June 2008, pp. 2–9.

Muller-Merbach, H., A System of Systems Approaches, *Interfaces*, 24, 4, July–August 1994, pp. 16–25.

Pisano, G.P., and W.C. Shih, Restoring American Competitiveness, *Harvard Business Review*, July–August 2009, pp. 2–14.

Pisano, G.P., and W.C. Shih, Does America Really Need Manufacturing? *Harvard Business Review*, March 2012.

Porter, M.E., and J.W. Rivkin, The Looming Challenge to US Competitiveness, *Harvard Business Review*, March 2012.

Skinner, W., The Focused Factory, *Harvard Business Review*, May–June 1974, pp. 113–121.

Stalk, G., E. Philip, and E.S. Lawrence, Competing on Capabilities: The New Rules of Corporate Strategy, *Harvard Business Review*, March–April 1992, pp. 57–69.

Wheelwright, S.C., and R.H. Hayes, Competing Through Manufacturing, *Harvard Business Review*, January–February 1985, pp. 99–109.

Source of Video Clips on Manufacturing

Alliance for Innovative Manufacturing (AIM): How Everyday Things Are Made.

Stanford University, through a cooperative venture with industry partners in the Alliance for Innovative Manufacturing (AIM), takes students on plant visits and guided tours of many kinds (see http://manufacturing.stanford.edu).

The material is exceptionally well designed and informative. Over 40 different products and processes with almost four hours of video are available to build an understanding of how things are made. For example, there is a 13-minute video on how to make glass bottles. Some other "how they are made" products include motorcycles and their engines, cars, chocolate, and clothes. Knowledgeable guides talk to the viewer through the stages and steps of processes as well as products. There are 12-minute clips on casting, 20-minute clips on molding plastics, 17-minute clips on forming and shaping, 5-minute clips on machining, and 4-minute clips of video on assembly operations.

Dr. Mark Martin hosts many other experts from industry in this enlightening excursion. Do not miss this chance to learn what successful P/OM students need to understand about both manufacturing and services. P/OM practitioners have to be conversant with every process detail.

Otherwise, the car will not move, the cake will not bake, the cost per unit will not be competitive.

Chapter 2

Strategy, Productivity, and History

Readers' Choice

Beinhocker, E., I. Davis, and L. Mendonca, The 10 Trends You Have to Watch, *Harvard Business Review,* July–August 2009, pp. 1–6. The authors state that organizations have to focus on all resources to be competitive.

Biema, M.V., and B.C. Greenwald, Managing Our Way to Higher Service-Sector Productivity, *Harvard Business Review,* July 1997. The authors argue that manufacturing techniques are applicable to service sector to improve productivity.

Drucker, P.F., Emerging Theory of Manufacturing, *Harvard Business Review,* May–June 1990, pp. 94–102. The theory of manufacturing proposed by Drucker includes statistical quality control, new developments in manufacturing accounting, modular organization, and systems approach.

Drucker, P.F., Knowledge–Worker Productivity: The Biggest Challenge, *California Management Review,* Winter 1991, 41(2), pp. 79–94. Drucker lists six major determinants of knowledge–worker productivity that include clear task definition, make workers responsible, continuous innovation, learning and teaching, focus on quality.

Hayes, R.H., and G.P. Pisano, Beyond World-Class: The New Manufacturing Strategy, *Harvard Business Review,*

January–February, 1994, pp. 77–86. The authors stress the importance of strategic flexibility in a turbulent environment.

Kumar, N., Strategies to Fight Low Cost Rivals, *Harvard Business Review,* December 2006, pp. 1–9.

Levitt, T., Production Line Approach to Service, *Harvard Business Review,* September–October 1972, pp. 2–11. Levitt argues that if customer service is consciously treated as "manufacturing in the field," it will get the same kind of detailed attention that manufacturing gets.

Ric, M., J. Calhoun, and D. Stevens, The Next Revolution in Productivity, *Harvard Business Review,* June 2008, pp. 1–10. The authors focus on the role of new web-based developments in improving productivity.

Skinner, W., Manufacturing—Missing Link in Corporate Strategy, *Harvard Business Review,* May–June 1969, pp. 113–121. Skinner advocates for a top-down approach for manufacturing operations. According to this approach, the manufacturing policy is developed starting with the company and its competitive strategy.

Skinner, W., The Focused Factory, *Harvard Business Review,* May–June 1974, pp. 113–121. Skinner states that a focused factory provides avenues to avoid sub-optimization of individual elements and promotes identification and building on competitive strength. A sharp focus results from good systems perception.

Starr, M.K., Modular Production—A New Concept, *Harvard Business Review,* November–December 1965, Vol. 43, No. 6, pp. 131–140. Starr noted the potential strategic impact of designing products for manufacture which share common components. Also see Starr, Martin K., Modular Production: A 45-year-old Concept, IJOPM, Emerald, Vol. 30, 2010.

This chapter illustrates the competitive edge organizations gain by using P/OM planning and decision-making. P/OM is the only function responsible for creating and running highly productive systems. If P/OM is not included in strategic planning, the odds are that productivity will be lower than it could be and lower than that of the most serious competitors. Less than best productivity is a severe handicap.

If P/OM knowledge and experience is included in strategic planning, productivity will be factored into all considerations. This kind of systems thinking is essential for maximum productivity. The history of operations management has been to continually invent methods that raise productivity. Some of these methods include better and better technology, off-shoring, and the return to on-shoring and training.

Several opinion leaders and thinkers in manufacturing field have put forward their ideas about manufacturing strategy, theory, and productivity. A few of these are described in the following paragraphs.

Skinner (1969), in his article, *Manufacturing—Missing Link in Corporate Strategy*, has advocated for a top-down approach for manufacturing operations. According to this approach, the manufacturing policy is developed starting with the company and its competitive strategy. Skinner states, "… that only when basic manufacturing policies are defined can the technical experts, industrial and manufacturing engineers, labor relations specialists, and computer experts have the necessary guidance to do their work."

Skinner (1974), in his article, *Focused Factory*, observes that "The conventional factory attempts to do too many conflicting production tasks within one inconsistent set of manufacturing policies." The lack of focus makes the plant noncompetitive. A focused factory provides avenues to avoid suboptimization of individual elements and promotes identification and building on competitive strength. A sharp focus results from good systems perceptions.

Drucker (1990), in his article, *Emerging Theory of Manufacturing*, puts forward four principles and practices for a new theory of manufacturing. These include statistical quality control (SQC), new developments in manufacturing accounting, modular organization, and a systems approach that "embeds the physical process of making things, that is, manufacturing, in the economic process of business, that is, the business of creating value."

Starr (1965, 2010) noted the potential strategic impact of designing products for manufacture that shared common components. Some components could be interchanged with others, for example, some (otherwise identical) laptops have black cases, while others have aluminum finishes. The manufacturing process is where the color and finish of the case is determined. Product and process designers must work together to achieve such manufacturing modularity. Such modular production is the achievement of predesigned component interchangeability. However, modularity is not always achieved by the production line. External modularity is a consumer option exemplified by a light fixture with a standard screw base that accepts a variety of light bulbs, for example, 40 W, 100 W, clear, frosted, yellow. Similarly, Michelin has many different kinds of tires that can be mounted on the same car. Both forms of modularity raise productivity.

Merrifield et al. (2008), in their article, *The Next Revolution in Productivity*, focus on the role of new web-based developments in improving productivity. According to the authors, the reengineering revolution is a thing of the past; the authors state that, "Thanks to the development of new technologies for using and sharing functions via the internet, the frontier is no longer the process but rather the business activities that make up every process—from pricing a product to issuing an invoice to assessing the risk of individual customers to prioritizing the potential features of a new product in development." However, the jury is out on this issue which seems to violate fundamentals of a proper systems approach.

After reading this chapter, you should be able to:

- Understand the strategic importance of productivity to an organization.
- Evaluate the applicability of various measures of productivity.
- Explain what contributes to good productivity.
- Discuss how to improve productivity.
- Explain why productivity is a systems issue.
- Relate productivity and price.
- Explain demand elasticity as a systems link between P/OM and marketing.
- Describe the effects of quality on elasticity.
- Explain why quality elasticity is critical to P/OM.
- Apply the concepts of economies of scale and division of labor to the management of operations.
- Relate productivity and CAD/CAM.
- Relate productivity and FPS/FMS.
- Explain mass customization and relate that capability to productivity.
- Explain how the history of P/OM shows continual improvement in the P/OM input–output transformation model with resultant increases in productivity.

2.1 The Systems Viewpoint

Attainment of productivity that is higher than, or at least equal to, the best that the competitors can achieve is essential to success of strategic plans. Put another way, inferior productivity is a disadvantage that must be overcome if a business is to be sustainable. However, the measurement of productivity creates quandaries because it can be defined in many different ways.

For adequate information, a firm may need to measure productivity in several ways. One thing is common to all definitions of productivity. Namely, productivity is always a measure of outputs over inputs. It is, therefore, a measure of the efficiency with which resources are utilized to create revenue and profit.

Productivity earns good grades when a high rate of output is obtained at a low cost. This is the case even when outputs are hard to measure, as in some service operations. Although P/OM is the custodian of the production input–output process and is responsible for achieving high productivity, *all employees at every level of the company are involved in attaining excellence in productivity.* Everyone has the ability to increase or decrease the organization's productivity. Those managers who have cultivated the systems point of view recognize that good productivity is contagious; so is poor productivity. Employees sense whether the company culture promotes high or low productivity, and they respond in accordance with the cultural norm. This makes the productivity condition a contagion factor with systems-wide implications.

Another strong systems-type interaction relates to price–demand elasticity, which links the price charged to the volume that can be sold. Competitive pressures to reduce prices lead to demands that P/OM improve productivity to decrease costs. Also, sales lost because of price competition decreases volume, which reduces capacity utilization, leading to larger overhead charges per unit. Productivity measures reveal the integration of all factors that are operating in the business system.

Kumar (2006), in his article, *Strategies to Fight Low-Cost Rivals*, foresees the continued growth of low-cost players and suggests strategies for the survival of regular value-added businesses. According to the author, the only options for the companies are "attack, coexist uneasily, or become low-cost players themselves." The appropriate framework can identify the best strategy to succeed. Attraction for cheap goods always increases in a recession. It is obvious that low cost is king when there is no sense of economic well-being. However, when the tide turns and prosperity returns, luxury items are the goal of an increasing number of consumers. Consequently, this is another article that reflects the economy of its time rather than taking a systems approach to market evaluation. The systems approach will ascertain the state of mind across various sets of potential consumers and develop the appropriate production and marketing strategy.

From a broader point of view, P/OM history shows that the overall trend in global economies is to increase productivity. Productivity growth has reflected the impact of a continuous stream of developments in technology and operations management methodology.

2.2 Strategic Thinking

A highly productive manufacturer of buggy whips in 1926 would be no better off today than an inefficient and lackadaisical one. The most productive maker of engineering slide rules during the 1950's was Keuffel & Esser. Although this company had the largest market share then, it is no longer in existence now. K & E did not have a strategy to cope with the advent of electronic calculators and then computers. Outmoded products would not be purchased even if the production process to make them had been highly productive. High productivity is necessary but not sufficient to assure competitiveness. Large market demand may thrive because of low price for a product that has elastic price–demand characteristics, but only when the product has true utility for some market segments. Collectors of buggy whips may have to deal with high prices because of nonelastic market demand based on the scarcity of an ancient or obsolete product. There will be low volume with a quite high price for such demand.

Adaptability is critical when product design technology (or style) changes. For example, the most efficient supplier of iceboxes had no advantage when refrigerators replaced the older technology. Additionally, some companies were really good at making vinyl records (collector's items now) but that did not help them survive rapid changes as the music industry moved from 8-track tapes to 8-mm audiocassettes to CDs, and now to DVDs and Blue Rays and eventually "the cloud." Regarding videotape formats, the Betamax system by Sony was the first successful commercial product. Matsushita aggressively pursued the VHS format, which replaced Betamax, but that is of little consequence now since DVDs are outmoding any tape system. Five-and-a-quarter floppy disks have disappeared, and 3.5-inch floppy disks have been largely superseded by memory cards, flash sticks, etc. What is coming next is a forecasting challenge.

Hayes and Pisano (1994) observe that strategic flexibility becomes the strategic goal in a turbulent environment and state that, "Being world-class is not enough; a company also has to have the capability to switch gears—from, for example, rapid product development to low cost—relatively quickly and with minimum resources. The job of manufacturing is to provide that capability."

The lesson to learn from all of these cases is that productivity excellence is useless without market acceptance. P/OM strategic thinking is "find the best product line for the marketplace, and then make it at the lowest cost, highest quality, and overall at max productivity levels." As Peter Drucker is quoted to have said, "Do the right thing, and then do the thing right." Strategy planning for operations is directly concerned with "Do the right thing."

Product design and product development lead, respectively, to process design and process development. Product and process are both components of the strategic imperatives for P/OM. Output as inventory that cannot be sold is unproductive and cannot be counted while measuring productivity.

2.3 Measurement of Productivity

Productivity is a critical business variable that directly impacts the "bottom line"; improved productivity raises net profits. P/OM is responsible for the productivity of the process. This is such a critical factor in a company's overall strategy that excellence in productivity achievement is a major P/OM issue.

Productivity is a system property that interacts with other system properties such as reliability and consistency, as well as customers' perceptions of quality. It interacts with the variable costs of goods and services, as well as with the fixed costs of facilities, training, and technology. It also interacts with the presence and the availability of management as a resource. It is often found that output rates and qualities (and therefore productivity) are better on regular work shifts as compared to "graveyard" shifts because managers are scarce resources for the latter.

Productivity measures the performance of the organization's processes for doing work. It is an important way of grading how well P/OM and the rest of the system are doing. It is a score like RBIs (runs batted in), your credit rating, or ROI (return on investment).

Productivity, according to The Association for Operations Management, which used to be called the American Production and Inventory Control Society (APICS), is defined as "an overall measure of the ability to produce a good or a service." It is the actual output of production compared to the actual input of resources. Productivity is a relative measure across time or against common entities (Blackstone and Cox 2004).

Operations management views the measurement of productivity as an essential tool for assessing the performance of an organization's productive capacity over a specific time period and in comparison to the competition. When outputs are high and inputs are low, the system is said to be efficient and productive. The APICS Dictionary definition of productivity can be converted into the terms of this text by recognizing that productivity is the ratio measure of the output divided by the input as given by

$$\text{Productivity} = \frac{\text{Outputs}}{\text{Inputs}} = \text{Measure of production efficiency.} \qquad (2.1)$$

This productivity measure compares the quantity of goods or services produced in a period of time (t) and the quantity of resources employed in turning out these goods or services in the same period of time (t) (Fabricant 1969).

Equation 2.1 is relatively easy to measure for physical goods. It is more difficult to find appropriate measures for some services outputs such as units of education or health care. Creative knowledge workers provide other instances of intangible outputs that are highly valued, but elusive to calculate. The effort has to be made to appraise the value of these outputs in a standardized way to provide a benchmark (or standard) for measurement. Several authors have proposed ways to measure service productivity. Some of these are described below.

Levitt (1972), in his article, *Production Line Approach to Service*, argues that, "if customer service is consciously treated as 'manufacturing in the field,' it will get the same kind of detailed attention that manufacturing gets. It will be carefully planned, controlled, automated where possible, audited for quality control, and regularly reviewed for performance improvement and customer reaction." In the article, *Managing Our Way to Higher Service-Sector Productivity*, Biema and Greenwald (1997) also argue that productivity improvement techniques used in manufacturing sector are applicable to service sector as well in spite of the complexity of the service sector.

According to Drucker (1999), improving *knowledge–worker* productivity is one of the biggest challenges in the twenty-first century. He lists six major determinants of *knowledge–worker* productivity. These include: offer a clear task definition, put responsibility for productivity on workers themselves and provide autonomy, use continuous innovation, continuous learning and teaching, focus on quality and not only quantity of output, and treat knowledge worker as an "asset" rather than a "cost."

Improving the productivity of specific service businesses has been the focus of attention of several other authors also—for example, hospitals and banks by Sherman (1984), Bell Labs by Kelley and Caplan (1993), and sales force by Ledingham et al. (2006).

Productivity measures are benchmarks for comparing how well the system is doing compared to other systems, or over time. The comparisons include: how A is doing over time; how A compares to B; how departments within A compare to each other; how A compares to an industry average; how A compares to the best of the industry, etc.

Sales perceives productivity as high customer sales volume (called an effective marketing system) accompanied by low producer costs (called an efficient producer system). Thus

$$\text{Productivity} = \frac{\text{Effectiveness}}{\text{Efficiency}} = \frac{\text{High customer sales volume}}{\text{Low producer expenses}}. \tag{2.2}$$

From a systems point of view, the inclusion of sales provides a correct measure of output for productivity measurement. It is never productive to make a lot of product that is not sold, even if the cost of making it is low.

At the same time, P/OM employs productivity measures to assess how well the production system is functioning. The kinds of questions that are being addressed are: how many units of resources are consumed to produce the output, and how many units of output can be made with a fixed amount of capacity? These are blueprints for productivity improvement.

Both ways of viewing productivity have benefits. They stem from different interests that need to be shared. The best interests of the company are served by merging what is learned about P/OM's efficiency and sales/marketing effectiveness.

2.3.1 Labor Productivity

Labor efficiency is an often-used measure of productivity where productivity is a ratio of output units produced to input labor resources expended per unit of time (t). The following equation gives the dimensions for output and input:

$$\text{Output} = \text{units per hour} \quad \text{and} \quad \text{Input} = \text{dollars per hour} = \text{hourly wages}. \tag{2.3}$$

Equation 2.4 measures labor productivity over a specific period of time, t, and gives the relationship between the dollars of labor resources required in period (t) to achieve the output rate (t). This is a measure of how much output is being obtained for each dollar spent.

$$\text{Productivity}(t) = \frac{\text{Output}(t)}{\text{Input}(t)} = \frac{\text{Units of output}(t)}{\text{Dollars of labor}(t)}. \quad (2.4)$$

Another way of stating the measure of productivity is to put a dollar value on the output volume per unit of time (t) as given in the following equation:

$$\text{Labor productivity}(t) = \frac{\text{Output}(t)}{\text{Input}(t)} = \frac{\text{Sales in dollars}(t)}{\text{Dollars of labor}(t)}. \quad (2.5)$$

The advantage of this method is that it can deal with the productivity of the system across different kinds of units. For example, the productivity of a paint company that puts paints of many colors in cans of many sizes could be measured. This approach is used for national accounting of productivity where there are many different kinds of units to be included (i.e., furniture, clothing, energy, and food). Dollars can standardize the output measure across diverse categories.

Labor productivity, often measured as the ratio of sales dollars to labor cost dollars, is called a partial measure of productivity. It is partial because it only looks at labor and does not include capital.

2.3.2 Capital Productivity

There are also other factors that can be used to measure productivity. For example, capital may be used as an input resource in place of labor, or both capital and labor may be used as discussed in Section 2.3.3. Capital productivity is the productivity of the invested capital in technology. Equation 2.6 measures capital productivity as the number of units of output per dollar of invested capital, whereas Equation 2.7 measures capital productivity as a pure ratio based on dollars of output per dollar of invested capital:

$$\text{Capital productivity}(t) = \frac{\text{Output}(t)}{\text{Input}(t)} = \frac{\text{Units of output}(t)}{\text{Dollars of capital}(t)}, \quad (2.6)$$

$$\text{Capital productivity}(t) = \frac{\text{Dollar value of output units}(t)}{\text{Dollars of capital resources}(t)}. \quad (2.7)$$

Such measures of capital productivity might help organizations in the service sectors to address the value of returns on investments in computers and telecommunications. By extending this reasoning, it is possible to develop other partial measures of productivity with respect to areas such as energy expended, space utilized, and materials consumed.

2.3.3 Multifactor Productivity

Multifactor productivity (MFP) can be measured in different ways. As shown in Equation 2.8, sales and finished goods are considered as outputs, but work-in-process is not. Also, labor costs and capital expenditures (amortized) are treated as inputs. MFP (see Equation 2.8) is measured as the ratio of dollars earned to dollars spent. Total factor productivity is another name associated with inclusion of more factors than labor. Multifactor (and total factor) productivity is measured on a regular basis to determine how well the US industrial base is doing and whether it is improving its competitive position in the world.

$$\text{Multifactor productivity}(t) = \frac{\text{All outputs of goods and services}(t;\$)}{\text{Total input resources expended}(t,\$)}. \tag{2.8}$$

> A change in multifactor productivity reflects the difference between the change in output (production of goods and services) and the change in labor and capital inputs engaged in the production of the output. Multifactor productivity does not measure the specific contributions of labor, capital, or any other factor of production. Instead, it reflects the joint effects of many factors, including new technology, economies of scale, managerial skill, and change in the organization of production. (USDL, July 7, 1994)

If it seems useful to include work-in-process in the numerator because WIP is certain to be sold, then that should be done. Similarly, it may be important to include all costs and expenses in the denominator (i.e., energy, material, and miscellaneous costs). This is called total productivity as distinguished from total factor productivity.

Beinhocker et al. (2009) state, “We believe that, in the years to come, ‘resource productivity’ (the output achieved from every unit of oil, power, water and other resource input) will become central to company competitiveness.” With this in mind, planning for resource price increases, volatility, and even shortages is important for strategists for future planning.

2.3.3.1 Trends in Multifactor (MFP) Productivity

Table 2.1 shows the average annual percent changes in MFP for the United States over different periods of time (USDL, February 1995). US productivity averaged just slightly above 2% per year from 1948 to 1973. That was considered to be healthy, though not robust, growth. Then came “the shrinking era,” which lasted for 22 years from 1973 to 1995. These were tough years for US manufacturing firms. The US auto industry observed that new Japanese plants in America (transplants in Tennessee, OH) in the 1970’s were twice as productive as traditional US auto plants in Detroit.

Table 2.1 Average Annual Percent Changes in USA MFP

Year	*Multifactor Productivity*	*% Change*
1948–1973	2.2	—
1973–1990	0.5	–77.2%
1990–1995	0.5	0.00%
1995–2000	1.3	160%
2000–2007	1.4	7.6%
2007–2012	0.6	–57%
2011–2012	1.4	133%

Source: This table is based in part on Table A in a Bureau of Labor Statistics News Release dated June 28, 2013, entitled *Preliminary Multifactor Productivity Trends—2012.*

There began a long learning process on the part of US auto manufacturers to increase their productivity. It took a dozen years (1995–2007), but it worked. US auto making reflected more productive processes. This improvement applied to more than automobiles. After many years of declining growth rates, and even some years of negative growth, US productivity had accelerated to levels equivalent to the post-World War II years.

The renewal of US productivity growth was not surprising. Great efforts had been made in the manufacturing sectors to improve efficiency. Subjects such as total quality management (TQM), reengineering, and turnarounds, discussed in the general press, were not fads. They heralded solid accomplishments. During the period from December 2007 to June 2009, there was a subprime mortgage crisis that led to the collapse of the housing bubble in the USA. This triggered a global financial crisis. In addition, productivity suffered as a result of sending abroad (off-shoring) manufacturing jobs which have always been more productive than service jobs. As seen in Table 2.1, productivity recovery began after 2011.

Adoption of new technology promises more developments yet to come. However, it is important to reiterate the comparison between manufacturing productivity and service-based productivity. In addition to measures of service productivity being much lower than those of manufacturing, they have not shown notable improvements over many years in all parts of the world.

There have been substantial investments in computers and telecommunications within the service sectors, but only recently are there real signs of improvement. At last, broad-based changes in productivity can be reported. Technological advances in voice recognition and intelligent service dialog between computers and people have begun to change the landscape. This will accelerate as P/OM managers learn how to apply the new technologies to service systems. This productivity differential constitutes a significant difference that currently exists between services and manufacturing. It is an area of great potential for P/OM.

Students and practitioners of business know that productivity improvements translate into more profits and greater profitability for organizations and improve the state of the economy. In the same vein, economists believe that productivity improvements translate into higher standards of living and greater prosperity. Productivity measures get factored into inflation calculations as well as other economic scenarios. Increases in productivity are generally regarded as a means of checking inflationary trends. Table 2.1 is well correlated with national prosperity.

This is because by working smarter—not harder—profit margins increase. There is more money available to invest in other business opportunities. The shortage of capital drives interest rates up. However, if there are more jobs available than people to fill them, labor becomes scarce, and rising salaries begin to fuel the expansion of inflation. High productivity is a counterforce to inflation. Fewer people are needed to produce more work. The cost of goods and services moves down with productivity improvements. The contribution of P/OM to the well-being of the national economy is widely recognized. The complaint that high productivity causes unemployment leads to the recognition that good jobs are not those which can be better done by machines. Good jobs are thinking jobs and creative activities. We have to redefine the beneficial occupations for a future in which machines do laborious work better than people.

Relative productivity compares the performance of competitive processes for which P/OM is accountable. This means that if two processes are under consideration for a new product, the productivity measures associated with each product should be derived. When quality problems exist, adjustments downward must be made to productivity. For example, the value of sales plus finished goods plus WIP must be reduced by the cost of defectives. Final decisions about the two processes will not be made on productivity advantages alone, but relative productivity will play a major part in planning.

2.3.4 Operational Measures of the Organization's Productivity

Productivity measures can be common sense ratios such as the value of pieces made in a factory divided by the cost of making them, or the number of documents produced by the typing pool divided by the number of people doing word processing. Such operational measures of productivity are valuable benchmarks to companies that are focused on continuous improvement. Relative productivity is such a benchmark.

A productivity measure used by restaurants is the daily dollars generated per table. Many fast-food restaurants use dollars generated per square foot. Airlines measure plane occupancy per flight and average occupancy per route, as well as for all routes flown. Department stores use sales dollars per square foot of space.

Mail-order companies measure sales dollars for categories (i.e., fashion, toys, and luggage) by type of illustration (i.e., color and size) on a percent of page basis. Trends in same-store sales can be followed as a benchmarking guide to retail productivity. Many companies measure the productivity of their complaint departments by the ratio of the number of complaints dealt with per day divided by the number of complaint handlers.

Productivity measures should be chosen to reflect strategic goals. For example, some schools can maximize instructor productivity by having large classrooms. The relevant input–output model is apparent. The total of student-in-class hours per instructor increases with large classrooms but personal contact with the instructor per student is reduced. With large class sizes, instructors must be entertaining to get good ratings. Student ratings of instructors may not be a satisfactory measure of educational productivity. Teachers' grading of students is especially difficult in large-size classes. Appropriate benchmarks for educational productivity exemplify the difficulties of measuring what counts.

Formulating appropriate productivity measures to capture the effectiveness of operations is always a P/OM benchmarking challenge. It takes insights and creativity to measure what matters in performance and what truly can be controlled and corrected.

2.4 System-Wide Issues Impacting Productivity

Every function in the company that has some measurable accomplishment can be evaluated with respect to productivity. The productivity of the company is the composite of the contributions of the individual productivity functions.

Volume of output sold is a measure of the productivity of the sales department. Cost of goods sold is a measure of the productivity of the process designers, the R&D department, and the operations managers. It also is a measure of the purchasing department's ability to find the best materials obtained at the lowest possible costs and highest qualities.

Output delivered to the customer is a measure of the ability of the distribution system to be on time with undamaged delivery. Output delivered is subject to warranties. Does the company listen to the "voice of the customers" with respect to difficulties in repairing or returning defectives for the entire product line? If not, future sales productivity can be impaired. Many companies lose track of the fact that warranties exercised represent productivity decreases.

Productivity issues are woven into all parts of the supplier–producer–customer value chain and the input–output transformation process. This further explains why productivity is a strategic systems issue. The supplier–producer–customer value chain, by definition, adds value at every step, which illustrates strategic impact operating on a global scale. The following section explores the relevance of international factors on productivity.

We discuss below some of the system-wide issues that influence the role of productivity and its measurement. These include: global issues, bureaucracy, size of firm, price–demand elasticity, quality economy of scale, and division of labor.

2.4.1 Global Issues

Although much knowledge exists about production systems and operations management, there have been and continue to be serious productivity problems in the world. These problems have afflicted many developing countries where capital to invest in new technology is scarce, and technical knowledge and training are lacking. There have also been productivity problems in industrialized countries where productivity growth has been cyclical.

Japan's phenomenal productivity growth rates of the 1980's could not be maintained. Nevertheless, Japanese productivity in a variety of industries continues to be formidable. Thus, with respect to auto parts, "On average, the plants in Japan were 18 percent more productive than ones in the United States, and 35 percent more productive than ones in Europe," the *New York Times* (November 5, 1994). Further, "Japanese parts makers surveyed increased their productivity by almost 38 percent from 1992 to 1994; the American companies made gains only in the mid-20's," the *New York Times* (November 5, 1994).

In 2008, Toyota was well on its way to overtaking GM as the leading auto producer in the world. Toyota has developed an astonishing ability to launch new products desired by the marketplace. Speed to market is phenomenal. The hybrid Prius and the inexpensive (at least, at first) Scion are two good examples. Scion, according to Forbes.com (July 26, 2007) "could be the new millennium version of the original Volkswagen Beetle." If this reference is not clear, look up Volkswagen Beetle in Wikipedia.

Toyota reclaimed the title of the world's largest auto maker in 2012 from General Motors Corp. For some reason, reports ignore the fact that China has been the largest producer of auto units since 2008. In fact, since 2009, annual production of automobiles in China exceeds that of the European Union or that of the United States and Japan combined (Wikipedia: Automotive industry in China). We have no information about the productivity of these production systems and know that they are subsidized by the Chinese government.

Toyota's productivity for all of its processes is outstanding. The Toyota Production System (called TPS) is referenced continually by businesses all over the world. TPS is P/OM-hub-centric, connecting all other business functions. Other spectacular global P/OM performers—from product development to process productivity—include Nintendo with its successful Wii and DS, and Apple's iPhone with sales of 525,000 units on the first weekend that the phone was available (Saturday, June 30, 2007, through Sunday, July 1, 2007). By 2013, the Wii system had been surpassed by Microsoft's Xbox with Kinect and Samsung's Galaxy phones with Android were way ahead of Apple's phone in share of market. As discussed in Chapter 11, the problem of protecting an advantage of being first to market with an innovative product requires total understanding and extreme commitment.

Productivity improvements can be registered by companies that reduce the number of direct employees, increase the use of part-time help, rely more on outsourcing and subcontracting of parts, and employ outside maintenance companies, among other things. In this regard, it is to be noted that Japanese firms typically rely heavily on their suppliers and part-time help. Other aspects of Japanese manufacturing methods also logically account for world-class productivity accomplishments.

Nevertheless, the phenomenal Japanese productivity records have diminished substantially in recent years. Other leading industrial nations have experienced productivity declines. Many hypotheses have been offered to explain the inability to sustain stable productivity growth in a turbulent era of new technological development. One suggestion, by Baumol et al. (1991), is that the productivity of industrial nations is converging to a global mean. Another, by Forrester (1978), is that old technology gets used up, and it is difficult to switch to the new technologies and make them profitable. It also takes smarter management. In spite of this, the US economy shows many signs of increased productivity while reducing the manufacturing labor force and increasing the service sector labor force (Table 2.1).

The outsourcing of services has become familiar. Outsourced call centers provide an excellent example of a global phenomenon. Calls originating in Bellevue, Iowa, may be answered by English-speaking operators in Bangalore, India. When outsourcing is used in a proper way, productivity does go up. Usually, input costs are reduced more than output rates. What is lost in the simple ratio measure is the damage to long-term loyalty.

Competitors keep jockeying and leapfrogging each other with the adoption of new technologies that initially reduce productivity. There are various causes for the reduction. The new technology is imposed on old processes by employees who lack training and experience with the new technology.

Product life is short. This allows little time to enjoy the advantages of new technology applied to evanescent and even obsolescent product lines. Quality deterioration is an almost hidden enemy of productivity when defectives are created global distances away from the managers who are responsible for carrying out strategic plans. Speed to remedy weaknesses is also impaired.

Increasingly, companies buy from suppliers located around the world and sell in markets that are equally dispersed. Production facilities, including fabrication, assembly, chemical and drug processes, and service facilities (e.g., call centers), are located globally. In the big picture, productivity performance is the result of international interactions. If productivity is being measured in dollars, or local currencies, exchange rate problems can distort the picture. Exchange rate imbalances can cause both gains and losses that relate to the productivity of investments.

2.4.2 Bureaucracy, Flexibility, and Productivity

A major factor that accounts for poor productivity is bureaucracy—the great inhibitor of flexibility. Bureaucratic systems are rampant worldwide. North America and Europe have more than their share, and they are equally prevalent in Asia, Latin America, and Africa. The definition of the term that is being used here is bureaucracy as institutionalized officialism, which has layers of red tape to cut through in order to conclude activities and operations. Bureaucracy spreads its reach as it proliferates through an organization placing controls over controls in the quest for low risk. It should be emphasized that all systems have an inherent risk aversion that results in their developing methods to protect the status quo.

The positive side to bureaucracy should not be overlooked. In the late nineteenth century, Max Weber was an advocate for legal domination (law administered by the state). See http://en.wikipedia.org/wiki/List_of_Max_Weber_works for Max Weber's works. Weber viewed bureaucracy as a force to counteract the Kafka-like (unpredictable intimidation and lack of civil rights) effects of traditional domination (feudal rules determined by monarchs and patriarchs).

In regard to positive aspects of bureaucracy, it can be noted that bureaucracy plays an important stabilizing role in organizations that are undisciplined and prone to accidents. Routines known to be safe are insurance against risk of catastrophic damage. The down side of bureaucracy is that the pendulum always swings too far. Once bureaucracy gains control, it strives to remove corrective counter swings and maintain existing conditions. The status quo often impedes progress and supports rigidity.

Flexibility is related to productivity in a number of ways. Conditions change and the ability to adapt to new situations is a function of the system's flexibility. New technology and the need to be global are among the most important changes in conditions that require flexibility. Product life is shorter and the need to modify product designs requires flexibility. The productivity advantages associated with producing large volumes of identical units is being replaced by mass customization methods that permit small volumes having greater variety to be produced. Design variations are being used for different countries and even for various regions of the same country. Carrying out strategic plans requires flexibility.

Bureaucratic organizations are dedicated to resisting change. What P/OM must deal with in fulfilling strategic plans is how to circumvent bureaucracy, which is, by intent, the protector and champion of the status quo. Bureaucracy is the opponent of operational change. When bureaucratic constraints are removed, often by decentralization, and more recently by reengineering, afflicted organizations regain some ability to rebound. However, there is always the danger of not having a carefully thought-out plan in place before un-constraining the system.

Japanese organizations that exhibited great resilience when they first started their major export drives also began to succumb to the problems of age and success. Age is correlated with circulatory insufficiencies in human beings and a lack of communication in organizations. This lack is also associated with zero empowerment of employees to do what makes sense instead of what the bureaucratic rulebook dictates.

Success leads to complacency and arrogance, even though it does not have to do so. Bureaucratic organizations are very successful at inhibiting innovation and change. It remains to be seen how successful organizations worldwide will be in learning to counter these inhibitors.

2.4.3 Size of Firms and Flexibility

It is worth noting that small- and medium-sized firms, and new businesses as well, tend to exhibit greater flexibility and adaptability to change than large, centralized organizations. That is why, under stress, AT&T, Dell, Disney, Ford, GM, IBM, Sony, Sears, and other giant corporations used different forms of decentralization to improve their chances of recovering market preeminence.

Organizational awareness of the need for flexibility has been recognized by many large organizations but solutions for big bureaucratic companies have been elusive. The Iacocca Institute at Lehigh University in the United States now supports its "Global Village" as well as the "Global Village on the Move." Both focus on leadership and have created uniquely interesting programs to develop future leaders for business and industry (see http://www.iacocca-lehigh.org).

As cited by the study, "Foreign-Affiliated Firms in America," conducted by the Center for the Study of Operations at Columbia University, 1991, small- and medium-sized firms are organizations with about 300 people. That number has been suggested by various managers, and there is also a great deal of unanimity that the upper limit should be no more than 500.

Given the usual proportions of administrative personnel and those of other functions, this suggests that a sensible limit for the size of an efficient production system is in the neighborhood of 100–200 people. By using divisional structures, it is reasonable to assume that a number of relatively autonomous divisions of sensible size can be related within the firm.

2.4.4 Price–Demand Elasticity and Productivity

Almost every company in the global market competes on price. It is one of the key decision factors for customers. Quality is another key factor but it is often hidden. When a company is competing on price, it means that it will lose some sales when a competitor offers a lower price that it cannot match. Everyone in the company looks to P/OM at this point. The CEO requests increased productivity to lower costs. That usually translates into attaining greater output volume. It is assumed that quality will remain unchanged.

To ask for increased productivity is a special way of asking for lower costs. Unions often take it to mean work faster for the same pay, which makes them reluctant to participate in productivity improvement. Speeding up production can compromise quality. Operations management should try to avoid supporting productivity increases gained in this way; the improvement is temporary, at best. Other ways of obtaining lower costs such as the use of cheaper components and raw materials may lower quality.

The CEO had something else in mind. When requesting increased productivity, the CEO meant using technology and good P/OM methods to improve the process without lowering quality. The CEO's call for increased productivity is in response to competitive strategies.

Decreasing quality to match lower prices is not a way to keep customers. Improved productivity, if it is to translate into greater customer satisfaction and loyalty, must come from working smarter, not harder. This means improving productivity by means other than asking people to work faster, which usually degrades quality.

This highlights the strong functional interaction between marketing and P/OM (which is emphasized in Chapter 11). The managers of these areas are associates working together to manage the effects of price–demand elasticity on production costs and on meeting quality standards. Price–demand elasticity is another example of a crucial relationship between systems partners (marketing and P/OM) required for successful strategic planning.

Elasticity is a rate-of-change measure that expresses the degree to which demand grows or shrinks in response to a price change. A product with high elasticity experiences large decreases (increases) in demand as price increases (decreases), whereas a product with low elasticity experiences small decreases (increases) in demand with the same degree of price increases (decreases). Low elasticity, called inelasticity, means that demand levels are relatively insensitive to price changes. Marketing managers frequently ask market researchers to study the price elasticity of products or services to determine how fast demand falls off as price is increased. Products that have no substitutable alternatives (as perceived by customers) usually have low elasticity. Product designers who strive for exceptional qualities and production managers who demand the highest feasible process qualities are creating barriers to substitutability (inelastic products).

Perfect inelasticity—when demand does not change, no matter what the price—is an accurate description of the situation when an industrial customer is dependent on one supplier for special materials. Most customers try to get out of such a constraining situation for obvious reasons.

Elasticity is a complex relationship. The rate of change between price and demand is not always smooth and regular. There can be kinks in the line or curve. These occur, for example, when an increase in price causes demand to increase, which might happen when price becomes high enough to have "snob appeal," which opens a new market. Despite difficulties, it is important to measure elasticity, thereby relating price and volume—which are critical factors for production planning.

The elasticity–productivity tie between operations management and marketing is attributed to the following:

1. Demand volume falls as price rises, but this is also relative to what prices competitors charge. When a competitor lowers prices, it is equivalent to a price increase for the customer who stays with a supplier who does not lower prices.
2. To be competitive, it is often necessary to find ways to match price decreases offered by competitors. This is a price–demand volume elasticity issue that assumes quality is unchanged.
3. If marketing lowers the price, then the profit margin will decrease.
4. P/OM is always trying to find a way to decrease total variable cost without degrading quality. For example, if a new material is developed that is as good as the old material but costs less, then P/OM shifts to the new material. This initiative is called value analysis.
5. The only way to achieve number 4 is to work smarter, and this is facilitated by means of technology-based or methodology-based productivity improvements.
6. Marketing tries to control demand volume through pricing. If competition drops the price, based on an improved process, emulation of the improvement is needed.
7. P/OM tries to match supply to demand through production scheduling and capacity planning. Marketing and P/OM must work together, combining their interactions by using the systems approach.

2.4.5 Elasticity of Quality and Productivity

The demand volume of goods or services sold as a function of price is the traditional focus of elasticity analysis. Years ago that simple model may have sufficed, but it no longer is valid.

For strategic planning, it is vital that marketing determines how quality levels of the product line affect competitive status (demand volume at a given price). In turn, P/OM must ascertain the unit costs for various process configurations operating at appropriate production volume levels. This subject brings together many systems factors that are of mutual concern to P/OM, marketing, and finance.

Customers in the marketplace take both price and quality into account. Customers' quality expectations often override price considerations. This applies to commercial and industrial customers as well as retail consumers of goods and services. For example, a gourmet restaurant cannot afford to serve meats and vegetables that might suffice for a neighborhood diner.

As previously noted, a product or service that has special qualities is said to have uniqueness, which means that other competitive products lacking those special qualities have a lower degree of substitutability. For example, a product with special features (such as the iPhone) or service rendered by a well-liked person (favorite waitress) has a competitive advantage. Competitive analysis will show when two products are competing head-on, as if they were identical products, being perfect substitutes for one another. Perceived highest quality renders a product less vulnerable to substitution. That means the product is less quality-elastic. What must production know in order to achieve this objective?

How can the effects of quality on demand levels be determined? This is equivalent to asking how to determine the quality–demand volume elasticity. There are ways in which market research can approach this issue that are similar to the way price–demand volume elasticity is determined. Essentially, it is necessary to establish how much extra money customers would be willing to pay for superior quality or for an added quality feature. By noting the distribution of the additional amounts of money that people would pay for superior quality or an added feature, it is possible to quantify the effects of quality and price on demand elasticity.

Throughout this discussion, it should be kept in mind that the achievement of quality standards is a direct responsibility of P/OM. Although this is a book about P/OM and not market research, these functions are highly interdependent. Market research enables P/OM to determine the kind of connections that link quality, price, and demand elasticity in the customer's mind.

These factors relate design and process decisions with the financial choices that are available to the firm. The system interaction includes the fact that quality varies with the kind of equipment that is used, the quality of the material used, and the amount of training that the employees receive.

2.4.6 Economies of Scale and the Division of Labor

Economies of scale are reductions in variable costs directly related to increasing volumes of production output. Economies of scale are driven by increases in production volume. Scale, as used here, is a surrogate for increasing volume. Total variable cost is both a P/OM and marketing responsibility. Since costs generally decrease with an increasing volume of production, this is also a matter of concern for finance. However, if overtime is used for increasing production volume, the costs may not decrease. The financial decision to use high-volume technology with greater fixed costs is made because the trade-off is lower variable costs.

Volume is a function of the total market size that exists, the number of competitors and their shares, and the controllable variables of the organization's price and quality. Although marketing and P/OM work together on the consequences of volume, P/OM is working on the reduction of variable cost which is also a function of volume—with no loss in quality.

Materials and labor are an important component of the variable per unit cost. The design of the product or service determines what materials are needed. It is common knowledge that greater purchase volumes generally are rewarded with discounts. That is only one of the interactions of variable cost and volume. The machines that can be used for high-volume outputs are significantly faster than machines that are economic for low-volume outputs. High volumes can sustain pre-engineering and improvement studies of the interactions between the design of jobs and the processes used, while low volumes cannot. High volumes generate learning about how to do the job better, whereas low volumes do not. The design of jobs determines the amount of labor and the skill levels required. The responsibility for low, unit variable costs leads P/OM to want high volumes so that it can take advantage of the resulting economies of scale.

For many reasons, including the material discounts previously discussed and a general learning effect, variable costs per unit decrease as volume increases. This result, called the economies of scale, is quite similar to the "Experience Curve" of the Boston Consulting Group (BCG) (1968), which yields a 20–30% decrease in per unit costs with each doubling of the volume. It is reasonable to consider the "doubling of volume" as a surrogate for the "doubling of experience," as in BCG terminology.

A coincident concept, proposed by Smith (1776) (a Scottish economist) in the 1700's, was for the division of labor. Labor was to be divided into specialized activities that could be honed to ever-greater skill levels. This notion follows from the theory that "practice makes perfect." To make division of labor worthwhile, the volume of production must be sufficient. Adam Smith said, "The division of labor depends on the extent of the market." With a large enough volume, activities could be segmented, and serialized process flows could be developed. Workers would be specialists in their assignments. Note that division of labor appears in the history section, which follows.

2.5 History of Improvements of P/OM Transformations

Literacy in P/OM requires an understanding of how the P/OM field has developed with respect to the transformation process and, thereby, productivity, quality, volume, and variety. The stages of history have moved production and operations capabilities from low-volume custom work through high-volume rapid and continuous output systems, modular production, and mass customization.

Attention shifts from custom crafts, which are art-based, to the theory of production, which has evolved over time. This theory consists of six established steps and a potential seventh one. There is emphasis on manufacturing because the theory evolved from the production of goods, but the theory has now transcended manufacturing and is applicable to service operations. Figure 2.1 depicts P/OM history in a timeline chart. Dates mentioned are approximate (but satisfactory). It is not possible to pinpoint exactly when each contribution was made.

The capability of P/OM processes to deliver goods and services has changed in steps or stages over time. The study of the history of P/OM production transformation processes allows us to determine which events triggered these stages of production theory. The ultimate goal is to learn the theory, understand it, and possess the advantages that accrue to literate managers.

2.5.1 Artisans, Apprentices, and Trainees—The Beginning

The Renaissance period (1300–1600's) signaled a surge of intellectual and productive vitality in Europe. That surge swept away the dark ages and fostered

Beginning
→ 1545: Benvenuto Cellini–goldsmith and sculptor

The beginning of the development of production transformation theory was in the Guildhalls by Mastercrafters. These artisans studied the transformation of materials using new tools and techniques.

Step 1
Interchangeable Parts (IP) and Division of Labor
→ 1776: Adam Smith–specialization of work activities
→ 1780: Eli Whitney–inventor of IP and facilitator of batch production

Step 2
Scientific Management (SM)
→ 1900: Frederick W.Taylor–father of scientific management

Step 3
Sequenced Assembly (SA)
→ 1912: Henry Ford–sequenced assembly and the serialized flow shop

Step 4
Statistical Quality Control (SQC)
→ 1930: Walter Shewhart–originator of statistical quality control
→ 1940's–1960: Batch Production and Sequenced Assembly Systems Predominate

In the same time frame:
The U.S. is supplier to the world and the leader in batch production systems-
in France, Servan-Schreiber wrote that American managerial dominance was a serious threat.
Quality control standards were established by the U.S. government for military acquisitions
Management Science (MS)-Solving complex problems–1950's-1990's: Operations Research (OR) and Management Science (MS) quantitative methods for solving resource allocation problems; OR and MS introduce major extensions of scientific management. The methods include linear programming, queuing theory, and inventory theory

Step 5
Lean Production Systems (LPS)
→ 1970's–1990's: Japanese manufacturing systems are emulated for low cost and high quality obtained with focus on statistical quality control (SQC) and other quality methods

In the same time frame: Product and process are designed together to minimize variability and maximize reliability, durability, and serviceability. Export industries gain economies of scale by using flow shop and serialized production processes. Development of time-based management concepts.

Step 6
Flexible Production Systems (FPS)
→ 1980–2000's: Growth of computer technology from the 1950's spurs new technology

Flexible manufacturing systems (FMS) are developed with extensions to flexible information systems and flexible office systems. Variety capabilities necessitate the systems approach to integrate marketing and OM. Flexibility and variety capabilities are added to the list of other "ilities" like durability, reliability, and reparability.

Step 7
Mass Customization (MC) and Global Competition (GC)
→ 2000's–2030's: The next steps are new energy sources, huge battery capacities, and eventually interplanetary travel

International linkages for sourcing, fabrication, assembly and marketing, as well as managing currencies; information systems based on networks of computers and global telecommunications alter the way that work is done. The systems approach is extended to all functions within the firm and to all partners along the supply chain, including suppliers and customers. The Internet homogenizes and polarizes. Same team members are more alike; opponents are increasingly different. Company mistakes are magnified by social media. New forces are unleashed that need to be understood and controlled. Big data systems increase the importance of P/OM.

Figure 2.1 The Histo-Map: a timeline of P/OM development.

accomplishments in the arts and sciences centered on artisans, apprentices, and craft guilds. Production transformations were by hand. Output volumes were very small.

Before the Industrial Revolution began (around 1770), craft guilds emphasized pride of workmanship and training for basic manual operations with appropriate hand tools. The shoemakers' children learned from their fathers and mothers. Process techniques were manual skills handed down from generation to generation.

From a transformation point of view, this was good management of the labor inputs. The use of apprentices improved productivity in the artisans' shops because the less skilled (and lower paid) apprentices did much of the preliminary work. This freed the master craftsmen to devote their time to the activities requiring higher skills. On-the-job training produced a continuous stream of greater skills.

Apprenticeship still has significance for many service functions. Great chefs almost always are the pupils of great chefs. The formula would seem to reside in the balance of art and science. When the important knowledge resides in the minds and hands of skilled workers, then the percent of art is high and the percent of science is low. Over time, this percentage has shifted in manufacturing so that engineering, technology, and computer programming play an increasing role.

The art element in manufacturing is disappearing. Computer know-how is replacing people know-how. It used to be that the tool and die department was crucial to the success of metal-working companies, and the best die makers were considered artists. (Tools and dies are the shape formers in the metal-working businesses.) Now, computer-aided design (CAD) and computer-aided manufacturing (CAM) are primarily science, and the old industrial arts are giving way to the new programming arts. This is also happening in service industries and is an effect that can be expected to accelerate in the future.

2.5.2 Interchangeable Parts (IP)—P/OM's First Step

Eli Whitney invented the concept of interchangeable parts for the fabrication of rifles around 1780, which coincides with the dates usually given for the beginning of the Industrial Revolution. The notion of interchangeable parts was the catalyst around which new methods for production transformation began to develop. These methods spawned and supported the Industrial Revolution.

Whitney was not the sole inventor of interchangeable parts. In France, Honoré LeBlanc had invented the same P/OM concept. Neither Whitney nor LeBlanc knew about each other's ideas. Whitney obtained a US government contract for "ten thousand stand of arms." The contract was awarded because of his newly developed production capabilities.

The concept of interchangeable parts is defined as follows: It allows batches of parts to be made, any one of which will fit into the assembled product. For example, headlights, fenders, tires, and windshield wiper blades are not specially made for each car. One 60-W bulb is like another and does not have to be fitted to each socket. The reason that the parts are interchangeable is that each one falls within the design tolerances. Designers are responsible for stating acceptable ranges, which are the design tolerances.

Machines that could produce parts to conform to the designer's tolerances were the keystone. Hand labor, better suited to custom work, began to be replaced by machinery. The effects of this change hastened the Industrial Revolution. Within a short time, IP was an accepted part of the production transformation process being applied to the manufacture of rifles, sewing machines, clocks, and other products.

In 1776, Adam Smith saw that the use of the division of labor as a means of increasing productivity was market-volume-dependent. The pin factory that he studied had sufficient production volume to warrant specialization. The production transformation process was revolutionized—combining worker specialization with interchangeable parts transformed all of the productivity standards. Expectations were raised to new levels.

2.5.3 Scientific Management (SM)—P/OM's Second Step

Frederick Winslow Taylor (1856–1915) introduced scientific management, the numerical measurement and analysis of the way work should be done. One of his landmark studies dealt with the speed and feed rates of tools and materials for metal cutting. Other studies focused on how to lay bricks and how to move iron castings. Taylor's testimony at hearings concerning the setting of rational railroad fees for shipments in interstate commerce brought national prominence to his analytic methodology.

This step in production theory added the idea that the transformation processes could be improved by studying and simplifying operations. This view required rationalizing the job, the workplace, and the workers. Strangely, it had been overlooked until the turn of the twentieth century. Finding economies of motion and putting materials near at hand were the kinds of improvements that Taylor addressed. In this step, the workplace and the design of the job were enhanced to improve the productivity of the transformation process. These industrial engineering ideas, often called methods engineering, have proved to be as useful for service applications as for manufacturing. They will continue to be totally relevant in the twenty-first century.

Called the father of "scientific management," Taylor was one of the key progenitors of industrial engineering (IE). There were others as well. Associated with this era are Henry L. Gantt, Frank, and Lillian Gilbreth, and other pioneers who were attempting to develop a theory of managing workers and technology in the United States. Henri Fayol was developing similar management theories in France.

Taylor started the process of systematizing all of the elements that are part of the manufacturing system. The same industrial engineering techniques developed by Taylor are still used by banks, insurance companies, and investment houses as well as by truckers, airlines, and manufacturers. They are well suited for repetitive operations such as those that characterize fast-food chains and information processing systems. Industrial engineering methods are recognizable as the forerunners of techniques that are currently applied in the search for continuous improvement.

Taylor and his associates developed principles and practices that led to a present-day backlash. He and other contributors to scientific management have been accused of dehumanizing the worker in pursuit of efficiency. Today, the accusation might hold, but it did not when judged by the value system of the early 1900's.

The criticism does not damage the case for benefits that can be derived from using the industrial engineering approach, which grew out of scientific management. Work simplification and methods engineering are both industrial engineering techniques for making jobs better for workers.

2.5.4 Sequenced Assembly (SA)—P/OM's Third Step

In 1912, Henry Ford developed sequenced assembly, which allows assembly to be a continuous flow shop process. Timing must be perfect so that what is needed for assembly arrives on time. Ford developed the sequenced assembly process as a continuous flow production line for automobiles, changing the pace from batch to continuous sequenced assembly.

The serialized flow shop was born over a hundred years ago. The key was learning to achieve synchronization and control of the process flows. The moving assembly line required a high level of component interchangeability. Ford succeeded in achieving complete synchronization of the process flows. Today, with computers facilitating smart supply chain management that reaches every corner of the world, synchronization takes a lot of P/OM training and knowhow.

By means of the principles of interchangeability, division of labor, and flow synchronization, Ford altered the production transformation process. He changed the perception of productivity standards and goals in a conclusive way. In so doing, he built an industrial empire that helped the United States become the world leader in productivity. The United States continues to maintain its lead, although other nations—especially those once considered to be less developed countries (such as China and India)—have been improving their productivity consistently.

Ford's contribution to production theory and to the revision of the transformation process had a major impact on the Japanese automobile industry. It also affected other industries of many kinds all over the world. There was a new rhythm to the transformation process.

Contrast US and Japanese production processes in the 1980's. The major portion of production and operations activities in the United States utilized batch processes. Batch work with small lots does not lend itself to the kind of synchronization that applies to the automobile industry or the continuous flows of chemical processes. Batch work costs per unit are much higher and its productivity is significantly lower than synchronized flow shop per unit costs.

When the Japanese export industry began to compete aggressively in global markets, they chose to shun batch-type production systems. Instead, they elected to specialize in highvolume, serialized flow shops, which extended the concept and application of assembly synchronization to manufacturing and assembly systems.

2.5.5 Statistical Quality Control—P/OM's Fourth Step

Interchangeable parts required manufacturing methods that made batches of parts conforming to tolerance limits. Shewhart developed the theory of SQC that enabled manufacturing to design and control processes that could achieve these objectives. SQC is discussed in detail in Chapter 8. SQC was focused on the producer's ability to control the variability of the process that was making the parts that had to fit within the specified tolerance limits. For the first time, the output of the transformation process could be stabilized and controlled. This was a major contribution to production theory.

Walter Shewhart's major work, which was published in 1930, described his concepts about why SQC works and how to apply it (see Shewhart 1939; Juran and Gryna 1980; Deming 1986) also participated in the development of SQC theory and later on played a crucial role in its implementation and dissemination.

The United States was the first country that consistently used SQC, which it did through the 1940's and the early 1950's, but by 1960 the majority of SQC users were in Japan. US organizations reported that they had dropped SQC to make cost reductions. Quality was considered good enough to replace costly staff departments with inspectors at the end of the production line. By the 1980's, however, under great competitive pressure from quality-driven Japanese organizations, many US companies restored SQC and enhanced it with broader concepts into TQM activities.

Organizations like Motorola, Toyota, and GE are considered to be pioneers leading the development of TQM and Six Sigma within the framework of the systems approach. The TQM approach applied to the production transformation system integrates the goals of productivity and quality. It represents a major step forward in the theory of production and an organizational feat to have gained broad acceptance at all levels. Six-Sigma^{SM}, a registered service mark of Motorola—which developed it—is a culmination of TQM. Motorola reported more than US $17 billion in savings from Six Sigma in the early days of application (see Six Sigma—Wikipedia). Many companies are now using Six Sigma, and certification programs are offered by dozens of schools.

2.5.6 Lean Production Systems—P/OM's Fifth Step

During the 1970's–1990's, Japanese organizations spearheaded by Toyota developed a new kind of production methodology called lean production systems (LPS; also called the Toyota Production System). These systems combine a deep understanding of quality with a desire to be fast (if not the fastest) and a fanatical distaste for all kinds of waste. LPS methodology is now a worldwide endeavor.

Time wasted is singled out. Every effort is made to use pre-engineering of products and process design to maximize quality achievements, minimize variability, and do it all as rapidly as possible. Part of being lean is being fast—in production. Many Japanese organizations were not fast in reaching decisions. Toyota substituted persistence for perfection and over time began to innovate at astounding speeds. Advocates of "lean" became lean producers with high-output volume targets, minimum cycle times, and rapid new product development.

By introducing time management and goals for short cycle times and rapid project development, other new factors were introduced into the production transformation system. The timing of transformations rose to a new level of importance, and, secondly, rapid project management began to mean that the transformation process could be changed from doing one thing to another very quickly. The idea of time management is consistent with the notion that bad scheduling causes delays in value-adding and wastes time, severely impairing productivity.

The Japanese auto industry has been leading in the development of lean and fast production systems. Toyota's production planners, who were architects of the revised production system, stated that Toyota's ideas were a continuation of the concepts that Henry Ford had been developing (Ohno 1978). In Europe, the notion of leanness was directly associated with speeding up cycle times and project development times. At one time, half-time systems were advanced by Saab. The goal was to cut in half the time currently required to do any operation. However, too little, too late, Saab was not lean enough. It ceased production in 2011 and petitioned the Swedish court for bankruptcy. Talks continue about restarting production.

Many US organizations have adopted at least some aspects of lean and rapid manufacturing methods. Six Sigma, when properly conceived and executed, is a means to LPS. Motorola's management originally set itself the goal of reducing defectives to less than 3.4 defects per million parts. This is called the six-sigma program. Having demonstrated that near-zero defect rates are attainable, Motorola now has set another radical target, which is to reduce existing cycle times by 90%. This means that a part that currently takes 10 minutes to produce eventually will be made in 1 minute. Note the Saab half-time objective mentioned earlier. Highlighting reduced cycle-time objectives brings new features to the production transformation process. However, wishing does not make it so. Goals that are unattainable can be more destructive than helpful.

The next two steps are in formative stages, and their impact on productivity cannot be fully evaluated at this time.

2.5.7 Mass Customization with CAD, CAM, and Flexible Production System—P/OM's Sixth Step

CAD that is able to program flexible production machinery represents powerful new technological capabilities. When design and programming are combined, opportunities develop for computers to instruct and control machines instead of needing hands-on command by human operators.

The purpose of using such equipment is different from the goal of "mass production" where one item is made in extremely high volume. The one color (black) Model T Ford epitomizes "mass production." However, with flexibility, the purpose is mass customization, where high levels of variety can be produced in great volume because the production line can be adjusted without incurring significant setup times and costs. Nano-times for changeovers is a desired goal. For example, many different colors and models can be made on the same production line at almost no additional cost when compared to traditional "mass production." Mass customization is discussed in more detail in Chapter 11.

CAD abilities alter traditional relationships of strategic planning. CAD and CAM work hand-in-hand. CAM is dovetailed and synchronized with CAD. Software for CAD is able to calculate strength of materials in specific configurations, which also ties-in with manufacturability. As a result of P/OM's responsibility for processes, P/OM must participate in strategizing for the product line design. CAM and flexible production systems are not the same, but they are related. Flexible processes are capable of switching production from one product to another with almost no time delay—in nano-times. CAD and CAM when used with flexible technology provide the ideal components for "mass customization."

The initial thrust was in manufacturing. Flexible manufacturing systems (FMSs) are designed to produce a high variety of outputs at low cost. Computer-controlled changeover is engineered into the system. Instead of human hands changing machine settings, electronics and mechanics provide the interface between the computer and machinery. Flexible technologies allow for fast and inexpensive changeovers. The equipment can be programmed to move from one product setup to another product setup in nano-times.

Using FMS, design and machine software talk to each other. CAD software creates new design drawings. It runs tests on all important reliability and durability characteristics, such as fatigue strength. CAD communicates design specifications to software that translates, instructs, and controls the production machinery, that is, CAM. CAD and CAM work together to determine the feasibility of manufacturing the new design and suggest improved design alternatives. CAD/CAM-type technology is used to design and manufacture many different products such as semiconductors, automobile grills, and aircraft parts. John Deere has invested millions of dollars in the creation of CAD/CAM systems for the manufacture of tractors. Boeing used CAD/CAM (specifically, CATIA V5 from Dassault Systèmes) to design all of the parts of its 787 Dreamliner. CATIA facilitated real-time meeting between design teams located all over the world with sophisticated audiovisual capabilities. Design at Boeing has moved from the draftsman's table to the computer.

Coordination problems can arise that are entirely new to airline designers. Airbus mega jet A380 fell two years behind schedule when preassembled cabin wiring bundles made in Germany did not fit the plane being assembled in France.

This global coordination problem (acknowledged by CEO of EADS, Christian Streiff, in October 2007) was caused by incompatible versions of CATIA software in Germany and France. The flexibility concept joins computers and equipment of many other kinds, including assembly-line processes and office machines. It is also possible and often desirable to include human beings in the network. Flexibility can be applied to information systems—flexible information systems and to flexible office systems as well as to FMS.

Flexibility in the office leads to consideration of telecommuting. That name is not properly descriptive of the effort to cut down on office space by having office employees work some part of the time at home or at a satellite office nearer to their place of residence. Thus, in addition to saving money by having smaller offices, people do not have to spend hours in traffic commuting to work. There are other advantages including having quiet time to concentrate. However, there is the opposite effect of not being able to focus on work because of distractions at home. Also the telephone is not a substitute for eye-to-eye contact with coworkers. The design of the office plays a part as well. The "newsroom" or "open room" setting is preferred by only 27% of 600 workers interviewed about office productivity. Use the following link to learn more on this issue: http://www.baselinemag.com/careers/slideshows/top-killers-of-office-productivity/?kc=BLBLBEMNL07092013STR3&dni=67553988&rni=25694712

Flexibility in the factory setting can apply to offices with repetitive functions. The concept of repetitiveness has been defined in different ways. For example, there is exact duplication which is characteristic of the flow shop in which statistically homogeneous replicas are turned out (one after another). On the other hand, generic repetitiveness is the property of flexible flow shops that allow certain changes to be made such as (three color varieties: green or blue or red stripes—depending on orders on hand) without requiring new setups of the system. This capability in the factory, called mass customization, can apply to many service systems including offices and restaurants.

Mass customization capabilities to produce small numbers of many varieties include extensive application of systems thinking to integrate marketing needs and P/OM scheduling abilities. Using marketing forecasts, P/OM managers decide what to make, but the decisions are constrained by the FMS menu, which was predetermined at the initial planning stage.

The need to produce increased variety is market-driven. The transformation process has to be able to change over from making one thing to another, quickly and inexpensively. A great deal of effort goes into altering the transformation process, so it can deal with the goal of increased variety. Technology and methodology must enable nearly instant setups and changeovers from one model to another to satisfy market demands. The payoff will be increased productivity of the joint production–marketing system, as exemplified by customized jeans.

Levi Strauss has put the customer directly in touch with the factory. "Sales clerks at an original Levi's store can use a personal computer and the customer's vital statistics to create what amounts to a digital blue jeans blueprint. When transmitted electronically to a Levi's factory in Tennessee, this computer file instructs a robotic tailor to cut a bolt of denim precisely to the customer's measurements," see *New York Times* (November 8, 1994). Ten years after this reference, Levi Strauss' "personal pair service," which manufactures and delivers made-to-measure denims, was recognized as an aspect of "mass customization."

Hellriegel et al. (2005) wrote, "Perhaps the most significant contribution of advanced manufacturing technologies is that of mass customization—that is the ability to produce a wide variety of a product by using the same basic design and production equipment but making certain modifications to the demand of a broader market. For example, Levi Strauss has successfully used computer-assisted design systems to help design customized leather outfits and jeans for customers." As of September 2012, Levi appeared to back off from mass customization and started offering hand customization at their Meatpacking District store in New York City. This exemplifies the difficulties of achieving mass customization and the dangers of announcing its availability before testing the capability and the consequences.

There is a dedicated strategic effort to achieve competitiveness through flexibility in America. It combines the goals of leanness (speed) and flexibility and is known by various names including "agile enterprise, agile business architecture, and agile project management." See Agile Alliance (2008) at http://agile2008.agilealliance.org/press.html. These names are meant to emphasize the ability to react quickly and with flexibility. The agile organization is supposed to be alert and nimble, keen, and lithe. Bureaucratic organizations (as we currently know them) cannot qualify.

2.5.8 Global Competition: Year 2010 Plus—P/OM's Seventh Step

It is conjectured that in the future the transformation process will continue increasing in complexity and productivity. On a worldwide scale, a broad range of goods and services should be within the spending capabilities of many people living in developing countries. More management will be needed to plan and control such systems. A greater number of operations managers will be required with far fewer workers on the production line. The global village will be sharing services—such as education and healthcare—that are mutually rewarding. Onerous service tasks will be relegated to service robots. Hopefully, people will have time to spend their money as they wish. The input–output production transformation model will be internationalized. There will be global competition at every link in the supply chain. International sourcing, fabrication, assembly, distribution, and marketing will prevail. The costs of the inputs and the values of the outputs will be affected by dozens, if not hundreds, of different currencies. Managing currencies will be part of the transformation process. The euro has simplified currency management and provides a good model for other regional currencies.

Information systems will be based on international networks of computers. Global telecommunication systems will transmit conversations that are spoken in 80 different languages.

Translation will be accomplished by language-capable computers with voice-language recognition. Voice response in the appropriate language will be expected.

The systems approach will extend to all functions within the organization and all partners along the supply chain, including suppliers and customers. Production and operations will develop transformation processes that require great management skills while decreasing burdensome labor components. Substantial productivity increases will be obtained. There have been many unexpected turns in the road, and there will be more as each new decade passes.

IBM provides a good case in point. In October of 2012, Mr. Palmisano was replaced as CEO and Chairman of the Board by Virginia (Ginni) Rometty who had been in charge of IBM's Global Business Services. This continued IBM's dedication to the improvement of productivity in the globalization of services. IBM and others are building networks for delivering technological services that are similar to the interdependent manufacturing networks that have evolved. For example, the 451 parts of the iPod are made by companies all over the world, but not by Apple.

We do expect that mass customization (MC) will become more economically feasible as a result of P/OM learning how to do MC effectively. There can be no doubt that global competition will become increasingly pervasive. Technology is going to be the drum major (leader of the marching band).

Summary

Productivity is a critical business systems variable because it strongly impacts the bottom line and it involves strategic planning among all of the business functions. Ultimately, the responsibility for being productive falls on the shoulders of P/OM, and not on marketing, or finance, or the other business functions. There are many ways to measure productivity, including those used to describe the national economy. Measures can be made in terms of labor productivity, capital productivity, or both. For business units, the measures relate more directly to the kinds of activities the company does. These are the operational measures of productivity used by an organization. Productivity measures should be adjusted to capture the relevant system. This challenge can only be met by using the systems approach.

Bureaucracy inhibits flexibility and, therefore, constrains productivity improvement. Small firms are less bureaucratic than large ones and therefore, are more flexible and able to pursue initiatives for greater productivity. Research and development (R&D) organizations are related to P/OM and should be studied in terms of their productivity, which is usually better for smaller organizations. These are strategic issues to consider.

Productivity and price–demand elasticity are interdependent. The importance to P/OM and marketing of the elasticity relationship for both price and quality are explained. Then, economies of scale and the division of labor are linked to productivity. The chapter concludes with the history of the improvement of P/OM input–output transformations. The resultant stages of increased productivity are presented in seven steps.

Review Questions

1. Why is productivity a crucial element of strategic planning?
2. Why is productivity measurement vital to P/OM?
3. Why is productivity measurement vital to national government economists?
4. What is the importance of productivity measurement to marketing management?
5. What role does the systems approach play with respect to productivity measurement?
6. What is good and bad about bureaucracy with respect to productivity? In 1922, the German sociologist, Max Weber wrote that bureaucracy is the most efficient and rational way to organize human activity (economy and society). He considered bureaucracy as a major organizational advance over what preceded it. Why is it now considered an impediment?
7. How does the division of labor concept help a market research firm put together a report for a client?
8. What is the value of knowing about the six historical steps in the development of production theory? Is the suggested seventh step likely to have an impact on the future of operations management?
9. How does the concept of interchangeable parts apply to
   a. vacuum cleaners? b. jigsaw puzzles? c. flashlights?
10. What is meant by lean (or agile) production systems?
11. How does business going global create new problems or opportunities for P/OM?
12. What relationship connects productivity and price–demand elasticity?
13. What relationship connects productivity and quality–demand elasticity? Is price implicitly included?
14. What are economies of scale? How do they relate to productivity and the systems approach?
15. How does telecommuting relate to productivity?
16. What are the good and the bad attributes of the "newsroom" or "open room" office setting?
17. Explain why consideration of flexibility in the office leads to evaluation of telecommuting.
18. Discuss the pros and cons of telecommuting.

19. What are two kinds of modularity?
20. How does modular production raise productivity?
21. Company X considers the ratio: number of complaints per day/number of employees needed to handle those complaints, as a productivity measure that is best when it is large.
    - What explains the company's point of view?
    - To a systems thinker, there is something wrong with this point of view. What is amiss?
22. Collectors of vinyl records are paying substantial sums for poor recordings of songs that can be purchased with much higher quality in digital form at far lower prices.
    - What is the anomaly that drives this situation?
    - Can an entrepreneur profit from this effect?
    - Does this opportunity have anything to do with P/OM?

Problems

Data for Problems 1–6:

In a 1-year period, the Productive Components Corporation (known as PCCorp) has shipped units worth \$1,200,000 to its customers. It produced units worth \$250,000 for finished goods (*FG*) inventory. PCCorp has \$50,000 of work-in-process (WIP) units. During the same 1-year period of time, PCCorp had a labor bill of \$140,000. Its capital expenses for the year are calculated to be \$430,000. Materials were purchased costing \$530,000. Energy expenses were \$225,000, and miscellaneous expenses were estimated to be \$75,000.

1. Calculate the labor productivity for PCCorp with respect to units shipped. In place of fixed cost, use sales; set variable cost (*VC*) to zero; in place of selling price (*SP*), use labor cost. Breakeven volume in solution is equivalent to labor productivity.
2. Calculate the MFP composed of labor and capital for units shipped plus finished goods for PCCorp. In place of fixed cost, use sales + finished goods; set variable cost to zero; in place of selling price, use labor cost + capital expenses. Breakeven volume in solution is equal to MFP.
3. What is PCCorp's total productivity? Abbreviating, in place of *FC* use sales + FG + WIP; $VC = 0$; in place of *SP*, use total of all costs. BEV in solution is equal to total productivity.
4. What is PCCorp's capital productivity? For *FC*, use sales + FG + WIP; $VC = 0$; SP = capital expenses. BEV is capital productivity.
5. The value of the units shipped must be reduced because \$350,000 worth of them have been returned as defective. Of the finished goods units (*FG*), \$150,000 worth been returned as defective. All WIP units are within

tolerances. Rework on the defective units reduces their value by 70%. What is the cost of the quality problem that has surfaced? Discuss what this means in terms of total productivity. Show that total productivity has dropped significantly (see Problem 3).

6. Calculate the MFP composed of capital, materials, and energy consumed for the total reworked output, which consists of units shipped plus FG plus WIP.
7. Productivity measures the ratio output values to the cost of inputs. Good productivity is shown by high ratio values; improved productivity is shown by higher ratio values over time. Does the following equation capture the meaning of productivity?
   Productivity = $pV/(vc)V$, where, p is the price per unit; V the volume sold per year; vc the variable cost per unit.
8. Because V is in both the numerator and the denominator of the productivity equation,

   Productivity = $pV/(vc)V$.

Reduce the equation to: productivity = p/vc and interpret the results.

Is productivity well described by the ratio of price per unit to variable cost per unit?

Readings and References

Anupindi, R., S. Chopra, S.D. Deshmukh, J.A. Van Meighem, and E. Zemel. *Managing Business Process Flows.* Upper Saddle River, NJ: Prentice-Hall, 1999.

Baumol, W.J., S. Anne Blackman, and E.N. Wolff. *Productivity and American Leadership: The Long View.* Cambridge: MIT Press, 1991.

Beinhocker, E., I. Davis, and L. Mendonca. The 10 Trends You Have to Watch, *Harvard Business Review*, July–August 2009, pp. 1–6.

Biema, M. Van, and B.C. Greenwald. Managing Our Way to Higher Service-Sector Productivity, *Harvard Business Review*, July 1997.

Blackstone, J.H., and J.F. Cox. APICS Dictionary, 11e. APICS Educational and Research Foundation, 2004.

Boston Consulting Group [BCG]. On Experience was published as part of the Perspectives series by the BCG company in 1968, with a third printing and preface by Henderson, Bruce then president of BCG, in 1972.

Britton, A. Economic Commission for Europe Discussion Papers. Economic Growth in the Market Economies 1950–2000, vol. 1, No. 1, 1991. New York: United Nations (GE.91-23315).

Deming, W.E. *Out of the Crisis.* MIT Center for Advanced Engineering Study, Boston, MA, 1986.

Drucker, P.F. Emerging Theory of Manufacturing, *Harvard Business Review*, May–June 1990, pp. 94–102.

Drucker, P.F., Knowledge–Worker Productivity: The Biggest Challenge, *California Management Review*, Winter 1999, Vol. 41, No. 2, pp. 79–94.

Fabricant, S. *A Primer on Productivity*. New York: Random House, 1969, p. 3. This definition adds "in a period of time (*t*)" to Fabricant's definition. He is considered to be the "father of productivity measurement."

Forrester, J.W. *Changing Economic Patterns*. Technology Review, August/September, 1978.

Hayes, R.H., and G.P. Pisano. Beyond World-Class: The New Manufacturing Strategy, *Harvard Business Review*, January–February 1994, pp. 77–86.

Hellriegel, D., S.E. Jackson, and J.W. Slocum Jr. Management: A Competency-Based Approach, 10th edition. Cincinnati, OH: Thomson South-Western Publishing Co., 2005.

Juran, J.M., and F.M. Gryna Jr. *Quality Planning and Analysis*, 2nd edition. New York: McGraw-Hill, 1980.

Kelley, R.E. and J. Caplan. How Bell Labs Creates Star Performers, *Harvard Business Review*, July 1993.

Kumar, N. Strategies to Fight Low Cost Rivals, *Harvard Business Review*, December 2006, pp. 1–9.

Ledingham, D., M. Kovac, and H.L. Simon, New Science of Sales Force Productivity, *Harvard Business Review*, September 2006, pp. 1–9.

Levitt, T, Production Line Approach to Service, *Harvard Business Review*, September–October 1972, pp. 2–11.

Merrifield R., J. Calhoun, and D. Stevens, The Next Revolution in Productivity, *Harvard Business Review*, June 2008, pp. 1–10.

Moser, K. *Mass Customization Strategies—Development of a Competence-Based Framework for Identifying Different Mass Customization Strategies*. Lulu.com, January 2007.

Ohno, T. *Toyota Production System: Beyond Management of Large-Scale Production*. Japan: Diamond Publishing Co. Ltd., London, UK, 1978.

Servan-Schreiber, J.-J. *The American Challenge*. Atheneum, New York, NY, 1968.

Sherman, H.D. Improving the Productivity of Service Businesses, *Sloan Management Review*, Spring 1984, Vol. 25, No. 3, pp. 1–10.

Shewhart, W.A. *Statistical Method from the Viewpoint of Quality Control*. Washington, DC: The Department of Agriculture, 1939. This is a classic in the P/OM field.

Skinner, W. Manufacturing—Missing Link in Corporate Strategy, *Harvard Business Review*, May–June 1969, pp. 113–121.

Skinner, W. The Focused Factory, *Harvard Business Review*, May–June 1974, pp. 113–121.

Smith, A. *An Enquiry into the Nature and Causes of the Wealth of Nations*, 1776 (a five-volume treatise).

Starr, M.K. Modular Production—A New Concept, *Harvard Business Review*, November–December 1965, Vol. 43, No. 6, pp. 131–140.

Starr, M.K. *Modular Production: A 45-year-old Concept*, IJOPM, Emerald, Vol. 30, 2010.

USDL, The U.S. Department of Labor, Bureau of Labor Statistics, News. USDL 94-327 (July 7, 1994). The Bureau of Labor Statistics public database, LABSTAT, is available on the Internet.

USDL, The U.S. Department of Labor, Bureau of Labor Statistics, Press release. (February 1995). Available at http://data.bls.gov/servlet/SurveyOutputServlet; June 2003.

Williams, F.P., D.E. D'Souza, M.E. Rosenfeldt, and M. Kassaee, Manufacturing Strategy, Business Strategy and Firm Performance in a Mature Industry, *Journal of Operations Management*, Vol. 13, 1995, pp. 19–33.

Chapter 3

Workload Assessment (Forecasting)

Readers' Choice: Forecasting methodology is a system of methods

Andrews, B.H., and Cunningham, S.M., L.L. Bean Improves Call-center Forecasting, *Interfaces*, 25(6), 1995, p. 1. The authors in this paper have developed models to forecast incoming calls to L.L. Bean's call center to prepare staffing schedules for telephone agents.

Basu, S., and Schroeder, R.G., Incorporating Judgments in Sales Forecasts—Application of the Delphi Method at American Hoist & Derrick, *Interfaces*, 7(3), 1977, p. 18. In this paper, the authors discuss the Delphi method for sales forecasting at the American hoist and derrick company.

Denton, B., Fitts Industries Enhances Forecasting and Production; Forecasting, *Interfaces*, 38(4), 2008, pages 338. This paper describes a forecasting model for Fitts Industries Inc. that manufactures a wide variety (2000 stock-keeping units) of wooden stair components.

Fader, P.S., and Hardie, B.G.S., Forecasting Repeat Sales at CDNOW: A Case Study, *Interfaces*, 31(3), 2001. This paper presents a model to forecast medium-term aggregate CD purchasing behavior.

Fildes, R., and Goodwin, P., Against Your Better Judgment? How Organizations Can Improve Their Use of

Management Judgment in Forecasting, *Interfaces*, 37(6), 2007, p. 570. The authors present the results of a survey of 149 forecasters to assess the effectiveness of using judgment in forecasting.

Helmer, F.T., Oppermann, E.B., and Suver, J.D., Forecasting Nursing Staffing Requirements by Intensity-of-Care Level, *Interfaces*, 10(3), 1980, p. 50. The authors present a series of regression models to estimate the nurse staffing needed to meet patient requirements in a 220-bed hospital that is affected by factors such as total census, intensity-of-care levels, and type of ward.

Kallina, C., Development and Implementation of a Simple Short Range Forecasting Model—A Case Study, *Interfaces*, 8(3), 1978, p. 32. This paper presents a simple short-range forecasting model for American Can Company. The model features include identification of trend components in a time-series, backward moving averages, feedback adjustment, and good accuracy.

Sanders, N.R., and Manrodt, K.B., Forecasting Practices in US Corporations: Survey Results, *Interfaces*, 24(2), 1994, p. 92. The authors document the forecasting practices at 500 US corporations.

Forecasts are necessary to describe the future. Examples of forecasting include the number of patients in a hospital, students in a college, customers in a grocery store, cars to be manufactured, and so on. The demand forecasts set the agenda for how the entire company will use its people, commit its resources, call on outside suppliers, and plan its work schedules. Forecasts provide information to coordinate demands for products and services with supplies of resources that are required to meet the demands. As such, the forecast is the platform for future planning. This is reactive planning. A good strategy also aims at modifying the forecast to influence what the future might bring rather than just accepting the forecast as an inevitable truth. In this way, management earns its rewards by better fitting production capabilities (short and long term) to marketing possibilities. This is proactive planning.

After reading this chapter, you should be able to:

- Describe the importance of forecasting.
- Explain various components of a time series.
- Choose an appropriate forecasting model.
- Perform regression analysis.

- Identify cause–effect relationships.
- Analyze and evaluate forecasting errors.
- Use the Delphi method.
- Pool information for multiple forecasts.
- Describe product life-cycle stages.

3.1 Introduction

Forecasting is not the term generally used to talk about the composition of a random draw of cards or how a roulette wheel performs. Those are statistical phenomena for which probabilities are known. Las Vegas and Atlantic City have built their gambling casino profits on the laws of probability, but customers still have to come and play. Customer attendance is not a known probability. The casinos try to forecast attendance.

Forecasting is to foretell sales demand volume even though the probabilities have never been formally studied. Business people often use their sense of what is happening to reach decisions that might be better made if someone had kept a record of what had taken place already. There is often some empirical basis for estimating what is likely to happen in the future.

Marketing models for predicting sales (lacking a contract) deal with levels of uncertainty that make forecasts of demand volumes, market shares, and revenues difficult, but not irrational. One of the best sources of information about the future is the past. For new products, there is no past, and so other methods can be tried.

The focus here is on using existing data to develop forecasts. The Rivet and Nail Factory has to forecast sales of products to develop departmental schedules for the next production period. The Mail Order Company has to forecast demand in order to have the right number of trained agents and operators in place. Ford Motor Company has to forecast car sales so that dealer stocks are of reasonable size for every model.

In every sales forecasting situation, the volatility of demand will determine how likely it is that a good forecast can be made. Stable patterns that persist for a long period of time make company forecasters confident that a credible job of forecasting can be done. Shaky estimates make company forecasters uneasy, so they search for new factors to correlate with the demand system.

Sales patterns are becoming less stable with increasing competition and information. Food sales that once were stable are now affected by medical reports about the food's effect on health. Auto and home sales are among many products that are strongly influenced by interest rates, which fluctuate more in a global environment. Exports and imports are sales that move around the globe in response to currency fluctuations that are increasingly unstable. Consequently, the best possible, economically feasible forecasting methods need to be used by the companies that are affected by increased volatility.

How well can one forecast the future? The answer will depend on the stability of the pattern of the time series for the events being studied. The underlying pattern can be hard to find, but not impossible. Even if a pattern is found, the question remains, how long will it persist? When will it change? Those willing to forecast should accept the challenge.

Mathematical equations are used for various kinds of forecasting. It is important to stress that equations do not make forecasts "the truth." Also, a great deal of good forecasting can be done without mathematics. Further, with or without mathematics, no forecast is ever guaranteed.

3.2 Time Series and Extrapolation

A time series is a stream of data that represents the past measurements. Each event (observation of demand) is time-tagged so that it is known where it is located in the series of data. The time series consists of data recorded at different time periods such as weekly or daily for the variable, which could be units produced or demands received. Forecasters attempt to predict the next value or set of values that will occur at a future time. In time-series analysis, external causes are not brought into the picture. The pattern of the series is considered to be time-dependent. That is why the definition of American Production and Inventory Control Society (APICS) states "the values of the variables are functions of the time periods." APICS now lists itself as The Association for Operations Management. Extrapolation is the process of moving from observed data (past and present) to the unknown values of future points. The extrapolation of time series is one of the main functions of forecasting.

The data in time series may consist of several different kinds of variations. Important among them are random variations, an increasing or decreasing trend, and seasonal variations. Random variations (see Figure 3.1) occur because the demand is seldom constant at a given level. Minor variations do occur from one

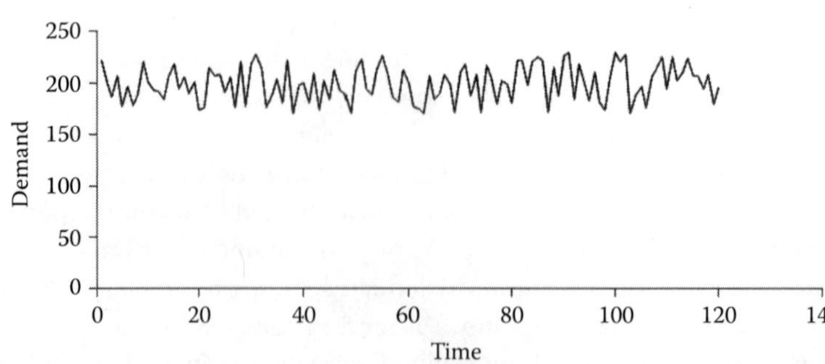

Figure 3.1 Time series with random variations.

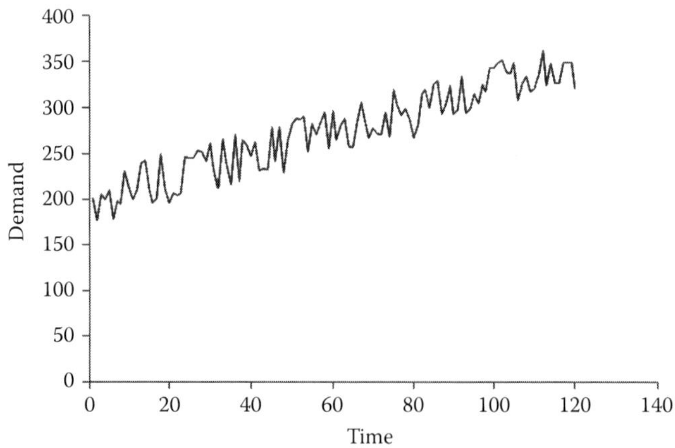

Figure 3.2 Time series with random variations and increasing trend.

period to another. In many instances, there are no specific assignable causes for these random variations. The random variations are a result of the economic environment and the marketplace within which an organization is operating.

In addition to random variations, the time-series data may exhibit an increasing or decreasing trend (see Figure 3.2 for an increasing trend and Figure 3.3 for a decreasing trend). If the time series shows an increasing linear trend, as in Figure 3.3, there is a constant rate of change (increasing) with the increase in time. Linear decreases can also occur. Nonlinear trend lines occur where the rate of change is

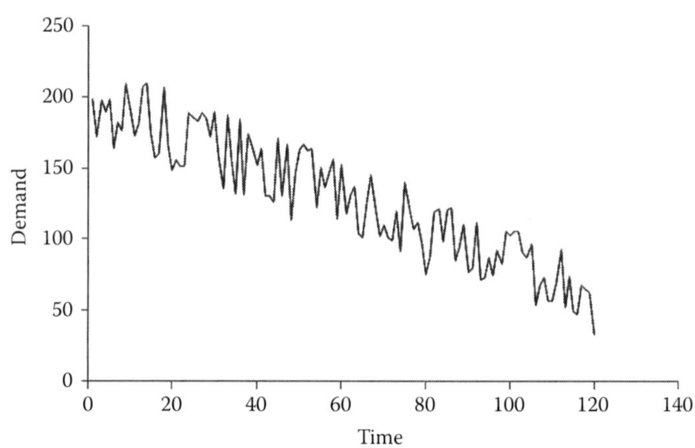

Figure 3.3 Time series with random variations and decreasing trend.

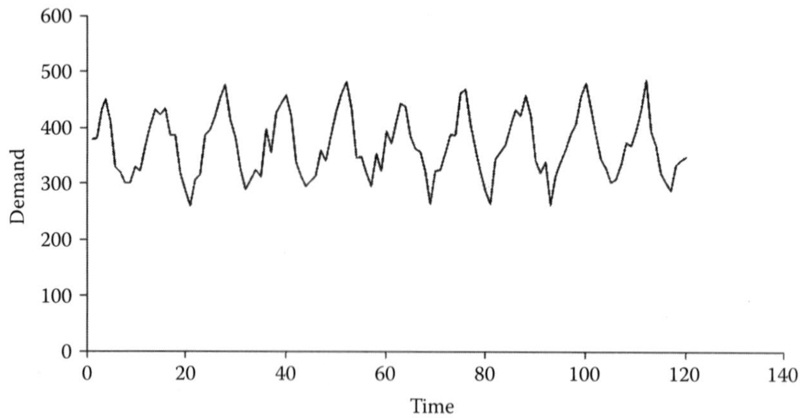

Figure 3.4 Time series with random variations and seasonal variations.

geometric. In such cases, it is possible to extrapolate the curve by eye, but the rate of change can also be calculated, and projections can be made mathematically.

The time series may also exhibit seasonal (or cyclical) variations. Figure 3.4 shows seasonal variations coupled with random variations, whereas Figure 3.5 combines all three components—random variations, an increasing trend, and seasonal variation. For example, one may observe seasonal variations in demand for resort hotels and home heating oil.

There are other categories of time series that are useful to know about. For example, step functions that look like stairs—going up or going down might also be present in a time series. It is hoped that the forecast can predict when the next

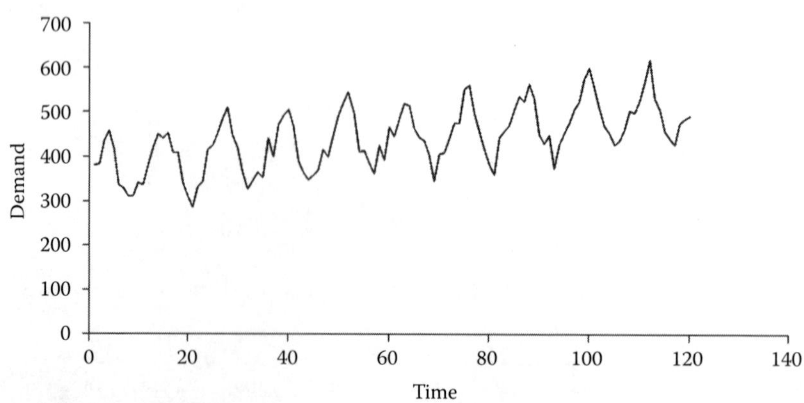

Figure 3.5 Time series with random variations, seasonal variations, and increasing trend.

step will occur and how far up or down it goes. If they occur regularly, a good estimate can be made of the period between steps and the timing of future steps. Step functions are characteristic of systems that can only change in given quantities. Thus, if sales must be made in lots of 100 units, then each change in demand will occur in hundred-unit steps. If December sales were 500, the sales for January could be 400, 500, or 600, that is, it could step up or down, or stay the same as the December sales of 500.

Time-series data can also reflect erratic bursts called impulses. If these spikes appear from time to time and some pattern can be associated with them, they could be added to the list of what can be extrapolated or projected into the future.

Short-term cycles can be piggybacked on long-term cycles. All kinds of combinations are possible. The key point is that cycles, trends, and steps are the basic pallet for the development of forecasting models.

The term "time-series analysis" deserves explanation. Time-series analysis is "analysis of any variable classified by time, in which the values of the variable are functions of the time periods." Time-series analysis deals with using knowledge about cycles, trends, and averages to forecast future events. It is always useful to try to find causal links between well-known cycles, such as the seasons, and demand patterns that are being tracked. Using symbols, a forecast is to be made from some series of numbers $x_1, x_2, \ldots, x_n$. The numbers might be monthly sales figures for the past year. The question is: What use can be made of these numbers to indicate the monthly sales figures for next year? How will the forecasted time series be used? Production schedules will be drawn up to satisfy demand. Orders will be placed for inventory to be purchased. Workers will be hired and trained, or let go as the production crew is downsized. Rented space will be increased or decreased, or it will stay the same. Many decisions that are interrelated require believable forecasts before they can be made.

Time-series analysis uses statistical methods, including trend and cycle analyses, to predict future values based on the history of sales or whatever the time series describes. In the following sections, we will discuss moving average (MA), weighted moving average (WMA), exponential smoothing (ES), and trend analysis methods for forecasting and extrapolation. How to make forecasts when seasonal variations are present is also discussed.

3.3 Forecasting Methods for Time-Series Analysis

We will study the following techniques for time-series analysis in this section:

- Moving average
- Weighted moving average
- Exponential smoothing
- Seasonal forecasting
- Trend analysis

3.3.1 Moving Average

The moving average (MA) method supplies a forecast of future values based on recent past history. MA is also called simple MA method. The latest n consecutive values, which are observations of actual events such as daily, weekly, monthly, or yearly demand, are used in making a forecast. These data are recorded and must be updated to maintain the most recent n values. With the passing of each time period, the most recent value is stored and the value of the earliest period is dropped off. For example, if you want to forecast demand for April using $n = 3$, then the demands for last three months, that is, January, February, and March, are needed. The forecast for April can then be calculated as

$$\text{Forecast for April} = \frac{\text{January demand} + \text{February demand} + \text{March demand}}{3}.$$

Example: Consider the demand (sales) data given in Table 3.1 for 10 periods (months in this example). The forecast for months 4–10 are given in the table. The numbers used in the formula for each month are given in the calculation column. For example, the forecast for the fourth month is

Table 3.1 An Example of the MA Method (*n* = 3)

Month	*Sales*	*Forecast*	*Calculation*
1	100		
2	80		
3	90		
4	110	90.00	= (100 + 80 + 90)/3
5	100	93.33	= (80 + 90 + 110)/3
6	110	100.00	= (90 + 110 + 100)/3
7	95	106.67	= (110 + 100 + 110)/3
8	115	101.67	= (100 + 110 + 95)/3
9	120	106.67	= (110 + 95 + 115)/3
10	90	110.00	= (95 + 115 + 120)/3
11	105	108.33	= (115 + 120 + 90)/3
12	110	105.00	= (120 + 90 + 105)/3

$$\text{Forecast}(4) = \frac{\text{Demand}(1) + \text{Demand}(2) + \text{Demand}(3)}{3}$$
$$= \frac{100 + 80 + 90}{3} = 90.$$

> In general, an n-month moving average is the sum of the observed values during the past n months divided by n.

It may be noted that the forecast cannot be made for the first three months, if $n = 3$, since the data for the latest three months is needed. Similarly, if $n = 4$, then the forecast cannot be made for the first four periods.

The term "moving average" is used because as each new observation becomes available a new average can be computed and used as a forecast. No forecast can be made until there are as many historical observations as are needed (n) for the MA. Further the method of MA is used to forecast only one time period in advance because of nonavailability of actual values needed to forecast more than one period in advance.

A decision must be made concerning how many data values should be included in the MA set (n). Is it better to use two, three, four, or more periods? The quest is to find the optimal number of prior periods to include in the MA series. How far back to go depends on the speed with which the series changes and the recency of events that tend to determine the future. "Very recent" means few values; "not so recent" means more values. That is a workable rule of thumb. The MA method is used when the demand for a product is fairly constant over time and does not have a rapid growth rate or seasonal characteristics. The method is primarily useful in removing the random fluctuations in the historical data. When the magnitude of the trend is great, and the pattern is very consistent, then the fewer the number of periods (n) in the set the better it is. If the trend is slow (up or down), and if fluctuations around the average are common, then having more periods of time in the set is better than having too few. In selecting the value of n, there are many conflicting effects. The larger the value of n, the greater is the effect of smoothing any random variation. Smoother value may be desirable if there is a lot of randomness in the historical data or if there is little change in the underlying pattern. On the other hand, if the underlying pattern in the data is changing, that is, there is a trend in the data either increasing the demand or decreasing it, then a small value of n may be desirable if we want to react to fluctuations more rapidly. In such cases, the MA based on a large value of n has the adverse characteristic of lagging the trend. With a short value of n, the follow up of the trend will be rapid. A shorter value of n may also be used where there is little randomness in the observed data. The large and small values of n vary around 36 and 6, respectively.

It may, however, be kept in view that a large value of n requires a larger data storage and when we are dealing with a large number of products, a large value of n will make the cost of data handling exorbitant. In such cases, we might choose a small value of n, even if we have to sacrifice some accuracy.

> Moving averages might be used to extrapolate next events if the following occurs: there is no discernible cyclical pattern, and the system appears to be generating a series of values such that the last set of values provides the best estimate of what will be the next value.

To find out the value of n that should be chosen we can calculate the errors for different values of n and the n which gives the least error may then be chosen. See more details in the section on error analysis. The value of n is specified by the manager and generally remains constant until there are reasons indicating a change in the time-series pattern. If the underlying pattern of the time series changes the current value of n may not give good forecast.

3.3.2 Weighted Moving Average

A way to make forecasts more responsive to the most recent actual occurrences (demand) is to use the weighted moving average (WMA) method. Just like the MA method, the most recent n period are used in forecasting. However, each period is assigned a weight between 0 and 1. The total of all weights adds up to 1. The highest weight is assigned to the most recent period and then the weights are assigned to the previous periods in the descending order of magnitude.

Example: Consider the data given in Table 3.2. Suppose the number of periods used in forecasting $n = 3$ and the weights are 0.2, 0.3, and 0.5. The highest weight (in this case 0.5) is assigned to the most recent period. For example, the forecast for period 4 will be calculated by using the weight 0.5 for period 3, 0.3 for period 2, and 0.2 for period 1. The forecast for period 4 will then be

$$\text{Forecast}(4) = 0.2 * (\text{Demand1}) + 0.3 * (\text{Demand2}) + 0.5 * (\text{Demand3}),$$
$$\text{Forecast}(4) = 0.2 * 100 + 0.3 * 80 + 0.5 * 90 = 89.$$

The demand forecast for all periods 4–10 are shown in Table 3.2. The comment column includes the values of the demand and the corresponding weights.

The WMA method allows the latest period to have the greatest impact, and later periods decreasing importance. Note that when all weights are equal ($1/n$), the WMA is the same as the MA.

The biggest weights are assigned to the most recent events when there is a continuing trend. In a rapidly changing system, w_t may be much greater ($\gg$) than w_{t-1}

Table 3.2 An Example of the WMA Method ($n = 3$)

Month	*Sales*	*Forecast*	*Calculation*	*Weights = 0.2, 0.3, 0.5*
1	100			
2	80			
3	90			
4	110	89.00	= 0.2 * 100 + 0.3 * 80 + 0.5 * 90	
5	100	98.00	= 0.2 * 80 + 0.3 * 90 + 0.5 * 110	
6	110	101.00	= 0.2 * 90 + 0.3 * 110 + 0.5 * 100	
7	95	107.00	= 0.2 * 110 + 0.3 * 100 + 0.5 * 110	
8	115	100.50	= 0.2 * 100 + 0.3 * 110 + 0.5 * 95	
9	120	108.00	= 0.2 * 110 + 0.3 * 95 + 0.5 * 115	
10	90	113.50	= 0.2 * 95 + 0.3 * 115 + 0.5 * 120	
11	105	104.00	= 0.2 * 115 + 0.3 * 120 + 0.5 * 90	
12	110	103.50	= 0.2 * 120 + 0.3 * 90 + 0.5 * 105	

and $w_t \gg w_{t-1} \gg w_{t-2},\ldots$, etc., where w_t is the weight for period t. This is equivalent to reducing the size of n, which was discussed in the section on MAs. Thus, WMAs can track strong trends more accurately than un-weighted moving averages. An error analysis using different values of n and different weights has to be done to find the best value of n and the corresponding weights. See section on error analysis.

3.3.3 Exponential Smoothing

The exponential smoothing (ES) method, like the WMA method, calculates an average demand (forecast). ES methodology remembers the last estimate of the average value of demand and combines it with the most recent observed, actual value to form a new estimated average. ES forecasts the demand for a given period t by combining the forecast of the previous period ($t-1$) and the actual demand of the previous period ($t-1$). The actual demand for the previous period is given a weight of α and the forecast of the prior period is given a weight of $(1-\alpha)$, where α is a smoothing constant whose value lies between 0 and 1. The equation for the forecast for period t is

$$\text{Forecast}(t) = \alpha * \text{Actual demand}(t-1) + (1-\alpha) * \text{Forecast}(t-1),$$
$$0 \le \alpha \le 1.$$

By rearranging the terms, this equation can also be written as

$$\text{Forecast}(t) = \text{Forecast}(t-1) + \alpha * \{\text{Actual Demand }(t-1) - \text{Forecast}(t-1)\}, \quad 0 \le \alpha \le 1.$$

Example: Consider the data given in Table 3.3. The ES method makes a forecast starting from period 2. The forecast for the first period is generally set equal to the actual demand in that period to get the forecasting process started. Forecasts for all periods 2–10 are given in Table 3.3. For example, the forecast calculations for periods 2 and 3 are shown below:

$$F(2) = F(1) + \alpha * \{A(1) - F(1)\} = 100 + 0.2 * (100 - 100) = 100,$$

$$F(3) = F(2) + \alpha * \{A(2) - F(2)\} = 100 + 0.2 * (80 - 100) = 96.$$

ES is a simpler method, requiring fewer calculations than WMA, which needs n weights and n periods of data for each forecast estimate. ES needs only three pieces of data. Also, it can be more effective because only one weight, alpha (α), has

Table 3.3 An Example of the ES Method ($\alpha = 0.2$)

Month	*Sales*	*Forecast*	*Comment and Calculation*
1	100	100	Forecast for period 1 should be available before starting the calculations. If it is not given, then set it equal to the sales of period 1
2		100.00	= (100 + 0.2(100 – 100))
3	90	96.00	= (100 + 0.2(80 – 100))
4	110	94.80	= (96 + 0.2(90 – 96))
5	100	97.84	= (94.8 + 0.2(110 – 94.8))
6	110	98.27	= (97.84 + 0.2(100 – 97.84))
7	95	100.62	= (98.27 + 0.2(110 – 98.27))
8	115	99.50	= (100.62 + 0.2(95 – 100.62))
9	120	102.60	= (99.5 + 0.2(115 – 99.5))
10	90	106.08	= (102.6 + 0.2(120 – 102.6))
11	105	102.86	= (106.08 + 0.2(90 – 106.08))
12	110	103.29	= (102.86 + 0.2(105 – 102.86))

to be chosen. This makes it easier to experiment with past data to see which value of α provides the least forecast error. Often, the use of the ES method for forecasting is preferred to the WMA method.

The response rate of the ES model is a function of the α value that is used. The response rate of a forecasting system is the speed with which the forecasting system makes adjustments to the forecasts if the data in the series show an upward or downward trend. A review of the two equations stated above provides a perspective on the response rate. Consider the first equation, Forecast $(t) = \alpha *$ Actual demand $(t-1) + (1-\alpha) *$ Forecast $(t-1)$. When α is large, the actual demand in the prior period is given great weight in the forecast for the next period. If α is 1, the forecast for the next period is the actual value of the prior period (same as $n = 1$ in MA). The forecast results are markedly affected by the α that is used.

The second equation, Forecast (t) = Forecast $(t-1) + \alpha *$ {Actual demand $(t-1)$–Forecast $(t-1)$}, provides a different perspective on the ES technique. The forecast in period t is equal to the forecast in the previous period plus a fraction (specified by α) of the error in the previous period. The error in the previous period is the difference between that actual demand and the forecast, which can be positive or negative. If α is close to 0, the new forecast does not show much adjustment for the error, and in the extreme case, if $\alpha = 0$, the new forecast is equal to the old forecast with no adjustments.

Small values of α (which will be analogous to large values of n in MA) are used for stable systems where there is, at most, a minimum amount of random fluctuation. Large values of α are used for changing and evolving systems where much reliance is placed on the last observation. New products, as they move through their life-cycle stages, start with a large α value, which gradually diminishes as the product enters its maturation stage. For most production scheduling systems in both the job shop and the flow shop, α is kept small, in the neighborhood of 0.050–0.150, to decrease the system's response to random fluctuations.

The ES method has a clear advantage over the MA method in terms of the amount of memory required for storing the data. In addition to the smoothing constant α, at any time we need to store only the exponentially smoothed average and the actual demand for the previous period. For the MA method, actual demands for the last n periods have to be stored.

The different values of α give different forecast just like different forecasts are obtained by changing the values of n in the MA method. A small value of α makes minor adjustments to the forecast results as discussed above—giving smoother forecasts. A smooth forecast is obtained in the MA method if n is large. Therefore, smaller values of α and large values of n tend to give similar forecasts. Similarly, large values of α and small values of n tend to give similar results. In fact, when $\alpha = 1$, or $n = 1$, the forecast in a given period is equal to the demand in the previous period and the forecasting system is extremely responsive to changes in the demand. In this case, the forecast fluctuates as actual demand fluctuates. However, forecasts are lagging behind the demand by one period.

If sudden change occurs, then the response will be sluggish irrespective of whether change is genuine or noise is present when α is small. On the other hand, if α is close to 1, there will be substantial adjustment for any error. With a large value of α, the bona fide changes will be reflected immediately in the new average.

It is difficult to say beforehand which value of α will serve the purpose best. Different values of α starting with a small value may be chosen to make a forecast. The actual observed data and the forecast using different values of α may be plotted on a graph and by examination that value of α which tracks the actual demand best is chosen. Alternatively, for different values of α, the errors—mean absolute deviation (MAD) and mean squared error—may be calculated and that value of α which gives a lower error may be chosen. See appropriate details in the section ahead on forecast errors.

As in the case of MAs, if there is a fundamental systems change, then prior history of patterns is not a help, and ES should be applied as if to a new startup system for the first time.

ES has been found to be very effective in a variety of situations. Many forecasting and control systems employ ES because it works better than the older methods of MAs and WMAs. It has proved effective for diverse applications. Fighter aircraft use ES to aim their guns at moving targets. In effect, they forecast the location of enemy jets during flying missions. This application shows how fast the ES method can track a consistent, but dynamically changing pattern. Manufacturers use ES to forecast demand levels, which experience the same kind of nonrandom but volatile shifts from time to time. In such cases, the recent past has the most information about the near future. ES can catch these shifts and make rapid adjustments to inventory levels. Other manufacturers and service organizations use it because it requires less computational work and is readily understood.

3.3.4 Forecasting with a Seasonal Cycle

Forecasting when seasonality is present can be done by assuming that what happened last year (or last month, etc.) will happen again. We call it the historical forecast method. The pattern is expected to repeat itself within the time period. This method works if a stable pattern (which is often seasonal) exists. The hotel and resort business is typically involved with historical forecasts, as is the agriculture business. In each case, special circumstances can arise that lead to the desire to modify the historical forecast. In the case of hotels and resorts, the state of the economy can modify the ups and downs of occupancy rates. Agricultural pursuits are modified by rainfalls and temperature fluctuations (see modifications to the historical forecast in the sections that follow).

When last year's monthly sales values are used to predict the next year's monthly sales values, the method is based on using past history to forecast. The same concept applies to weekly or daily sales, which can be forecast on a semiannual, quarterly,

Table 3.4 An Example of Seasonal Cycles (Historical Forecast with Seasonal Data)

Month	*Actual Sales Last Year*	*Forecast of Sales Next Year*
January	1500	1500
February	1600	1600
March	1800	1800
April	2000	2000
May	2300	2300
June	2500	2500
July	2350	2350
August	2100	2100
September	1850	1850
October	1650	1650
November	1550	1550
December	1400	1400

or monthly basis. Table 3.4 shows the historical forecast technique applied to data that exhibit a seasonal cycle. No modification has been made for an overall change in annual sales, which is the basis of the historical forecast.

When cycles are stable, they can provide insights that are very important for P/OM tactical planning. When they work, historical cycles allow P/OM to excel at capacity planning and production scheduling for mature manufactured products and services. The historical forecast is not appropriate for a new product introduction unless there is similarity to some other product that is already on the market.

Historical seasonal cycles can become apparent by studying the calendar. In the case of P/OM, there are also other cyclical regularities and patterns that lend themselves to capacity planning, production scheduling, and purchasing decisions for both manufacturers and service systems.

If the time-series pattern remains fixed, but the demand level has increased overall, then a base series modification can be used. Assume that in 2013 the quarterly demands were 10, 30, 20, and 40. This gives a yearly demand of 100 units. Further, assume that in 2014 the yearly demand is expected to increase to 120 units. Then the quarterly forecasts would be adjusted:

Forecast for 2014

Quarter 1: 120 (10/100) = 12

Quarter 2: 120 (30/100) = 36

Quarter 3: 120 (20/100) = 24

Quarter 4: 120 (40/100) = 48

The adjusted quarterly demands total 120 units. The cyclical patterns are matched. In the year 2015, assuming the pattern continues and the base level increases to 150, the time series would be 15, 45, 30, and 60 with the sum of 150.

The above reasoning can be extended to those situations where the demand data exist for several years with similar demand patterns.

Consider the data in Table 3.5 that gives the quarterly demand for last 4 years. There is a clear pattern of demand variation among the four quarters. Based on these data, we can forecast the demand for the four quarters of year 5. Assume that the annual demand is expected to be, 2800, in year 5. This demand has to be estimated using other techniques. The model discussed in this section will divide the yearly demand into quarterly demands based on the past quarterly demand pattern. The following five-step process will be used to obtain the quarterly demand.

Step 1: Find average quarterly demand for each year.

The total demand for each year is divided by four (number of quarters in each year) to obtain the average demand in each year. Table 3.5 gives the average demand for each quarter.

Step 2: Compute seasonal index (SI) for each quarter for each year.

The seasonal index for a quarter for a given year is obtained by dividing the demand in that quarter by the average quarterly demand for that year. For

Table 3.5 Demand

Quarter	*Year 1*	*Year 2*	*Year 3*	*Year 4*
Fall	2530	2690	2790	2860
Winter	2300	2420	2410	2600
Spring	1900	2000	2105	2175
Summer	1510	1775	1875	1945
Average	2060	2221	2295	2395
Calculation	= (1510 + 1900 + 2300 + 2530)/4)	= (1775 + 2000 + 2420 + 2690)/4)	= (1875 + 2105 + 2410 + 2790)/4)	= (1945 + 2175 + 2600 + 2860)/4)

Table 3.6 Step 2: Compute Seasonal Index (SI) for Each Quarter for Each Year

Quarter	*Year 1*	*Calculation for SI Year 1*	*Year 2*	*Calculation for SI Year 2*	*Year 3*	*Calculation for SI Year 3*	*Year 4*	*Calculation for SI Year 4*
Fall	1.228	= 2530/2060	1.211	= 2690/2221	1.216	= 2790/2295	1.194	= 2860/2395
Winter	1.117	= 2300/2060	1.090	= 2420/2221	1.050	= 2410/2295	1.086	= 2600/2395
Spring	0.922	= 1900/2060	0.900	= 2000/2221	0.917	= 2105/2295	0.908	= 2175/2395
Summer	0.733	= 1510/2060	0.799	= 1775/2221	0.817	= 1875/2295	0.812	= 1945/2395

example, the seasonal index for the winter quarter of year 2 is 1.090 which is obtained by dividing 2420 (demand for the winter quarter in year 2) by 2221 (average quarterly demand for year 2). The seasonal indices for each quarter of each year are given in Table 3.6. The formulas to calculate the seasonal indices are also given in this table.

Step 3: Calculate the average SI for each quarter.

The next step is to find the average seasonal index for each quarter which is simply the average of the seasonal indices calculated in step 2. Table 3.7 gives the average seasonal index for each quarter. For example,

$$\text{Average SI for the summer quarter} = 0.790 = \frac{0.733 + 0.799 + 0.817 + 0.812}{4}.$$

Step 4: Calculate the average quarterly demand for next year.

In this step, we find the average quarterly demand for the next year. The yearly demand estimate for next year is 2800. Therefore, the average quarterly demand = 2800/4 = 700. This step is shown in Table 3.8.

Step 5: Forecast demand for quarters of next year.

The quarterly demand for each quarter of next year is obtained by multiplying the average demand by the SI for each quarter. Table 3.9 gives the forecast

Table 3.7 Step 3: Calculate the Average SI for Each Quarter

Quarter	*Average SI*	*Calculation*
Fall	1.212	= (1.228 + 1.211 +1.216 + 1.194) /4
Winter	1.086	= (1.117 + 1.09 +1.05 + 1.086) /4
Spring	0.912	= (0.922 + 0.9 +0.917 + 0.908) /4
Summer	0.790	= (0.733 + 0.799 +0.817 + 0.812) /4

Table 3.8 Step 4: Calculating Average Demand per Quarter for Next Year

Total demand for next year[a]	2800
Average demand per quarter	700

[a] The total demand for next year is given or determined by using an appropriate forecasting method.

Table 3.9 Step 5: Forecast Demand for Quarters of Next Year by Multiplying the Average Demand by SI for Each Quarter

Quarter	*Year 5*	*Calculation*
Fall	848	= (700 * 1.212)
Winter	760	= (700 * 1.086)
Spring	638	= (700 * 0.912)
Summer	553	= (700 * 0.79)

of the quarterly demand of next year. For example, forecast for the Spring quarter = 638 = 700 * 0.912, where 700 is the average quarterly demand and 0.912 is the average SI for the Spring quarter.

In the above example, we have used year as the time period in which the seasons are represented by quarters. The seasons could have been represented by months if the time series shows variations by month. In some instances, the week could be the time period and the days could be the "seasons" if the demand fluctuates on a daily basis. Restaurants are a good example of where the demand fluctuates on a daily basis. Even hour of the day can be the "season" if the demand fluctuates hourly in a given day, for example, the power generation requirement in a power utility or the number of calls in a telephone company vary on an hourly basis.

3.3.5 Trend Analysis

If the time series exhibits an increasing or decreasing trend, then the techniques discussed above (MA, WMA, and ES) may not be appropriate for making a forecast. We perform a trend analysis to make a forecast in this case. Consider the graphs given in Figure 3.6. The zigzag line shows the demand for 10 periods. There is clearly an increasing trend. The trend analysis will fit a trend line through these data to make forecast. The equation of the trend line is, $Y = a + bX$, where Y is the demand forecast and X is the time period. X is the independent variable and Y is the dependent variable since the demand depends on the time period. As X increases Y increases in an increasing trend. Y will decrease as X increases in a decreasing trend.

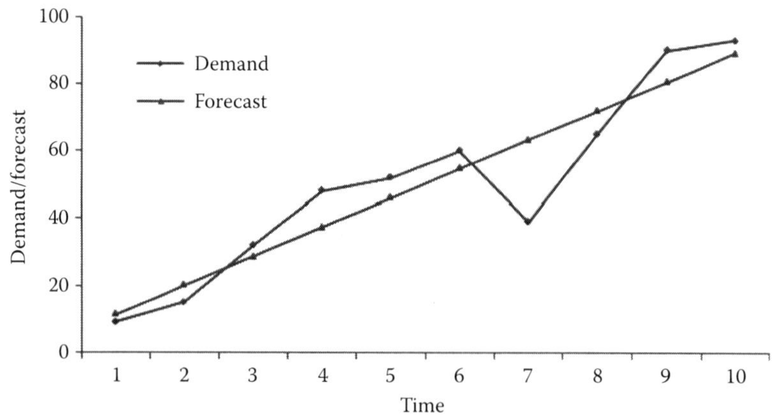

Figure 3.6 Time-series analysis.

In the trend line equation, *a* is the intercept on the *Y*-axis which is the value of the demand (variable *Y*) when $X = 0$. The slope of the line is represented by *b* which gives the change in the value of demand (variable *Y*) for a unit change in the value of *X*. That is, *b* is the amount by which demand will change if the time period changes by 1.

There are mathematical formulas for finding the values of *a* and *b*. However, we omit listing those formulae because of their complexity and also because the Excel computer program has built-in functions to find the values of *a* and *b*. The "intercept" function in Excel calculates *a* and the "slope" function is used to find the value of *b*. Excel functions are employed in this text to find the values of *a* and *b*.

Example: Consider the demand data for 10 periods given in Table 3.10. The Excel function gives $b = 8.65$ and $a = 2.73$. The forecast can now be made for any time period using these numbers. For example, the forecast for period 11 ($X = 11$) will be $F(11) = 2.73 + 11 * 8.65 = 97.87$. Similarly, the forecast for period 12 will be $F(12) = 2.73 + 12 * 8.65 = 106.52$. The forecast can be made for any future period.

The straight line in Figure 3.6 gives the forecast for periods 1–10 using the calculated values of *a* and *b*. For any given time period, the difference between the forecast (values on straight line) and the actual demand (values on zigzag line) gives the error in that period. The trend analysis method minimizes the sum of the squares of these errors in calculating the values of *a* and *b*.

Table 3.10 Data for Time-Series Analysis

Time: Independent Variable (x)	1	2	3	4	5	6	7	8	9	10
Demand: Dependent Variable (y)	9	15	32	48	52	60	39	65	90	93

3.4 Regression Analysis

Another valuable method that assists forecasting is regression analysis. This method is useful for establishing a relationship between two sets of numbers that are time series. For example, when a series of *Y* numbers (such as the monthly sales of cameras over a period of years) is causally connected with the series of *X* numbers (the monthly advertising budget), then it is beneficial to establish a relationship between *X* and *Y* in order to forecast *Y*. The first assumption that is generally made in regression analysis is that the relationship between the correlate pairs is linear. It is the easiest assumption to check. However, if nonlinear relations are hypothesized, there are strong, but more complex methods for doing nonlinear regression analyses. In regression analysis, *X* is the independent variable and *Y* is the dependent variable.

Example: Consider the data given in Table 3.11. This table gives 10 observations for the values of *X* (independent variable) and the corresponding values of *Y* (dependent variable). The data are plotted in Figure 3.7.

We can fit a straight line (called the regression line) to this data. This line will be an estimate of the way in which the value of *X* affects the value of *Y*. For this example, a linear trend is assumed. Once the line is determined, it can be used to extrapolate future demand—the value of *Y*. The equation of the regression line is

Table 3.11 Data for Regression Analysis

Observation number	1	2	3	4	5	6	7	8	9	10
Independent variable (x)	10	12	11	9	10	12	10	13	14	12
Dependent variable (y)	400	600	700	500	800	700	500	700	800	600

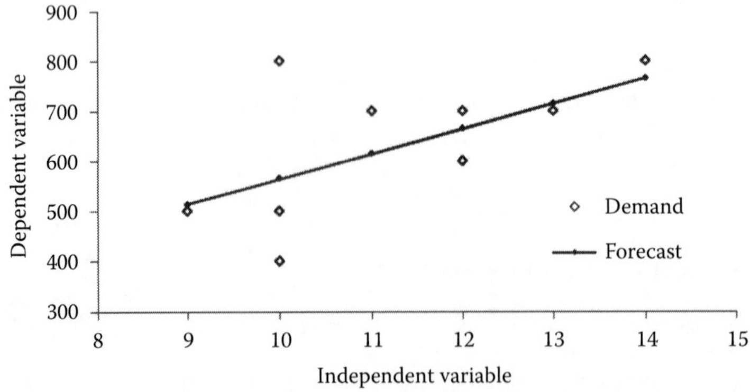

Figure 3.7 Regression analysis.

$Y = a + bX$. As in trend analysis, a is the intercept on the Y-axis which is the value of the demand (variable Y) when $X = 0$. The slope of the line is represented by b which gives the change in the value of demand (variable Y) for a unit change in the value of X. That is, b is the amount by which demand will change if the independent variable changes by one unit.

We can use the Excel program's built-in functions to find the values of a (intercept) and b (slope). The Excel function gives $a = 62.44$ and $b = 50.23$. A forecast can now be made for any value of X. suppose $X = 15$, then the forecast will be, Y (for $X = 15$) $= 62.44 + 50.23 * 15 = 815.84$.

3.5 Coefficients of Correlation and Determination

An important prerequisite to use regression analysis is the existence of a causal relationship between X and Y. Therefore, before we even determine the regression equation, it is important to find out whether there exists a relationship between X and Y. The scatter diagrams visually displays the relationships between X and Y. Scatter diagrams are useful visual aids to find whether there is a relationship between X and Y. Figures 3.8 and 3.9 are the scatter diagrams for two different data series. It is easy to see that there is a relationship between X and Y in Figure 3.8, whereas Figure 3.9 shows very little relationship. These relationships can also be established in quantitative terms by calculating the correlation coefficient. A correlation coefficient (r) shows the extent of correlation of X with Y, where r can take on values from −1 to +1. At −1, X and Y are perfectly correlated—going in opposite directions. As X gets large, Y gets small, and vice versa. When $r = 0$, there is no correlation and when $r = +1$, X and Y are perfectly correlated going in the same direction. The correlation coefficient can be found by using Excel's built-in function "Correl." For the series

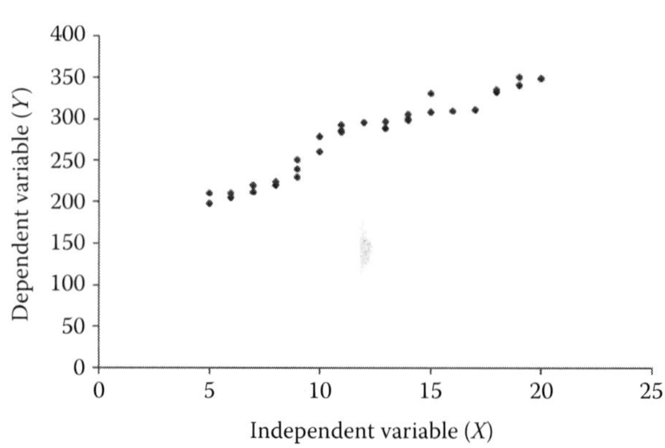

Figure 3.8 Scatter diagram with high coefficient of correlation.

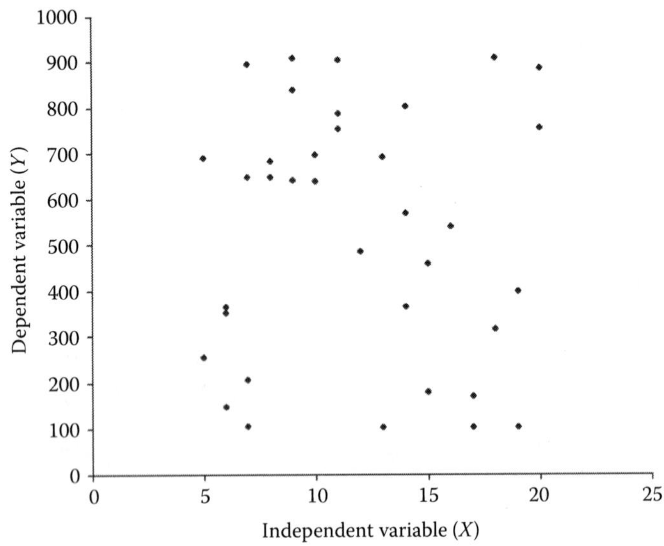

Figure 3.9 Scatter diagram with low coefficient of correlation.

shown in Figure 3.8, the coefficient of correlation is 0.97 showing almost a perfect relationship, whereas the coefficient of correlation is −0.04 for the series in Figure 3.9, indicating an absence of any relationship. We should not use regression analysis for the time series represented in Figure 3.9, whereas the time series represented in Figure 3.8 is a good candidate for regression analysis.

It is not only the relationship that exists between *X* and *Y* before doing the regression analysis, but it is important that a causal relationship also exists between the independent (*X*) and the dependent (*Y*) variables meaning changes in *Y* can be explained due to changes in *X*. A causal factor that is common to both *X* and *Y* and which operates as an unknown link may be responsible for whatever relationship is found between *X* and *Y*. When a series of *X* numbers is causally connected with the series of *Y* numbers, then it is beneficial to collect the information about *X* in order to forecast *Y*. Often by chance, correlation appears to be strong although the relationship is purely random.

The coefficient of determination (r^2) is a measure of the variability that is accounted for by the regression line for the dependent variable. The coefficient of determination always falls between 0 and 1. For example, if $r = 0.8$, the coefficient of determination $r^2 = 0.64$ meaning that 64% of the variation in *Y* is due to variation in *X*. The remaining 36% variation in the value of *Y* is due to other variables. If the coefficient of determination is low, multiple regression analysis may be able to account for all variables affecting the independent variable *Y* (not covered in this chapter).

3.6 Forecasting Errors

It is very rare to get 100% accurate forecasts. There are always some errors in forecasting. A good forecasting technique tries to minimize the errors. The forecasting errors are computed by comparing the *actual demand* for time period t with the *forecast* for that same period, that is, Error (t) = Demand (t) – Forecast (t). Two kinds of forecasting errors can occur. First, actual demand is greater than the forecast. This is a forecasting underestimate. By convention, when actual demand is greater than forecast demand, the error term is positive. Second, actual demand is less than the forecast. This is a forecasting overestimate and in this case the error is negative.

To choose a forecasting method, it is necessary to be able to compare the errors that each method generates for particular circumstances. There are several different ways of measuring errors. The choice is dependent on the situation and what kind of comparison is wanted. Thus, absolute measures are used to count all errors conservatively. Absolute measures mean that positive and negative errors are treated the same way. The reason for doing this is to prevent positive and negative errors from canceling each other out.

In this chapter, we will study a most commonly used method that calculates mean absolute deviation (MAD) of the error terms. To calculate MAD, take the sum of the absolute measures of the errors and divide that sum by the number of observations. MAD treats all errors linearly and is a more conservative measure.

Example: Consider the data given in Table 3.12. The demand and forecast are given for 10 periods in columns 2 and 3. Column 4 gives the error for each period, and column 5 gives the absolute error. The sum of the absolute errors for 10 periods is 171. Therefore, MAD = 171/10 = 17.10.

To select a particular forecasting method, say ES, we will calculate the values of MAD by using different values of α. The value of α that minimizes MAD will be selected. Similarly, applied to the MA method, we will calculate MAD for different values of n. The n that minimizes MAD will be selected. A similar procedure will be used with the WMA method to find the best combination of weights.

Managers change forecasts that are derived by numerical methods because they know about other factors that are not included in the calculations. This results in modifications of forecasts that can be called predictions and estimates. It is, therefore, wise to maintain a history of all forecasting errors. This should be done for all methods that are used and for all personnel who are making the forecasts, predictions, and estimates. Some people are good at forecasting and others are not. Also, some people are good at forecasting only under certain circumstances and not good in all circumstances.

Historical records of pertinent information are essential. Companies gain major advantages by finding out who can make good estimates under what circumstances. Further, there are situations where people can learn to make better forecasts, predictions, and estimates as a result of feedback about how well they have done in the past. When a record is not kept, all such advantage potentials are lost.

Table 3.12 Forecasting Errors

Period	*Demand*	*Forecast*	*Error*	*Absolute Error*
1	212	206.0	6.0	6.0
2	224	207.0	17.0	17.0
3	220	210.0	10.0	10.0
4	211	212.0	– 1.0	1.0
5	198	205.0	– 7.0	7.0
6	236	209.0	27.0	27.0
7	219	224.0	– 5.0	5.0
8	296	238.0	58.0	58.0
9	280	249.0	31.0	31.0
10	252	261.0	– 9.0	9.0
		Total	127.0	171.0

3.7 The Delphi Method

Delphi is a forecasting method that relies on expert estimation of future events. In one of its forms, the experts submit their opinions to a single individual who is the only one that knows who the participants are and what they have to say. The person who is the Delphi manager combines the opinions into a report, which, while protecting anonymity, is then disseminated to all participants. The participants are asked whether they wish to reevaluate and alter their previous opinions in the face of the body of opinion of their colleagues. Gradually, the group is supposed to move toward consensus. If it does not, at the least, a set of different possibilities can be presented to management.

The regression results could be shown to the Delphi panel of experts with the questions: "Do you think sales will be higher, lower, or the same as the regression results? Why?" Consensus might lead to modification of the managers' targets that were based on the regression results. Why does this method not let these "experts" talk to each other and discuss their opinions? If one of the experts is the CEO, or a Nobel prize winner, the dialogue might not be unbiased. People with greater debating capabilities are not necessarily people with more insights. The Delphi method is meant to put all participants on an equal footing with respect to getting their ideas heard.

There is no evidence that the Delphi method provides forecasts (and/or predictions) with smaller errors than other techniques. It is apparent that managers gain greater perspective about forces that should be considered when they are contemplating possible outcomes. That is a positive benefit of Delphi.

3.8 Pooling Information and Multiple Forecasts

Methods for pooling information to provide stronger forecasts should be explored. It is critical that all parties share their forecasts as much as possible and try to find ways to combine them. Usually, stronger forecasts can be obtained if both data and experience are pooled.

One of the keys to success in combining forecasts is trial and error. What seems to work is retained and what fails is discarded. As an example of pooling, the results of a regression analysis could be augmented by a Delphi-type estimation. In this way, experience and data analysis might be combined to provide a stronger conclusion.

Formal methods can be used to evaluate how well different forecasting techniques are doing. At each period, the method that did best the last time is chosen for the forecast that will be followed. Forecasts are still derived from the alternative methods, but they are recorded and not followed. When the actual demand results are known, the various forecasting methods are evaluated again, and the one that is most successful is chosen to make the prediction for the next period.

Averaging of forecast results also is used. The results of taking forecasts from more than one method and averaging these results to predict demand has been successful in circumstances where choosing the best method (as described above) produces frequent alteration of the chosen method.

3.9 Product Life-Cycle Stages and Forecasting

Throughout the company, the planning function marches to the drumbeat of product life-cycle stages. Operations managers need to be aware of the timing and stages that drive the development schedules of new products (goods and services), as well as the production and delivery schedules of the company's mature products.

The product life cycles are composed of four stages that appear in a regular way over time. All products and services go through the following stages:

- Introduction to the market
- Growth of volume and share
- Maturation, where maturity is the phase of relative equilibrium
- Decline occurs, because of deteriorating sales; decline leads to restaging or withdrawal

These life-cycle periods are discrete stages in each product's life that need to be understood in order to manage that product. Marketing is responsible for using different pricing, advertising, and promotion activities during appropriate stages. P/OM is responsible for intelligent management of the transformation system, which changes in various ways according to the life-cycle stage. The changes or transitions between stages require the knowledge of knowing how to adjust the production system's capabilities.

3.9.1 Introduction and Growth of the New Product (Goods and Services)

There are two new product early phases of life-cycle stages. These are initial introduction and sales growth of the product. The "idea" for the product and its development precedes the introduction. The entire team works on ascertaining the marketing feasibility of the idea, as well as the feasibility of making it and delivering it. Research and Development (R&D) may have made sample product so that market research can test its customer acceptability. When it is approved, P/OM and engineering swing into action to create the production system that can make and/or assemble it. During this process of bringing the idea to reality, there are many "make or buy" analyses.

None of this is easy. It takes a lot of effort and attention to detail. There is much time and talent needed to conceptualize the product, design its specifics, organize the process for making it, cost it out, pilot test it, and so forth. When the product is accepted, it is released for production and marketing. All of these take place in the introductory stage.

3.9.2 Maturation and Decline of the New Product (Goods and Services)

When the new product or service stops growing, it is considered mature. This means that its volume is stabilized at the saturation level for that brand. The competitors have divided the market, and only extraordinary events, such as a strike at a competitor's plant, are able to shift shares and volumes.

During the introductory phase, P/OM had to deal with producing larger and larger quantities of product. Now the marketing–production relationship reaches equilibrium. Marketing takes specific actions during this phase to maintain the product's share of the market. Prices often are lowered. Coordination between P/OM and sales is essential to meet delivery schedules on time. Finally, the product begins to lose share, volume drops, and, depending on the strategy, the product is either restaged or terminated.

It is expected that a new product will have been introduced previously that has grown to a reasonable extent. It is the replacement for the other product and has been assigned the capacity of the product that has been replaced. The cycle of the new product is similar to that of the one that is replaced in that it goes through

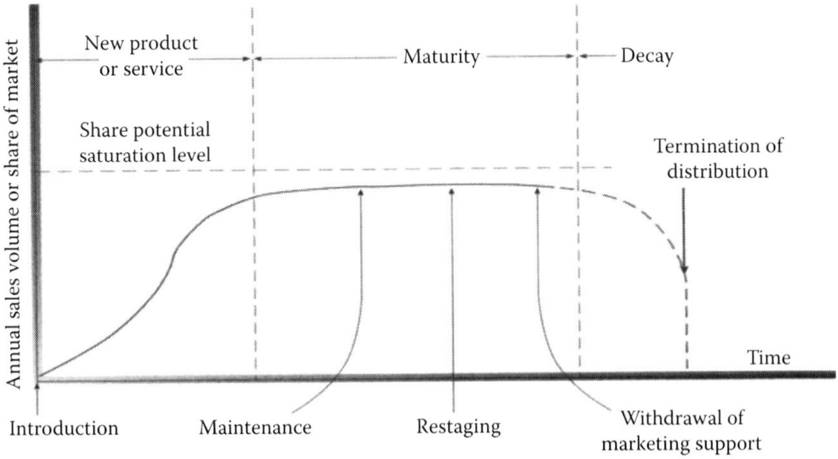

Figure 3.10 Trajectory representing the evolution of new product or service life-cycle stages.

introduction, growth, maturation, and decline. Figure 3.10 illustrates the trajectory that represents the evolution of new product or service life-cycle stages.

3.9.3 Demand Prediction in Life-Cycle Stages

Life-cycle stages provide a classification for understanding the kinds of trends that can be expected in demand. During the introduction, the demand is led by the desire to "fill the pipeline." This means getting product into the stores or into warehouses or wherever it must be to supply the customers.

When growth starts to occur, there is a trend line of increasing sales. The trick is to estimate how fast demand will increase over time and for how long a period growth will continue. There are good methods used by market research to find out what is happening. This information is fed back through sales to P/OM for purposes of production scheduling.

The revenue-generating capacity of a product varies over its lifetime. (The word "product" is applicable to goods and/or services, so product should be read as either or both.) During the introduction, expenditures usually are high to bring the new product to the distributors, retailers, and customers. The cash flow is negative. More is being spent on the introduction of the product and promotion of growth than the product revenue returns. Later, when the customer base has been established, net revenue increases, and expenditure decreases. During the stable period of product maturity, if the product enjoys a loyal following, promotions of any kind may be unnecessary. Eventually, competitive moves are likely to lead to price cuts and the withdrawal of the product from the market. The extent of the stable interval has been decreasing in many categories because of rapid technological changes as well as aggressive competitive behaviors.

Product life cycles have been speeding up, which means that growth has to occur faster. The product does not stay in the mature stage long and has to be replaced more often. More new product introductions are required to replace waning products. The challenge for P/OM is the rapidity of adjustment that is required to all of the life-cycle stage changes and to go through them frequently, that is, plant expansion and contraction, new supply chain sources, inventory obsoleted by immediate demands for new products (e.g., iPhone 5).

P/OM thinking is crucial to each of the four stages. The new design that is about to enter production has been finalized. There still are many questions to be answered. How much product must be made for distribution at start-up? How much product will be needed to keep the supply lines filled to meet the demand during growth? How much capacity is to be used at each of the stages? How much training will the evolving work configurations require? Making or buying decisions can change. What should be bought at first, at the lower volumes, might be better made in the plant, at the higher volumes that are associated with maturity. The global supply chain is a dynamic factor which demands up-to-the-minute information and rapidly changing plans.

3.9.4 Protection of Established (Mature) Products (Goods and Services)

Organizations that have been successful with their new product introductions can bank on having established products or services that generate cash flow. The "bank" is not as good as it used to be though, because the competitive rate of new product introductions has increased markedly in recent years. P/OM requires the cash to pay labor and buy materials from suppliers. Generally, P/OM spends more cash than any other department in the company. These cash accounts for operating expenses may total 70% of all cash spent by the company.

To meet the cash flow requirements and to have the right plans in place requires excellent forecasting abilities of the marketplace dynamics. This is the phase where many big companies failed to forecast changes and they lost their edge or went bankrupt. Examples include, Sony's Betamax, Blockbuster, Borders, Circuit City, Digital Equipment, Eastman Kodak, Enron, Global Crossing, Lehman Brothers, Nortel, Polaroid, Pan Am, Sharper Image, Sun Microsystems, and Wang Laboratories. Each and every one of these situations might have been avoided if forecasting had called attention to the cliff that was in front of their fast moving trajectory.

Because of global organizations, there are more competitors in the game, and because there are more competitors, each organization competes harder. Increased competition has led to higher levels of market volatility. Nevertheless, for most product categories, there is still plenty of opportunity to benefit from the mature product life-cycle stage. P/OM benefits from the security of established products because they have stable material and labor skill requirements. They also provide the opportunity to improve the productivity of these processes.

However, it is essential to obtain good forecasts about timing when the market equilibrium is deteriorating. Market share deteriorates slowly at first and then severely when the tipping point is reached. Suddenly, the production system is using materials and labor to make a product or deliver a service that no one wants. What has triggered this state of affairs? Was the handwriting on the wall? Could Polaroid have forecast the digital camera? The interesting and conclusive answer is "yes." In every one of the cases previously mentioned, with proper methods and clear vision, forecasting could have seen what was coming even though the cause of the problem could not have been anticipated and the timing prediction might have been flawed.

The Pan Am case provides an interesting example. Pan Am flight 103 on Wednesday, 21 December 1988 was destroyed by a bomb over Lockerbie, Scotland killing all 259 people on board the plane. The effect on bookings was immediate and the continued publicity barrage kept negative associations uppermost in future flyers' minds. It cannot be known if Pan Am could have reversed the revenue decay but many have supposed that creative efforts on a grand scale might have succeeded. What is known is that Pan Am's management tried to wait it out without forecasting the inevitable bankruptcy that would occur when load factors fell below breakeven. Production planning for a period of very low occupancy might have conserved enough cash to weather the storm. Forecasting played no part in Pan Am's planning after the catastrophe.

By current definitions, this bombing was a *black swan* event which is an occurrence that is very rare, unexpected, and even impossible. In the sixteenth century, black swans were presumed not to exist. Later, it was realized that black swans are very rare but they do occur in nature. The Black Swan theory of forecasting is relevant to many situations (such as the Chicago Tylenol unsolved murders—these poisonings occurred in 1982). Johnson & Johnson made a major P/OM decision to redesign the package and to publicize with extensive marketing expenditures their new quality control (safety) measures. The Tylenol case is often cited because J&J used forecasts to correctly redirect marketing efforts and P/OM strategies. Disaster was avoided and crisis management was totally successful. When established product designs and brands (such as Keuffel & Esser slide rules, RCA 45 rpm vinyl records, 3.25 floppy disk drives, and Kodak 35-mm films) disappear in the market, it is essential to note that many producers were surprised but very few consumers were surprised.

Summary

Strategic planning requires teamwork. P/OM should be part of the team. Chapter 3 discusses P/OM's role in helping to develop strategies and illuminates P/OM's assignment to carry them out. P/OM is responsible for seeking out and adopting "best practice." Life-cycle stages are developed. This includes introduction and growth of new products, followed by their maturation and decline. P/OM is expected to protect established products. The challenge is difficult when competitors introduce disruptive innovations. Recognition of problems and fast action to deal with disruptions are needed.

Forecasting is essential for proper management of life-cycle stages. Extrapolation and correlation are vital methods used with time-series analysis. Other important forecasting methods that are covered include MAs, WMAs, regression analysis, correlation coefficients, coefficients of determination, multiple forecasts, and the Delphi method. Also, the means for evaluating the different forecasting methods by comparing the kinds of errors they generate is explained.

Review Questions

1. What is suboptimization and why is it a concern of P/OM?
2. What is meant by the statement "P/OM is a scarce resource"?
3. What is a time series? How would it be used to make predictions about emerging technological developments?
4. Is there a basis for predicting periods of prosperity and of economic slowdowns on the basis of cycles?
5. What are life-cycle stages? Why does the concept of life-cycle stages unite P/OM with other members of the organization in using the systems approach?
6. Detail the life-cycle stages of new products and explain them for both goods and services.
7. What is extrapolation and how is it used?
8. What is correlation?
9. What is an historical forecast? When is it used?
10. Why are long life-cycle stages considered desirable? Does this apply to start-up and growth as well as the mature stage of new products?
11. When are multiple forecasts desirable?

Problems

1: The larger the number of periods in the simple MA forecasting method, the greater the method's responsiveness to changes in demand. (True or False)?

2: The coefficient of correlation can never be negative. (True or False)?

3: Most forecasting techniques assume that there is some underlying stability in the system. (True or False)?

4: The percent of variation in the dependent variable that is explained by the regression equation is measured by which one of the following? (a) MAD, (b) slope, (c) coefficient of determination, (d) correlation coefficient, or (e) intercept.

5: Time-series data may exhibit which of the following behaviors? (a) Trend, (b) random variations, (c) seasonality, (d) all of these.

6: Weekly sales of bread at a local Whole Foods Market are given in the table below. Forecast sales for week 7 using a 3-week MA.

Week	1	2	3	4	5	6	7	8
Sales	402	385	420	382	410	432	385	411

7: Weekly sales of ten-grain bread at the local Whole Foods Market are given in the table below. Forecast demand for week 6 using a 4-week MA.

Week	1	2	3	4	5	6	7	8
Sales	415	389	420	382	410	432	380	410

8: What is the forecast for May based on a WMA applied to the following past demand data and using the weights: 0.7, 0.2, 0.1 (largest weight is for most recent data)?

Nov.	Dec.	Jan.	Feb.	Mar.	April
37	36	40	42	47	43

9: What is the forecast for April based on a WMA applied to the following past demand data and using the weights: 0.6, 0.3, 0.1 (largest weight is for most recent data)?

Nov.	Dec.	Jan.	Feb.	Mar.	April
37	40	36	47	42	43

10: What is the forecast for May based on a WMA applied to the following past demand data and using the weights: 0.5, 0.3, 0.2 (largest weight is for most recent data)?

Jan.	Feb.	Mar.	April	May	June
137	140	136	145	138	144

11: Given an actual demand of 110, a previous forecast value of 120, and an $\alpha = 0.3$, the ES forecast for the next period would be ______.

12: Given an actual demand of 103, a previous forecast value of 99, and an $\alpha = 0.4$, the ES forecast for the next period would be _______.

13: For a given product demand, the time-series trend equation is $25 + 3.2X$. What is the demand forecast for period 10?

14: The demands for an item for the 12 months of year 2014 are given below.

Period (T)	1	2	3	4	5	6	7	8	9	10	11	12
Month	Jan.	Feb.	Mar.	April	May	June	July	Aug.	Sep.	Oct.	Nov.	Dec.
Demand	151	152	132	161	182	192	174	159	183	169	175	181

The trend projection equation is $Y = 148.97 + 2.864T$ where T is the number of the month. Make a forecast for June of 2015. (Note that the value of T to be used in the trend equation is 18. June is the 6th month of 2015.)

15: A forecasting method has produced the following data over the past 5 months. What is the MAD?

Actual	*Forecast*	*Error*	*|Error|*
10	11	−1	1
8	10	−2	2
10	8	2	2
6	6	0	0
9	8	1	1

16: For the data given below, is the forecasting system overestimating or underestimating the demand? (a) Overestimating, (b) Underestimating.

Actual	*Forecast*	*Error*	*|Error|*
10	11	−1	1
8	10	−2	2
10	8	2	2
6	6	0	0
9	8	1	1

17: A forecasting method has produced the following data over the past 6 months. What is the MAD at the end of 6 months?

Month	Actual	Forecast	Error	|Error|
1	10	9	1	1
2	8	10	−2	2
3	10	8	2	2
4	6	6	0	0
5	9	8	1	1
6	11	12		

18: If the coefficient of correlation is −0.8, what is the percent of variation in the dependent variable that is explained by the regression equation?

19: If you were selecting a forecasting model based on MAD, which of the following MAD values reflects the most accurate model?
(A) 0.2, (B) 0.8, (C) 2.0, (D) 4.5, (E) 100.

Past actual demand (in units) for a company is given below. Answer the next three questions.

Month	1	2	3	4	5	6	7	8	9
Actual demand	700	645	660	648	655	760	682	670	756

20: Compute a weighted 3-month MA for month 7 by using the following weights: 0.5, 0.3, and 0.2. The weight is highest for the latest month.

21: The demand forecast for month 7 by using a simple 3-month MA will be ___.

22: Find the forecast for month 9 by using ES with $\alpha = 0.3$.

23: Find the value of MAD at the end of month 8 by using the demand and forecast data in the table given below.

Month	1	2	3	4	5	6	7	8	9
Actual demand	700	645	660	648	655	760	682	670	756
Forecast	720	675	635	587	702	718	740	753	

24: Which of the following statements is true on the basis of the data given in the above table?

a. The demand is being overestimated.
b. The demand is being underestimated.

25: In regression analysis, you determine the intercept value of the model to be 1000 and the slope value to be 50. What is the resulting forecast value using this model if the value of the independent variable X is 12.

26: The following trend projection, based on past 36 months (years 2014, 2015, and 2016) data is used to predict monthly demand:

$Y = 250 + 1.5t$, where $t = 1$ is the first month of year 2014 (January 2014)

What is the seasonally adjusted forecast for March 2017? It is given that the seasonal (monthly) index for March is 0.8.

27: The following trend projection, based on past 12 quarters (years 2014, 2015, and 2016) data is used to predict quarterly demand:

$Y = 250 - 2.5t$, where $t = 1$ for the first quarter of 2014.

Seasonal (quarterly) indices are Quarter 1 = 1.5; Quarter 2 = 0.8; Quarter 3 = 1.1; and Quarter 4 = 0.6.

What is the seasonally adjusted forecast for the second quarter of 2017?

28: The demand data for an item for 3 years is given in the table below. Find the forecast for the Winter quarter of the fourth year using the seasonal forecasting model. Total demand in the fourth year is expected to be 2212 units.

Quarter	*Year 1*	*Year 2*	*Year 3*
Fall	110	135	180
Winter	240	355	500
Spring	600	750	830
Summer	450	540	610

29. The Highway Department is considering using a 6-month MA to forecast crew hours needed to repair roads month by month for the coming year. To test whether this method is accurate, use last year's data (shown in the following table) to predict the last 6 months' crew hours to compare with the observed values. Start with the average of January through June to compare with the actual value of 200 for July and so on through December.

Month	*Crew Hours*
January	110
February	120
March	140
April	180
May	250
June	200
July	200

Month	*Crew Hours*
August	220
September	280
October	120
November	100
December	80

30. Referring to Q29, prepare an analysis of the errors in using the 6-month MA as a predictor.
31. Referring to Q29 and Q30, use a 3-period MA to forecast the last 6 months and compare directly to the 6-month forecast developed there. Here, the first forecast is the average of April, May, and June to compare with the actual value of 200 for July and so on through December.
32. Referring to Q29, Q30, and Q31, use a 6-period WMA to forecast the last 6 months and compare directly to the un-weighted forecasts for 6-month and 3-month periods. Use the weights 0.1, 0.1, 0.1, 0.2, 0.2, and 0.3 with the 0.3 weighting the most recent datum point.
33. The following table presents data concerning the sales of minivans and minivan tires for 10 years. It is believed that the demand for minivan replacement tires is highly correlated with the sales figures of minivans for the previous years. Based on these data which is a better predictor of minivan tire sales—sales of minivans three years prior or four years prior?

Year t	*Minivan Sales in Year t (in millions)*	*Minivan Tire Sales yt in Year t (in millions)*
1	10	4
2	12	6
3	11	7
4	9	5
5	10	8
6	12	7
7	10	5
8	9	7
9	8	8
10	7	6

Readings and References

Andrews, B.H. and Cunningham, S.M., L. L. Bean Improves Call-Center Forecasting, *Interfaces*, 25(6), 1995, p. 1.

Basu, S. and Schroeder, R.G., Incorporating Judgments in Sales Forecasts—Application Of the Delphi Method at American Hoist & Derrick, *Interfaces*, 7(3), 1977, p. 18.

Campbell, O., Forecasting in Direct Selling Bussines:Tupperware's Experience, *The Journal of Business Forecasting*, 27(2), Summer, 2008, pp. 18–19.

Cederlund, J.P., Kohli, R., Sherer, S.A., and Yao, Y., How Motorola put CPFR into Action, *Supply Chain Management Review*, October, 2007, pp. 28–35.

Chambers, J.C., et al., How to Choose the Right Forecasting Technique, *Harvard Business Review*, 49(4), July–August, 1971, pp. 45–74.

Clemen, R.T., Combining Forecasts: A Review and Annotated Bibliography, *International Journal of Forecasting*, 5, 1989, pp. 559–583.

Denton, B., Fitts Industries Enhances Forecasting and Production; Forecasting; *Interfaces*, 38(4), 2008, p. 338.

Fader, P.S. and Hardie, B.G.S., Forecasting Repeat Sales at CDNOW: A Case Study, *Interfaces*, 31(3), 2001.

Fildes, R.G., Against Your Better Judgment? How Organizations Can Improve Their Use of Management Judgment in Forecasting, *Decision Sciences*, 37(6), Nov.–Dec. 2007, pp. 570–576.

Galbraith, C.S. and Merrill, G.B., Politics of Forecasting: Managing the Truth, *California Management Review*, 38(2), 1996, pp. 29–43.

Gardner, E.S., Exponential Smoothing: The State of the Art, *Journal of Forecasting*, 4, March 1984, pp. 1–28.

Gardner, E.S. and Dannenbring, D.G., Forecasting with Exponential Smoothing: Some Guidelines for Model Selection, *Decision Sciences*, 11(2), 1980, pp. 370–383.

Georgoff, D.M. and Murdick, R.G., Manager's Guide to Forecasting, *Harvard Businesss Review*, 64, Jan.–Feb. 1986, pp. 110–120.

Gilliland, M. and Leonard, M., Forecasting Software—The Past and the Future, *The Journal of Business Forecasting*, 25(1), Spring 2006, pp. 33–36.

Helmer, F.T., Oppermann, E.B., and Suver, J.D., Forecasting Nursing Staffing Requirements by Intensity-of-Care Level, *Interfaces*, 10(3), 1980, p. 50.

Kallina, C., Development and Implementation of a Simple Short Range Forecasting Model—A Case Study, *Interfaces*, 8(3), 1978, p. 32.

Kimes, S.E. and Fitzsimmons, J.A., Selecting Profitable Hotel Sites at La Quinta Motor Inns, *Interfaces*, 20(2), 1990, pp. 12–20.

Li, X., An Intelligent Business Forecaster for Strategic Business Planning, *Journal of Forecasting*, 18(3), 1999, pp. 181–205.

Lim, J.S. and O'Connor, M., Judgmental Forecasting with Time Series and Causal Information, *International Journal of Forecasting*, 12, 1996, pp. 139–153.

Makridakis, S.G., The Art and Science of Forecasting, *International Journal of Forecasting*, 2, 1986, pp. 15–39.

Makridakis, S.G., Accuracy Measures: Theoretical and Practical Concerns, *International Journal of Forecasting*, 9(4), 1993, pp. 527–529.

Melnyk, S., Forecasting Software Product Listing, *APICS—The Performance Advantage*, April 1997, pp. 62–65.

Moon, M.A., Mentzer, J.T., and Thomas, D.E., Customer Demand Planning at Lucent Technologies: A Case Study in Continuous Improvement Through Sales Forecast Auditing, *Industrial Marketing Management*, 29(1), 2000.

Onkal, D., Gonul, M.S., and Lawrence, M., Judgmental Adjustments of Previously Adjusted Forecasts, *Decision Sciences*, 39(2), May 2008, pp. 213–238.

Rowe, G. and Wright, G., The Delphi Technique as a Forecasting Tool: Issues and Analysis, *International Journal of Forecasting*, 15(4), October 1999.

Sanders, N.R., Measuring Forecast Accuracy: Some Practical Suggestions, *Production and Inventory Management Journal*, 38(1), 1997a, pp. 43–46.

Sanders, N.R., The Status of Forecasting in Manufacturing Firms, *Production and Inventory Management Journal*, 38(2), 1997b, pp. 32–36.

Sanders, N.R. and Manrodt, K.B., Forecasting Practices in US Corporations: Survey Results, *Interfaces*, 24(2), 1994, p. 92.

Sanders, N.R. and Manrodt, K.B., Forecasting in Practice: Use, Satisfaction, and Performance, *Interfaces*, 33(5), 2003, pp. 90–93.

Sanders, N.R., and Ritzman, L.P., The Need for Contextual and Technical Knowledge in Judgmental Forecasting, *Journal of Behavioral Decision Making*, 5(1), 1992, pp. 39–52.

Sanders, N.R. and Ritzman, L.P., Bringing Judgment into Combination Forecasts, *Journal of Operations Management*, 13, 1995, pp. 311–321.

Shah, P., Techniques to Support Better Forescating, *APICS Magazine*, Nov.–Dec. 2008, pp. 49–50.

Smith, L., West Marine: A CPFR Success Story, *Supply Chain Management Review*, March 2006, pp. 29–36.

Steerman, H., A Practical Look at CPFR: The Sears-Michelin Experience, *Supply Chain Management Review*, July–August 2003, pp. 46–53.

Tabatabai, B., Improving Forecasting, *Financial Management*, October 2008, pp. 48–49.

Tersine, R.J. and Riggs, W., The Delphi Technique: A Long-Range Planning Tool, *Business Horizons*, 19(2), 1976.

Urs, R., How to Use a Demand Planning System for Best Forecasting and Planning Results, *The Journal of Business Forecasting*, 27(2), Summer 2008, pp. 22–25.

VICS, *Collaborative Planning, Forecasting, and Replenishment*, Version 2.0, www.cpfr.org, 2002.

Yurkiewicz, J., Forecasting 2000, *OR/MS Today*, 27(1), 2000, pp. 58–65.

Yurkiewicz, J., Forecasting Software Survey, *OR/MS Today*, February 2003.

Chapter 4

Capacity Management and Aggregate Production Planning

Readers' Choice: 12 roses, 9 tulips, and 3 lilies aggregate to yield two dozen flowers

Alden, J.M., Burns, L.D., Costy, T., and Hutton, R.D., General Motors Increases Its Production Throughput, *Interfaces*, 2006, 36(1), p. 6. The following three activities increased GM's revenue and saved \$2.1 billion: (1) better estimate of throughput, identifying bottlenecks, and better buffer allocation, (2) real-time data collection, and (3) establishment of common processes.

Caixeta-Filho, J.V., van Swaay-Neto, J.M., and de Padua Wagemaker, A., Optimization of the Production Planning and Trade of Lily Flowers at Jan de Wit Company, *Interfaces*, 2002, 32(1), p. 35. The model developed by the authors produced the following benefits to the company. "Between 1999 and 2000, company revenue grew 26 percent, sales increased 14.8 percent for pots of lilies and 29.3 percent for bunches of lilies, costs fell from 87.9 to 84.7 percent of sales, income from operations increased 60 percent, return on owner's equity went from 15.1 to 22.5 percent, and best quality cut lilies jumped from 11 to 61 percent of the quantities sold."

Denton, B., An Integrated Outbound Logistics Model for Frito-Lay: Coordinating Aggregate-Level Production and Distribution Decisions, *Interfaces*, 2009, 39(5), p. 460. The authors developed a model to simultaneously optimize Frito-Lay's inventory and transportation decisions and demonstrated that the company can reap substantial benefits by implementing this model.

Fleischmann, B., Ferber, S., and Henrich, P., Strategic Planning of BMW's Global Production Network, *Interfaces*, 2006, 36(3), p. 194. The authors developed a strategic-planning model for BMW to allocate various products to global sites that led to more transparency and flexibility of the strategic planning process.

Gerald, B., Keegan, J., Vigus, B., and Wood, K. The Kellogg Company Optimizes Production, Inventory, and Distribution, *Interfaces*, 2001, 31(6), p. 1. Kellogg's operational planning system reduced production, inventory, and distribution costs by approximately $4.5 million in 1995, whereas its tactical system helped in consolidating production capacity with an expected savings of $35 million to $40 million per year.

Kuchta, M., Newman, A., and Topal, E., Implementing a Production Schedule at LKAB's Kiruna Mine, *Interfaces*, 2004, 34(2), p. 124. The authors developed a production plan for Kiruna's mining operations. The model identifies the production blocks to mine and the time to mine them. The plan minimizes variations from monthly production plans within the operational constraints.

Capacity management relates to how the existing system is used. The systems viewpoint broadens the scope of inquiry to include questions about the existing arrangement and whether alternative configurations might not provide superior alternatives. The discussion of options cannot be pursued without consultations between marketing, finance, R&D, and P/OM.

Capacity is always limited by the bottleneck operations in both manufacturing and service departments, or in the entire supply chain. Supply chain management is discussed in detail in Chapter 9. In this chapter, we discuss capacity management from the viewpoint of an operations department, but references to the entire supply chain are made when relevant. The viewpoint must be systems-wide to scan across facilities with excess capacity in order to spot the overloaded resources that are struggling to keep up with the rest of the system.

Delays are another source of capacity problems that are often caused by factors that are outside the organization's boundaries, for example, vendor postponements.

Thus, suppliers must also be viewed as part of the system to understand swings in demand caused by delays that create above peak and below normal requirements throughout the entire supply chain.

After reading this chapter, you should be able to:

- Define and measure capacity.
- Make clear what aggregation does and why that is important.
- Explain the function of aggregate planning (AP).
- Discuss the systems nature of AP in terms of classes of resources and product-mix families.
- Explain standard units of work.
- Relate the importance of forecasting to AP.
- Compare constant (or level) production with a chasing policy—where supply chases demand.
- Detail the cost structure for aggregate planning.
- Develop aggregate plans using overtime and subcontracting.
- Distinguish between planning for manufacturing and service industries.

4.1 Definitions of Capacity

Actual capacity of the supply chain, or that of a manufacturing or service department, is the greatest throughput rate that can be achieved with the existing configuration of resources and the accepted product or service mix plans. Altering the product or service mix can (and usually will) change actual realizable capacity to produce output. Modifying the existing configuration of resources, equipment, and people in the supply chain workforce alters real capacity. The systems point of view includes cash as part of the resources because cash can be converted into new machines, which alter real deliverable throughput capacity. The systems viewpoint includes good ideas, which can increase supply chain capacity with minimum expenditures.

The formula for actual measured capacity of a manufacturing or service department is

$$C = T \times E \times U,$$

where

C = actual measured capacity (in units converted to standard hours)

T = real time available

E = efficiency
U = utilization

T is determined by calculating the amount of time that is available when fully utilizing the resources that are already in place to make and deliver product throughput. Doubling the number of machines, trucks, etc., doubles T (the amount of available time). E is the efficiency with which time T can be utilized to make and deliver different kinds of products. Then, $T \times E$ is equivalent to standard hours available to make and deliver the products. U is how much of the available throughput capacity can be (or is) utilized. Lack of orders or breakdowns of productive systems diminish U. When T and E and U are multiplied, the product is C, the actual capacity that is being (or has been) utilized. Table 4.1 illustrates the calculations.

Assume that the rated maximum throughput capacity of the plant for the data given in Table 4.1 is 150 standard hours. It never achieves the maximum, but comes closest to doing so on Wednesday. As briefly explained before, there are a variety of reasons why the system does not achieve maximum throughput. Systems with bottlenecks and flow disruptions can never achieve maximum product throughput.

E, the efficiency, is a proportional factor used to convert units of throughput to standard times. Systems of machines and people that work slower have lower efficiency than those that have a higher productive output. Often, the best in the class is given an efficiency of 1. It is to be expected that variations in efficiency will occur. Sometimes the source of the variation can be traced. If it is significant variation, then it should be corrected. Unexplained fluctuations in capacity are unpleasant and unprofitable.

If a supply chain is operating at 90% of the standard time because a supplier (somewhere along the supply chain line) has delivered a defective product, remedial action must be taken with supplies on hand and the problem must be corrected with future deliveries. Marketing may have assured deliveries, so customer notification and promises of remedial action reflect the urgent nature of the systems problem so often involved in capacity management.

Table 4.1 Capacity Utilized (C)

Product	*Monday*	*Tuesday*	*Wednesday*	*Thursday*	*Friday*
T101	25	20	65	18	30
T102	65	10	25	40	0
MW11	40	90	50	70	80
Total	130	120	140	128	110

Note: The numbers are product throughput.

U, the utilization, is applied as a proportional correction to standard time when there are supply chain disruptions. Even when everything is running "as planned", the value of *U* is often less than 100%. If the system is operating even faster, the value of *U* can exceed 100%. There are pros and cons related to running operational systems above the maximum rated capacity. How long maximum capacity is exceeded also counts.

U is a measure to be wary of when it becomes an objective of management to keep *U* as close to 100% as possible. There are sound economic reasons not to operate a production department above capacity. For example, it is economically sensible to stop production when the planned output quota has been met and safety stock is sufficient. Shutting down the production system for 2 h of an 8-h day means that the *U* measure goes to 75%. Actual measured operational capacity is going to be reduced by a fourth. For various reasons, that may be a good thing.

Management must establish the fact that supply chains are composed of complex sub-systems some of which cannot function above capacity for very long. The costs of unsold throughput must be analyzed. How long will stored output remain as inventory? Are there fluctuations in demand that they will buffer or is the buffer in place already? The costs of arbitrary utilization of supply chain capacity to eliminate less than 100% utilization should be recognized for what it is—waste of time and money due to fear of seeming to waste (unutilized) capacity. An appropriate decision model can be constructed to set proper supply chain throughput based on maximizing the whole systems performance.

As an example, assume that *U* is 0.963. This is likely to be viewed as a more reasonable utilization factor than 0.750. P/OM, in general, will not tolerate a permanent situation where utilization factors are below 0.900. Still, the target numbers will be dependent on the situation. For example, in service organizations, a high value of *U* might be required to keep queues short. Cyclical supply chain demand systems can be expected to cycle between utilization factors in the 70% range and then up to more than 100% capacity. Cyclical industry companies prefer having excess capacity in reserve and expect to operate effectively below the misleading ideal of 100% utilization.

Using data in Table 4.1, the value of 130 actual standard hours of throughput on Monday might have been obtained as follows:

$$C = T \times E \times U = 150 \times 0.9 \times 0.963 = 130 \text{ actual standard hours.}$$

The ratio of actual standard hours to maximum standard hours is equal to 130/150 = 0.87 or 87%.

Suppose in a bank, there are six tellers' windows. If the bank is open for 8 hours (h) per day, the maximum designed capacity is 48 h. This is a long-term decision because the number of windows cannot be changed every day. If there are only four tellers on a given day then the operating capacity is 32 h. The operating capacity can be changed on a daily basis (even every hour) by changing the number of tellers. The operating capacity cannot be more than the designed capacity. The actual utilization of the four tellers may only be (say) 27 h—based on the number of customers that visit the bank.

The same situation applies to grocery stores with one cash register and supermarkets with dozens of checkout counters. The number of operating rooms in a hospital is the maximum design capacity. The number of seats in a restaurant is the maximum design capacity but the service rate and the length of time that customers stay in their seats determines throughput capacity.

In that regard, maximum operational capacity can also be defined as maximum sustained throughput of goods or services. Operations capacity describes how many units can be supplied per unit of time. For services, a bank might compare the maximum number of people the bank teller can process per hour with the maximum number of people the ATM can process per hour. This is a supply chain service capacity comparison.

For a manufacturing supply chain example, compare the maximum number of hot dogs Oscar Meyer (OM) can make and ship per hour with the maximum number of hot dogs Hebrew National (HN) can make and ship per hour. The maximum number is likely to be much bigger than the breakeven number (see Appendix A). In this supply chain the ingredients required to make the products have a flow through rate that must match the producers' rates. It is not possible to determine if OM is better off than HN without the entire picture of the system.

Still, this comparison is one that both companies would like to make to compare their PMCs (productivities at maximum capacity). PMC makes an excellent benchmarking measure. In supply chain terms, benchmarking is a systematic comparison of fundamental measures with those of contestants performing similar supply chain functions.

Many other aspects of capacity planning are discussed in Chapter 9, Supply Chain Management. In the following sections, we will discuss capacity planning for an operational department what is popularly known as aggregate production planning.

4.2 Introduction to Aggregate Production Planning

Aggregate planning (AP) moves the focus onto the shop floor, inside a manufacturing plant, or in a service facility, where work is done. Operations managers must deal with workforce planning. The number of employees is strongly related to the outputs required from the input–output transformation process. Workforce planning is directly related to the control of inventories based on forecasts of demand.

AP is the process to develop a generalized production plan for all job types in an organization. For example, the job types may include: the water-based, oil-based, and acrylic paints for a paint manufacturer; different kinds of services offered by a market research company; or a hospital with tests and treatments for all kinds of cases. The drivers of AP are forecasted customer demand, specific customer orders, or the workload determined through any other rigorous planning process.

The forecasts and workloads are translated into operational plans. The goal is to be prepared to make the products and deliver them when needed. The same applies to delivering services to the customers at the right time. Aggregate plans outline the strategies to meet the fluctuating demand for products and services for a planning horizon spanning over 6–18 months. These are called medium-term plans. These plans are made after the demand forecasts are known for all periods covered by the planning horizon. The plans are developed for the total demand (total of demands for all products) and not separately for the demands for individual products—giving it the name "aggregate planning." As an example, three oranges, two bananas, five apples, and seven grapes are in the aggregate 17 pieces of fruit.

Aggregate plans specify the schedule of work, that is, the number of units to be produced or services to be provided in every period. The plans also specify the resource requirements. The investment in machines and other long-term facilities bound the production levels and generally do not change during the planning horizon. The resources that can be altered and changed include permanent full-time workers, temporary workers, and part-time workers. The plans are revised after periodic intervals—most frequently every quarter. The planning period, "the time bucket," is generally a month but any time bucket can be used. We are going to use "month" as the planning time bucket in this chapter.

AP starts a chain reaction in the supply chain of suppliers–producers–customers. After the internal assignments that relate to the factory and the office are scheduled, the external flows begin to function. These are flows of materials from the factory to the warehouses by trucks and other transport systems, which must be coordinated. Transport from the warehouses to the customers must be organized. External flows from suppliers to the production transformation system must be activated. Equipment may have to be rented or bought and people hired and trained, or workforce reductions may be initiated. Material flows inside the company can be scheduled in detail for specific items or in categories of specific items that have been aggregated.

Before trying to do detailed, tactical scheduling (discussed in Chapter 6), AP (which is generic) should be used to avoid costly mistakes arising from not being prepared with the proper resources at the right time. It is important to note that AP is an internal production management function that leads to detailed internal production scheduling in later stages. We will illustrate the nature of the AP problem and the strategies for solving the problem through the following three examples.

4.3 Example 1: Aggregate Production Planning

Consider the case of a company that is manufacturing a single product to understand the nature of the problem. Table 4.2 gives the "expected demand" for this

Table 4.2 Expected Demand

Period	1	2	3	4	5	6	7	8	9	10	11	12	Total	Average
Expected Demand	210	440	600	300	480	610	560	800	200	400	380	540	5520	460

product for the next 12 months. However, as we studied in the chapter on forecasting, the actual demand could be different from the expected demand. While solving the AP problems, we make an implicit assumption that the actual demand is equal to the expected demand. The discrepancy between the expected and actual demands can be incorporated during the revisions of plans. The plans are generally revised after a certain interval like a month or quarter. The provision for safety stocks can also be included in the demand forecast to take care of these discrepancies. However, we are not considering safety stocks in this chapter to reduce the complexity of the problem.

The demand in Table 4.2 varies from a high of 800 in period 8 to a low of 200 in period 9. How do we meet these fluctuating requirements? We will study three different plans and discuss each of them in the following section.

4.3.1 Production Plans

We examine three different plans to meet the production requirements. In all three plans, the total production in 12 months is equal to the total demand in Table 4.2, namely 5520. However, these plans result in different production quantities for each month. Two of the basic (or "pure") plans include: (a) Level plan, in which the same number of units are produced in every period, that is, produce at an average rate throughout the planning horizon, and (b) Chase plan, in which the number of units produced in each period is equal to the demand in that period. That is, produce what you need in every period. A third strategy, Mixed or Hybrid strategy,

INSET 1

Production Plans

Level plan: The number of units produced in each month is constant at a chosen level.

Chase plan: The number of units produced in a given month is equal to the exact demand in that month.

Mixed or hybrid plan: The number of units produced is constant at a given level for a few months and then changes. The level may change several times.

is a mixture of these two pure strategies. These production plans are listed in Inset 1 and are discussed below.

Level plan: In this plan, the production quantity is constant at the average demand level 460 (= 5520/12). The details of this plan are given in Table 4.3. The plan results in inventory in some months and shortages in other months. The inventory/shortage amounts can be obtained by comparing the demand and production. For example, in the first month the demand is 210 and the production is 460. Therefore, 250 (= 460 – 210) units go in stock as inventory. In the second month, the demand is 440 and the production is 460. Production exceeds demand by 20 units and therefore the inventory increases to 270 (= 250 + 20). In the third period, the demand (600) exceeds production (460) by 140 units. Therefore, 140 units will be withdrawn from the inventory to meet the demand and the inventory decreases by 140 units to a total of 130 units.

This is how the inventory levels are calculated. The inventory levels can also be calculated by simply taking the difference between cumulative production and cumulative demand in each period. The third column which is derived from the

Table 4.3 Level Production Plan

Month	*Expected Demand*	*Production Plan*	*Cumulative Demand*	*Cumulative Production*	*Inventory*
1	210	460	210	460	250
2	440	460	650	920	270
3	600	460	1250	1380	130
4	300	460	1550	1840	290
5	480	460	2030	2300	270
6	610	460	2640	2760	120
7	560	460	3200	3220	20
8	800	460	4000	3680	–320
9	200	460	4200	4140	–60
10	400	460	4600	4600	0
11	380	460	4980	5060	80
12	540	460	5520	5520	0
Total	5520	5520			
Average	460				

Note: Inventory = cumulative production – cumulative demand.

second column gives the cumulative demand. The cumulative demand in a given month is the total of all demands up to and including that particular month. For example, the cumulative demand for period 3 will be 210, 440, and 600, for a total of 1250. In a given period, there will be inventory if the cumulative production exceeds cumulative demand and there will be a shortage if the cumulative production is less than the cumulative demand. As shown in Table 4.3, there are shortages in months 8 and 9. There is a shortage of 320 units in month 8 because by the end of eighth month we have produced a total of 3680 units but the demand is 4000 units. In the next month, we produce 460, whereas the demand is 200 units. So the shortage goes down to 60 units. This is how the calculations are done from period 1 to the last period. In months 10 and 12, the inventory or shortages are zero.

Figure 4.1 gives a graphical representation of the Level plan. The cumulative demand (solid line) and cumulative production (dotted line) are plotted in this figure. These graphs depict that in some periods the dotted line is above the solid line (indicating inventory) and below the solid line (indicating shortages) in other periods. The solid line is above the dotted line in months 8 and 9, indicating shortages as discussed earlier. The inventory (shortages) levels are zero in periods 10 and 12. It may also be noted that this plan may not work in service industries because the services cannot be inventoried. The Chase plan, discussed next, is more appropriate for service industries.

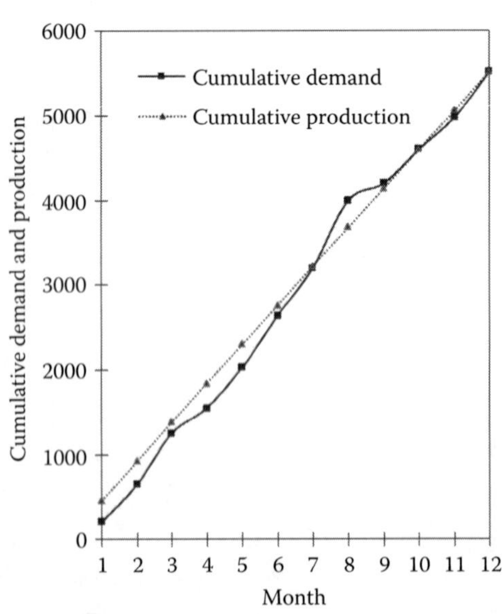

Figure 4.1 Graphical representation of level plan.

Chase plan: In this plan, production is equal to the demand in every period: 210 in the first month, 440 in the second month, 600 in the third month, and so on for a total of 5520 at the end of the year. This is called the "chase plan" because production "chases" demand. The inventory and shortages are calculated just like these were calculated in Plan 1. However, there are no inventories and shortages in this plan. The details of the plan are given in Table 4.4. The cumulative production and cumulative demand graphs overlap as shown in Figure 4.2.

Hybrid plan: The Hybrid plan, also known as "Mixed plan," is a mixture of the Level and Chase plans. The Level and Chase plans are at two extremes because in Level plan the production level is constant and in Chase plan it changes in every month. In the mixture of these plans, we can produce at a selected level for a few months, then change the production level for a few months, and may change it again. There can be many mixed plans. One such plan is given in Table 4.5. In this plan, 440 units are produced per month for the first 6 months; then the production level is changed to 680 per month for the next 2 months; and finally 380 units are produced per month for the last 4 months. In this plan, 440 represents the average

Table 4.4 Chase Production Plan

Month	*Expected Demand*	*Cumulative Demand*	*Production Plan*	*Cumulative Production*	*Inventory*
1	210	210	210	210	0
2	440	650	440	650	0
3	600	1250	600	1250	0
4	300	1550	300	1550	0
5	480	2030	480	2030	0
6	610	2640	610	2640	0
7	560	3200	560	3200	0
8	800	4000	800	4000	0
9	200	4200	200	4200	0
10	400	4600	400	4600	0
11	380	4980	380	4980	0
12	540	5520	540	5520	0
Total	5520				

Note: Inventory = cumulative production − cumulative demand.

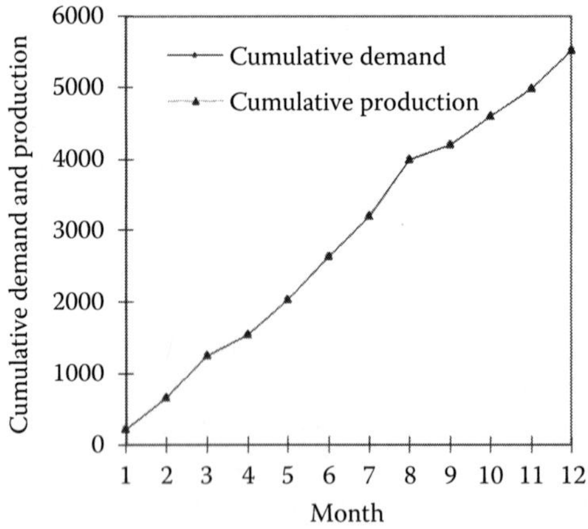

Figure 4.2 Graphical representation of chase plan.

Table 4.5 Hybrid (Mixed) Production Plan

Month	*Expected Demand*	*Cumulative Demand*	*Production Plan*	*Cumulative Production*	*Inventory*
1	210	210	440	440	230
2	440	650	440	880	230
3	600	1250	440	1320	70
4	300	1550	440	1760	210
5	480	2030	440	2200	170
6	610	2640	440	2640	0
7	560	3200	680	3320	120
8	800	4000	680	4000	0
9	200	4200	380	4380	180
10	400	4600	380	4760	160
11	380	4980	380	5140	160
12	540	5520	380	5520	0
Total	5520				

Note: Inventory = cumulative production – cumulative demand.

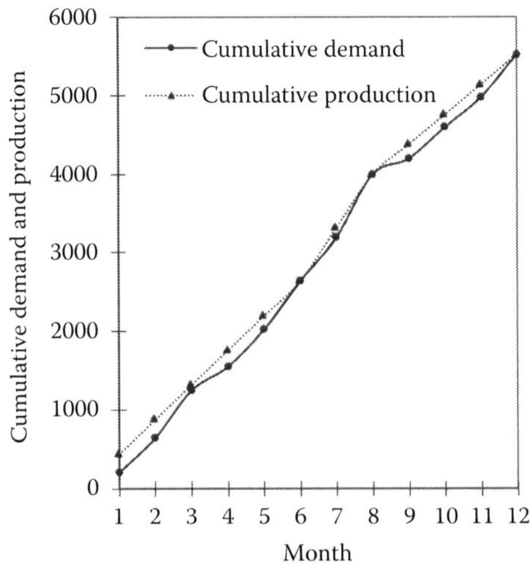

Figure 4.3 Graphical representation of hybrid (mixed) plan.

of the first 6 months, 680 is the average of months 7 and 8, and 380 is the average of the last 4 months. Inventories and shortages are calculated as we did earlier in the Level plan. Table 4.5 shows the inventory/shortage levels. The plan is depicted graphically in Figure 4.3. This plan is based on judgment, and different hybrid plans will give different inventory/shortage levels.

In discussing these plans, we did not consider the constraints on production capacity which is the subject matter of the following section.

4.3.2 Production Capacity

The feasibility of the plans discussed in the previous section depends on the productive capacity of the plant. A plant might be operating in a single shift or two shifts; or some plants may even work round the clock for three shifts also. In this chapter, we consider only the case of those plants that operate in a single shift and may work overtime if necessary. The production capacity is specified as the number of units that can be produced in a single shift during regular time. This capacity can be increased by working overtime; and overtime capacity is generally specified as a percentage of regular time. Outsourcing (subcontracting) can also augment the in-house production capacity.

Every plant is designed for a specific regular time capacity and investment in machines and facilities is made to achieve this capacity. In AP, we assume that the investment in facilities will not be changed during the planning period because these are long-term investments. Marginal adjustments to the productive capacity can be made by changing the work force. Large adjustments in productive capacity, by changing the workforce level, can be made in those cases where the output is largely labor-dependent, particularly in service organizations. See Example 3 in Section 4.5.

In Example 1, the plant could have been designed to produce, say, 460 units per month during regular time in a single shift representing the level production of 460 in every period. This is appropriate for the Level plan.

However, in the Chase plan, production varies from a low of 200 in period 9 to a high of 800 in period 8. For what level of demand should the plant be designed if the Chase plan is being used? Suppose the plant is designed at a level of 500 units per month. In that case, the plant will remain idle for some part of those periods where production is less than 500; and on the other hand, if the production requirements are more than 500, then overtime and/or subcontracting will have to be used. The plant could even be designed to produce 800 units per month during regular time. In this case, the plant will remain idle for some part of all months except month 8 when the demand is 800. The plant could also be designed at 200 units per month during regular time which is the lowest demand in any month. In that case, overtime and/or subcontracting will be used in all months except in month 9 when the demand is 200.

Let us do some more analysis of the mixed plan in which the production level is 440 per month for the first 6 months, then it increases to 680 per month for the next 2 months, and finally the production level is 380 per month for the last 4 months. Suppose the plant is designed for a production capacity of 480 units per month during regular time. In this plan, for the first 6 months, we will produce 440 units every month and there will be an idle capacity of 40 (480 – 440) units in each month. Similarly, during the last 4 months the idle capacity will be 100 (480 – 380) units per month since the production level is 380 per month.

In months 7 and 8, the production level is 680; but only 480 units can be produced during regular time. The additional 200 units have to be obtained by working overtime and/or through subcontracting. Suppose overtime is limited to 20% of regular time production, meaning 96 (= 20% of 480) units can be produced during over time. So we produce 480 units during regular time per month and 96 units during over time per month. However, total production should have been 680. The remaining 104 (680 – 480 – 96) units will be obtained through subcontracting. The maximum number of units that can be subcontracted depends on the availability of appropriate suppliers.

Each of these plans has cost implications. Various components of aggregate plans as discussed above include: regular time work, overtime work, subcontracting, changing regular time production levels, inventories, shortages, and idle capacity. Changing production level can be accomplished by changing work force levels up or down. The costs of these components are added together to find the total cost of a plan. The plans are compared based on total costs to identify the best plan.

We are going to describe the calculations of various costs using Examples 2 and 3 directly ahead. Example 2 is a manufacturer making products for sale which can be inventoried when productive capacity is greater than demand. Example 3 studies AP in a blood-testing laboratory which is a service-type situation. When the laboratory has more testing capacity on hand than requests for testing, some testers are idle and testing equipment is underutilized. The main difference between Examples 2 and 3 is that with services, inventory cannot be built up during periods when demand for that service falls below supply of that service.

4.4 Example 2: Aggregate Manufacturer's Production Planning

Consider the data given in Table 4.6. This table gives the expected aggregate demand for a manufacturer's products. A similar set of data will be used for Example 3 which deals with aggregate service production planning for a blood-testing laboratory. The service case and the manufacturing case use the same set of numbers for demand levels, production capacity, and applicable costs, but they are used to develop different strategies for matching supply and demand.

Table 4.6 also provides the number of working days for each month over a 12-month planning horizon. The total yearly demand is 5856 product units (henceforth called units) with an average of 488 units per month. The demand varies from a high of 650 units in August to a low of 350 units in October. The total number of working days in the year is 251. The demand per day for each month and the average demand for the 12-month planning period are also included in this table. Inset 2 shows the calculations for the units of product demanded per day.

INSET 2

Demand per day: Unit demand per day in a given month is a function of the demand in that month and the number of working days in that month.

Demand per day = Demand per month/working days per month.

Suppose in a given month, say March, the demand is 600 units and the number of working days is 22. Then,

Demand per day = 600/22 = 27.27 units.

The average demand per day for the entire 12-month period can be calculated in the same way.

The average demand per day for the 12-month period is 23.33 (= 5856/251).

Table 4.6 Data for AP Problem for the Manufacturer

Month	*Jan.*	*Feb.*	*Mar.*	*Apr.*	*May*	*Jun.*	*Jul.*	*Aug.*	*Sep.*	*Oct.*	*Nov.*	*Dec.*	*Total*	*Average*
Expected demand	400	440	600	396	480	610	560	650	450	350	380	540	5856	488
Working days	22	20	22	21	20	22	19	22	22	21	20	20	251	
Expected demand per day	18.18	22.00	27.27	18.86	24.00	27.73	29.47	29.55	20.45	16.67	19.00	27.00		23.33

Note: This example employs fractional numbers for expected demand per day. These fractions result from dividing expected demand by working days.

How does the manufacturer plan to meet the fluctuating requirements? That is the question that AP addresses, and the answer depends on various costs and the production capacity that are influenced by the planning strategies. Various costs and capacity calculations are discussed in Section 4.4.1. It may be noted that there are a different number of working days in every period due to holidays such as Thanksgiving, Christmas, etc. Also each month will have a different number of days, for example, January (31), February (28 or 29), March (31), and April (30) etc.

4.4.1 Costs in AP

The costs incurred in AP are listed in the Inset 3.

INSET 3

Costs in Aggregate Planning

- Inventory carrying (or holding) cost—(this cost is applied to the number of units in stock). This is the case where supply (productive capacity) is greater than the demand.
- Stock-out or shortage or back order cost. The backorder costs apply when demand for units is greater than the supply (productive capacity).
- Regular time labor cost.
- Overtime labor cost.
- Subcontracting or outsourcing cost.
- Cost of increasing production level (hiring workers).
- Cost of decreasing production level (firing workers).
- Material costs.

Inventory carrying (holding) costs usually include the cost of capital, storage, insurance, etc. A more detailed description of inventory costs is given in the chapter on inventory management. This cost is specified as per unit per period. It arises when production supply is greater than demand.

Stock-out or shortage or backorder cost is the cost incurred for not meeting the demand on time. When shortages occur, the demand is either lost or backlogged. Backlogging of demand means that the customer waits for the order to be fulfilled in a future period. Lost orders and backlogged orders can have a substantial cost. Lost orders sometimes translate into the loss of the lifetime value of customers.

In manufacturing situations, shortage costs are usually based on an estimate of lost revenue, possibly ameliorated by giving discounts to customers for late deliveries. There is also the potential damage to the company's reputation for reliability which can adversely affect future business and revenues. In this example, we assume that the demand is backlogged. The shortage cost is specified as per unit outage per period of

Table 4.7 Cost Data for Manufacturer's Example of Aggregate Production Planning

Shortage cost per unit per period ($)	100
Inventory carrying (holding) cost per unit per period ($)	25
Cost of increasing production level per unit* ($)	200
Cost of decreasing production level per unit* ($)	300
Regular time labor cost per hour ($)	20
Overtime labor cost per hour ($)	30
Subcontracting (outsourcing) cost per unit ($)	650
Material cost per unit ($)	100

* Increase and decrease in production levels are normally achieved by hiring and firing workers, respectively.

time. When a flyer is bumped from a flight, the airline compensates with hundreds of dollars. That is a fair representation of the effect of being out of seats on a flight.

Regular time labor cost is a function of the normal hourly wage rate and the number of hours worked during regular time in each month. Similarly, the overtime labor cost is a function of the hourly overtime wage rate and the number of overtime hours worked. Subcontracting cost is the cost of buying the product from outside suppliers and is specified as cost per unit purchased.

The material cost per unit is the cost of the material from which the item is produced. The production level can be changed up or down by hiring or firing (layoff) workers, respectively. The costs are incurred in adjusting production levels. For this example, the estimates used are shown in Table 4.7. Actual costing requires a true understanding of all elements of the specific system's characteristics and cost data. P/OM is responsible for providing guidance and wisdom in assigning costs.

4.4.2 Production Capacity

The feasibility of a production plan depends on the production capacity of the plant. The information that establishes production capacity (in general) includes the number of working days in each month, the number of operating shifts, the number of workers, amount of overtime allowed, and the maximum number of units that can be bought through subcontracting (outsourcing). A plant might operate in a single shift or in two shifts; some plants may even work round the clock for three shifts.

In this chapter, all of our text examples are for single-shift operations. Regular production capacity is specified as the number of units that can be produced in a single shift during regular time by a given number of workers. This capacity can be increased by working overtime; and the overtime capacity is generally specified as a percentage of regular time capacity. Outsourcing (subcontracting) can augment the in-house production capacity. Adjustments to the production capacity are made by changing the number of workers.

For this example, the production time per unit is assumed to be 4 h and each worker is on hand for 8 h per day. Also, we assume that the number of workers remains constant for each entire month. We assume that no defective items are produced. Alternatively, the demand figures can be inflated to allow for delivery of the required number of good units. This is often done in the job shop where shorter runs occur. In this regard, we will solve our AP problems with the simple assumption that actual demand is equal to the expected demand and production has no defectives. Discrepancies that arise between the expected and actual demands can be adjusted at the time of the periodic revisions of plans.

The first step in developing alternate production plans is to establish a production level per day for each month. This production level remains constant within a given month but may change from one month to the next. The calculations for the production level per day and the number of workers required to meet that level are given in Inset 4.

INSET 4

Monthly production level per day: The production level per day in a given month is a function of the production in that month and the number of working days in that month.

Production level per day = production per month/working days in that month

Suppose in a given month, say March, we expect to produce 600 units and the number of working days is 22. Then to produce 600 units in 22 days,

Production level per day = 600/22 = 27.27.

The fractional result can be interpreted in various ways. It can be rounded up and treated as inventory. Alternatively, it can be completed using overtime. We deal with a production level of 27.27 units as production of 27 units on that given day; then, production continues to work on the unfinished product on the next day.

Number of workers required: The number of workers required on any given day is a function of the production demand for that day, the number of hours worked on that day, and the production time (time required to produce one unit which is equal to 4 h in our example).

Number of workers = (production × production time per unit)/hours worked

In March suppose each production employee works for 8 h per day and required production time is 4 h per unit. To achieve a production level of 27.27, the required number of workers is: (27.27 × 4)/8 = 13.64. A work force of 13.64 is considered as 13 full-time workers and 0.64 part-time workers.

We will now develop several production plans and discuss them in the next section.

4.4.3 Production Plans

A production plan specifies the number of units to be produced per day. For this example, the number remains constant within a given month but may change from one month to the next month. Table 4.6 gives the expected demand per day for each month in the planning horizon. The calculation of expected demand per day was explained in Inset 2. The expected demand per day fluctuates between a minimum of 16.67 in October to a maximum of 29.55 in August. The average demand per day for the 12-month period is 23.33 (see Inset 2).

We will now develop different plans that specify the number of units produced in each month. We are going to develop the following four production plans: (1) Level, (2) Chase, (3) Hybrid, and (4) a plan that uses overtime. Three of these were also discussed in Example 1.

We assume that the plant is currently set up to produce 21 units per day.

4.4.3.1 Level Plan

In the Level plan, the production per day is constant at the average demand level of the 12-month period which is 23.33 units per day. Table 4.8 gives the calculations for this plan. The production in any given month is calculated by multiplying the number of days in that month by the production rate per day (23.33). The production numbers are rounded to the nearest integer. For example, in January, 513 is used for the number of units produced whereas 23.33 × 22 = 513.26.

However, the demand in January is only 400 units. Therefore, 113 (513 – 400) units go into inventory. In February, 467 (23.33 × 20) units are produced, whereas the expected demand is for 440 units. Therefore, 27 units (467 – 440) are added to inventory and the inventory level at the end of February is 140. In March, the production is 513 (23.33 × 22) which is 87 units less than the demand (600 units). Therefore, 87 units are withdrawn from inventory to meet the demand for March. The inventory level at the end of March becomes 53. In this way, the calculations continue.

In some months, the production is less than the demand and there are not enough units in inventory to meet the demand. In such situations, shortages occur. For example, in July we produce 117 units less than the demand (443 – 560). Only ...nits are available in inventory at the end of June. These 37 units are used to off... shortage of July. However, there still will be a shortage of 80 (117 – 37) units. ...st, the shortage increases to 217, then decreases to 154 units in September, ...decreases to 14 units in October. At the end of November, there are 73 ...ntory. Finally, we see zero inventories at the end of December. Total

Table 4.8 Level Plan: Production per Day = 23.33 units for a Manufacturer of Storable Product

Month	*Jan.*	*Feb.*	*Mar.*	*Apr.*	*May*	*Jun.*	*Jul.*	*Aug.*	*Sep.*	*Oct.*	*Nov.*	*Dec.*	*Total*
Expected demand (units)	400	440	600	396	480	610	560	650	450	350	380	540	5856
Working days	22	20	22	21	20	22	19	22	22	21	20	20	251
Production per day (units) avg	23.33	23.33	23.33	23.33	23.33	23.33	23.33	23.33	23.33	23.33	23.33	23.33	
Total production (units)	513	467	513	490	467	513	443	513	513	490	467	467	5856
Inventory addition/shortages (units)	113	27	–87	94	–13	–97	–117	–137	63	140	87	–73	
Ending inventory/shortages (units)	113	140	53	147	134	37	–80	–217	–154	–14	73	0	
Ending inventory (units)	113	140	53	147	134	37	0	0	0	0	73	0	697
Ending shortages (units)	0	0	0	0	0	0	80	217	154	14	0	0	465
Inventory cost ($)	2825	3500	1325	3675	3350	925	0	0	0	0	1825	0	17,425
Shortage cost ($)	0	0	0	0	0	0	8000	21,700	15,400	1400	0	0	46,500
Change in production level	2.33	0	0	0	0	0	0	0	0	0	0	0	
Change in production level—up	2.33	0	0	0	0	0	0	0	0	0	0	0	2.33
Change in production level—down	0	0	0	0	0	0	0	0	0	0	0	0	0
Cost of changing production level up ($)	466	0	0	0	0	0	0	0	0	0	0	0	466
Cost of changing production level down ($)	0	0	0	0	0	0	0	0	0	0	0	0	0

The number of workers required in the Level plan is constant in each month and is calculated as follows: number of workers = (total demand in 12 months × production time per unit)/(number of days in 12 months × hours worked per day) = (5856 × 4)/(251 × 8) = 11.67. Alternatively, number of workers = (production in units per day × production time per unit)/hours worked per day = (23.33 × 4)/8 = 11.67.

production is equal to demand. Nevertheless, with the Level plan, both overstock and out-of-stock situations occur in spite of ending the year with zero inventories.

The inventory and shortage costs in each month are calculated by multiplying the end of month inventory or shortage by the respective costs. Alternatively, we can total the end-of-month inventories (sum of all end-of-month inventories = 697) and also total the end-of-month shortages (sum of all end-of-month shortages = 465). Multiply these numbers by their respective costs. Therefore, the total inventory cost is \$17,425 (697 × 25) and the total shortage cost is \$46,200 (462 × 100). The total production during regular time is 5856. Therefore, the regular time labor cost is \$468,480 (5856 × 4 × 20). The total material cost is \$585,600 (5856 × 100). The production level changes from 21 to 23.33 in January. After that the production level remains constant at 23.33. Therefore, the cost of changing production level is \$466 (2.33 × 200). *The total annual cost of this plan is \$1,118,171. Note: rounding errors can make small changes in the various totals alluded to in this paragraph.*

4.4.3.2 Chase Plan

An alternative to the Level plan is the Chase plan in which the production level per day in each month is equal to the demand per day in that month. The production per day changes from 1 month to the next. This is called "Chase plan" because production chases (matches) demand. The details of this plan are given in Table 4.9. In this plan, there are no inventories or shortages. However, the production level has to be adjusted in each month. In January, the production level is 18.18 units per day. Therefore, the production level is reduced by 2.82 (21 − 18.18) units, assuming that the plant is currently set up to produce 21 units per day. In February, the production level increases to 22 and the level has to be adjusted upward by 3.82 (level in February − level in January = 22 − 18.18) units.

As shown in Table 4.9, the production levels have to be adjusted upwards in March (5.27), May (5.14), June (3.73), July (1.75), August (0.07), November (2.33), and December (8.00). The numbers in parentheses show the amount of upward adjustments. In January, April, September, and October, the production levels have to be adjusted down by 2.82, 8.42, 9.09 and 3.79, respectively. The total upward adjustment is 30.11 with a cost of \$6023 (30.11 × 200). The total downward adjustment is 24.11 with a cost of \$7234 (24.11 × 300). The total production during regular time is 5856. Therefore, the regular time labor cost is \$468,480 (= 5856 × 4 × 20) and the total material cost is \$585,600 (= 5856 × 100). *The total annual cost of this plan is \$1,067,336. Note: rounding errors can make small changes in the various totals alluded to in this paragraph.*

4.4.3.3 Mixed or Hybrid Plan

The Level and Chase plans are at two extremes. The Level plan keeps the workforce intact. There are no firings and employees feel secure. With chasing, it is important

Table 4.9 Chase Plan: Production per Day is Variable Matching Expected Demand for a Manufacturer of Storable Product

Month	*Jan.*	*Feb.*	*Mar.*	*Apr.*	*May*	*Jun.*	*Jul.*	*Aug.*	*Sep.*	*Oct.*	*Nov.*	*Dec.*	*Total*
Expected demand (units)	400	440	600	396	480	610	560	650	450	350	380	540	5856
Working days	22	20	22	21	20	22	19	22	22	21	20	20	251
Production per day (units)	18.18	22.00	27.27	18.86	24.00	27.73	29.47	29.55	20.45	16.67	19.00	27.00	
Total production (units)	400	440	600	396	480	610	560	650	450	350	380	540	5856
Inventory addition/shortages (units)	There are no inventories or shortages in this plan												
Change in production level (units)	−2.82	3.82	5.27	−8.42	5.14	3.73	1.75	0.07	−9.09	−3.79	2.33	8.00	
Change in production level—up (units)	0.00	3.82	5.27	0.00	5.14	3.73	1.75	0.07	0.00	0.00	2.33	8.00	30.11
Change in production level—down (units)	2.82	0.00	0.00	8.42	0.00	0.00	0.00	0.00	9.09	3.79	0.00	0.00	24.11
Number of workers	9.09	11.00	13.64	9.43	12.00	13.86	14.74	14.77	10.23	8.33	9.50	13.50	

Chase plan: Production rate = demand rate in each month. The number of workers is calculated using the formula given in Inset 4. For example, in May the required number of workers is $(24 \times 4)/8 = 12$.

to explain to the workforce how its interests are represented by this strategy which hires and fires according to the customer demand system. These considerations lead to the idea that we could have a "Mixed or Hybrid" plan in which the production rate is kept constant at a given level for a few months and then changed to another level. This adds some stability for the benefit of the employees while producing savings on seasonal shifts in demand.

The production level can be changed several times during the planning horizon. In the Hybrid plan proposed in this section, the production level is constant at 23.04 units per day for the first 6 months which is the average demand for the first 6 months (January–June); then it changes to 26.35 which is the average demand for the next 3 months (July, August, and September) and finally the production level changes to 20.82 which is the average demand for the last 3 months. This plan entails inventories, shortages, and changing production levels up and down. The calculations are done in the same way as for the Level and Chase plans. This plan is shown in Table 4.10. *The total annual cost of this plan is $1,093,359* which includes the regular time labor cost ($468,480), material cost ($585,600), inventory carrying cost ($17,650), shortage cost ($18,900), increasing production level ($1070), and decreasing production level ($1659). *Note: rounding errors can make small changes in the various totals alluded to in this paragraph.*

4.4.3.4 Overtime in a Chase Plan with Level Production and Overtime

In this plan, we show the usage of overtime for aggregate production planning. We use overtime in a combination of the Level plan and the Chase plan. In the Level plan, the production is kept constant at a specified level. In Example 2, the Level plan used 23.33 as the production rate per day. It was the average production rate per day over the planning horizon. However, when overtime is used, we can set the production level at a lower number. For this problem, we have set the level production at 16.67 units per day. It could have been any other reasonable number. The number 16.67 is equal to the minimum of the production per day in 12 months. The minimum production per day, 16.67, is for October (see Table 4.6).

Regular time production in a given month is obtained by multiplying 16.67 by the number of days in that month. The production numbers are rounded to the nearest integer. The difference between the demand and production in any month gives the number of units that have to be produced during over time. For example, the production in January is 367 (16.67×22) units. Therefore, overtime production will be 33 ($400 - 367$). The calculations for this plan are given in Table 4.11.

In this plan, 4184 units are produced during regular time and 1672 units are produced during overtime for a total of 5856 units. Therefore, the regular time production cost is $334,720 ($4184 \times 20 \times 4$) and the overtime cost is $200,640 ($1672 \times 30 \times 4$). The total material cost is $585,600 which is the same as in the other plans. The production level per day had to be decreased by 4.33 which is the

Table 4.10 Mixed (Hybrid) Plan

Month	*Jan.*	*Feb.*	*Mar.*	*Apr.*	*May*	*Jun.*	*Jul.*	*Aug.*	*Sep.*	*Oct.*	*Nov.*	*Dec.*	*Total*
Expected demand (units)	400	440	600	396	480	610	560	650	450	350	380	540	5856
Working days	22	20	22	21	20	22	19	22	22	21	20	20	251
Production per day	Average Demand per Day (Jan. to Jun.)							Average Demand per Day (Jul. to Sep.)			Average Demand per Day (Oct. to Dec.)		
	23.04	23.04	23.04	23.04	23.04	23.04	26.35	26.35	26.35	20.82	20.82	20.82	
Total production (units)	507	461	507	484	461	507	501	580	580	437	416	416	5856
Inventory addition/ shortages (units)	107	21	–93	88	–19	–103	–59	–70	130	87	36	–124	
Ending inventory/ shortages (units)	107	128	35	122	103	0	–59	–130	0	87	124	0	
Ending inventory (units)	107	128	35	122	103	0	0	0	0	87	124	0	706
Ending shortages (units)	0	0	0	0	0	0	59	130	0	0	0	0	189
Inventory cost ($)	2675	3200	875	3050	2575	0	0	0	0	2175	3100	0	17,650
Shortage cost ($)	0	0	0	0	0	0	5900	13,000	0	0	0	0	18,900
Change in production level (units)	2.04	0	0	0	0	0	3	0	0	– 6	0	0	

continued

Table 4.10 (continued) Mixed (Hybrid) Plan

Month	*Jan.*	*Feb.*	*Mar.*	*Apr.*	*May*	*Jun.*	*Jul.*	*Aug.*	*Sep.*	*Oct.*	*Nov.*	*Dec.*	*Total*
Change in production level—up (units)	2.04	0.00	0.00	0.00	0.00	0.00	3.31	0.00	0.00	0.00	0.00	0.00	5.35
Change in production level—down (units)	0.00	0.00	0.00	0.00	0.00	0.00	0.00	0.00	0.00	5.53	0.00	0.00	5.53
Cost of changing production level—up ($)	407.87	0.00	0.00	0.00	0.00	0.00	661.97	0.00	0.00	0.00	0.00	0.00	1070
Cost of changing production level—down ($)	0.00	0.00	0.00	0.00	0.00	0.00	0.00	0.00	0.00	1658.86	0.00	0.00	1659
Number of workers	11.52	11.52	11.52	11.52	11.52	11.52	13.17	13.17	13.17	10.41	10.41	10.41	

Table 4.11 Overtime in a Chase Plan with Level Production

Month	*Jan.*	*Feb.*	*Mar.*	*Apr.*	*May*	*Jun.*	*Jul.*	*Aug.*	*Sep.*	*Oct.*	*Nov.*	*Dec.*	*Total*
Expected demand (units)	400	440	600	396	480	610	560	650	450	350	380	540	5856
Working days	22	20	22	21	20	22	19	22	22	21	20	20	251
Production per day (units)	16.67	16.67	16.67	16.67	16.67	16.67	16.67	16.67	16.67	16.67	16.67	16.67	
Regular time production (units)	367	333	367	350	333	367	317	367	367	350	333	333	4184
Overtime production (units)	33	107	233	46	147	243	243	283	83	0	47	207	1672
Total production (units)	400	440	600	396	480	610	560	650	450	350	380	540	5856
Ending inventory/shortages (units)	0	0	0	0	0	0	0	0	0	0	0	0	
Ending inventory (units)	0	0	0	0	0	0	0	0	0	0	0	0	0
Ending shortages (Units)	0	0	0	0	0	0	0	0	0	0	0	0	0
Inventory cost ($)	0	0	0	0	0	0	0	0	0	0	0	0	0
Shortage cost ($)	0	0	0	0	0	0	0	0	0	0	0	0	0
Change in production level (units)	−4.33	0	0	0	0	0	0	0	0	0	0	0	
Change in production level—up (units)	0	0	0	0	0	0	0	0	0	0	0	0	0
Change in production level—down (units)	4.33	0	0	0	0	0	0	0	0	0	0	0	4.33
Cost of changing production level up ($)	0	0	0	0	0	0	0	0	0	0	0	0	0
Cost of changing production level down ($)	1299	0	0	0	0	0	0	0	0	0	0	0	1299
Number of workers	8.34	8.34	8.34	8.34	8.34	8.34	8.34	8.34	8.34	8.34	8.34	8.34	

difference between the current production level of 21 and the production level in January of 16.67. The cost of changing the production level is $1299.

The total cost of this plan is $1,122,259. Note: rounding errors can make small changes in the various totals alluded to in this paragraph.

4.4.4 Selection of a Production Plan

Table 4.12 gives the cost comparison of the four production plans discussed above.

The Chase plan is the least expensive plan and is a top candidate for being selected. However, other nonfinancial considerations must also be taken into account when choosing the final plan. The aggregate plans are developed within the overall constraints imposed by corporate policies and management directives. The final selection of the plan depends on executive judgment. A least expensive plan may not be selected because of other qualitative considerations. For example, the least-cost plan may not find favor with management if that plan requires much hiring and firing of workers. Management may prefer having a stable workforce to project the image of a good employer within its community. Similarly, management may not choose a plan that allows shortages because the company wants to provide a 100% service level to its customers. Alternatively, a plan that uses overtime production may be preferred to one using subcontracting even if the overtime plan may be more expensive. Overtime work puts money in workers' pockets.

4.4.5 Beginning Conditions

Beginning conditions specify the existing inventory level and the existing production level per day when planning starts. The same beginning conditions must be used in each plan for comparing the costs.

Table 4.12 Cost Comparison of the Four Production Plans

	Level Plan	*Chase Plan*	*Hybrid Plan*	*Level Plan with Overtime*
Regular time labor cost ($)	468,480	468,480	468,480	334,720
Over time labor cost ($)				200,640
Inventory carrying cost ($)	17,425		17,650	
Shortage cost ($)	46,200		18,900	
Increasing production level ($)	466	6023	1070	
Decreasing production level ($)		7234	1659	1299
Material cost ($)	585,600	585,600	585,600	585,600
Total cost ($)	1,118,171	1,067,336	1,093,359	1,122,259

The beginning inventory is used to reduce the production requirements of the first period while solving the problem. For example, if the beginning inventory for the example discussed above was 100 units, then the demand in the first month will be reduced to 300 (= 400 – 100) because the 100 units in stock can be used to meet the demand of the first month. We assumed a beginning inventory of zero in the above example.

The existing production level affects the cost of changing the production level (up or down) in the first month. In the example above, we assumed the existing production level per day to be 21 units. Therefore, we had to increase the production level by 2.33 (= 23.33 – 21) in the Level plan. The production level was decreased by 2.82 (= 21 – 18.18) in the Chase plan, increased by 2.04 (= 23.04 – 21) in the Hybrid plan, and decreased by 4.33 (= 21 – 16.67) in the Level plan with overtime.

4.5 Example 3: Aggregate Production Planning in a Service Industry

This example is built around Example 2 with the purpose of illustrating some of the differences that distinguish AP for making goods from providing services. The healthcare service sector is very large and growing. It requires a great deal of testing for diagnostic reasons. A blood-testing laboratory provides a good illustration of a service system with testing capacity that cannot be fully used when there is not enough demand. Technicians' time cannot be stored when technicians are idle. On the other hand, when demand is greater than the technical capacity to supply tests, the tests must be postponed or subcontracted or done on overtime. The big difference with the manufacturer's example is the inability to create a useful inventory when capacity to test is greater than demand for testing. That point will now be illustrated by Example 3 using the data in Table 4.13.

This table gives the expected demand for blood tests in a laboratory. Blood tests are the product of the lab. Because all tests are individual and unique, they can only be made on demand. Demand for tests of different kinds are aggregated—some take longer than others and use different kinds of materials and equipment. The aggregation is based on averaging across the normal mix of blood test procedures. Demand, in this example, is stated in units (occasionally as test kit units).

We repeat that this service case and the prior manufacturing case use the same set of numbers for demand levels, production capacity, and applicable costs applied to different strategies for matching supply and demand. The difference to note is that the manufacturer can store units in inventory to meet later demand but the blood-testing laboratory cannot.

Table 4.13 provides the number of working days for each month over a 12-month planning horizon. The total yearly demand is 5856 blood tests with an average of 488 testing units per month. The demand varies from a high of 650 test kit units in August to a low of 350 units in October. The total number of working days in the year is 251. The demand per day for each month and the average demand for the

Table 4.13 Data for AP Problem for the Blood Test Laboratory

Month	*Jan.*	*Feb.*	*Mar.*	*Apr.*	*May*	*Jun.*	*Jul.*	*Aug.*	*Sep.*	*Oct.*	*Nov.*	*Dec.*	*Total*	*Average*
Expected demand	400	440	600	396	480	610	560	650	450	350	380	540	5856	488
Working days	22	20	22	21	20	22	19	22	22	21	20	20	251	
Expected demand per day	18.18	22.00	27.27	18.86	24.00	27.73	29.47	29.55	20.45	16.67	19.00	27.00		23.33

**Note:* This example employs fractional numbers for expected demand per day which result from dividing expected demand by working days. Demand for different kinds of blood tests are aggregated (as different manufactured products would be) into units (of blood tests required).

12-month planning period are also included in this table. Inset 5 shows the calculations for the blood tests demanded per day.

> **INSET 5**
>
> Demand per day: Blood test demand per day in a given month is a function of the demand in that month and the number of working days in that month.
>
> Demand per day = demand per month/working days per month
>
> Suppose in a given month, say March, the demand for blood tests is 600 units and the number of working days is 22. Then,
>
> Demand per day = 600/22 = 27.27 test units.
>
> The average demand per day for the entire 12-month period can be calculated in the same way.
>
> The average demand per day for the 12-month period is 23.33 (=5856/251).

The AP addresses the question: How does the laboratory plan to meet the fluctuating requirements? The answer depends on the costs and the production capacity set by the planning strategies. Costs and capacity calculations applicable to the blood-testing laboratory are discussed in Section 4.5.1.

Different numbers of working days occur in each period because of holidays like New Years and Labor Day, and the different number of days in a month. We multiply the number of workers by the number of hours per day (eight in this case) to find out how many regular time work hours are available per day for production. Production rate per day is based on the number of workers available in the month and the hours required for one unit blood test.

4.5.1 Costs in AP for the Blood-Testing Laboratory

The costs incurred in AP for the service system are differentiated in Inset 6.

We do not have inventory carrying (holding) costs in this service system example. In this case, the cost of supply being greater than demand is associated with idle workers (paid by the hour) and blood-testing equipment sitting around unused. While this cost is difficult to determine, it can be calculated and it is specified as dollars wasted per unit per period. The inability to create inventory that can be used later on when demand is greater than supply makes this a typical service example of aggregate scheduling.

Stock-out or shortage or backorder cost is the cost incurred for not meeting the demand on time. When shortages occur, the demand is either lost or backlogged. In this case, it is a cost for backlogging of demand for testing which means that the patient (customer), doctor and nurse, wait for the blood test order to be fulfilled in a

INSET 6

Costs in Aggregate Planning

- Instead of an inventory carrying (or holding) cost—(in this service example)—the cost of idle workers and underutilized facilities must be calculated. This occurs when supply of units by technicians is greater than the demand for units of blood tests.
- Stock-out or shortage or back order cost. In this case, the cost of backlogging is used. Demand for blood tests is greater than the supply (of testing capacity). We do not permit this lab to use an outside vendor, that is, no subcontracting or outsourcing cost.
- Regular time labor cost is the time of a technician to do a standard blood test on regular time.
- Overtime labor cost—in this example, no overtime is used.
- Subcontracting or outsourcing cost—in this example, neither cost is allowed.
- Cost of increasing production level (hiring workers and more equipment).
- Cost of decreasing production level (firing workers).
- Material cost.

future period. There can be a *very* high cost for this lack of information if the patient has an unknown critical situation that must be diagnosed on the basis of blood testing. In reality, the doctors could indicate a priority for testing urgency. Blood tests are being postponed because there is a shortage of testing personnel or of testing supplies, or of testing equipment. These are the main reasons that a shortage cost is incurred.

Also, when an urgent situation exists, there is the opportunity to subcontract with another laboratory or employ overtime.

In service situations, shortage costs are usually based on estimates of lost revenue if delay causes the hospital to switch to another laboratory. There is certainly some damage to the laboratory's image. Losing a grade-A reputation can affect adversely future business and revenues. In this example, we assume that the demand is backlogged. Neither patients nor doctors are happy. The shortage cost is specified as per unit outage per period of time.

Regular time labor cost is a function of the normal hourly wage rate for blood-testing technicians, and the number of hours worked during regular time in each month. Similarly, the overtime labor cost is a function of the hourly overtime wage rate and the number of overtime hours worked. Subcontracting costs can be quite substantial if, for example, outside testing is used when "our" laboratory does not have the proper equipment or skill sets.

The material cost per unit is the cost of the chemicals used for testing. As before, production levels can be changed up or down by hiring or firing (layoff) workers

Table 4.14 Cost Data for Blood Testing Using Aggregate Production Planning

The backorder cost arises because the laboratory is not able to provide the blood test on time in the given month. Probably, in this situation, planning should be done on a daily basis ($)	100
Idle time costs occur when there is too much capacity for testing. This is higher than regular time labor cost due to idle apparatus ($)	25
Cost of increasing production level per unit* ($)	200
Cost of decreasing production level per unit* ($)	300
Regular time labor cost per hour ($)	20
Overtime labor cost per hour—not used ($)	30
Subcontracting (outsourcing) cost per unit—not used ($)	650
Material cost per unit ($)	100

* Increase and decrease in production levels are normally achieved by hiring and dismissal, respectively. Changes are costly because blood-testing involves responsibility for proficiency.

respectively. Costs are incurred in adjusting production levels. For this example, the estimates used are shown in Table 4.14. Actual costing requires a true understanding of all elements of the specific system's characteristics and cost data. P/OM is responsible for understanding the total structure of blood testing.

4.5.2 Production Capacity

The quality of a production plan depends on proper use of the production capacity of the plant, or in this case, the blood-testing capabilities of the laboratory. The production capacity includes the number of working days in each month, the number of operating shifts, the number of workers, amount of overtime allowed, and the maximum number of units that can be bought through subcontracting (outsourcing). This laboratory might operate in a single shift or in two shifts; some laboratories work three shifts because expensive equipment justifies higher labor costs.

In Example 3, we employ a single shift without overtime. Regular production capacity is the number of units that can be produced in a single shift during regular time by a given number of workers. Especially in job shops, such as the blood-testing laboratory, adjustments to the production capacity are made by changing the number of workers.

The production time per testing unit is assumed to be 4 h and each technician works for 8 h per day. We want to point out that in blood testing there is a need to check and double check every test procedure used. For example, there are two measures of cholesterol (LDL and HDL). Both must be double-checked. Some blood tests take little time, while others are very time-consuming. Also, we assume that the number of workers remains constant for each entire month and that no defective items are produced. Quality control is important. The consequences of incorrect results can produce both physical and mental damage. To avoid such problems, checking is essential. The costs for doing that are included in the time required to produce a unit.

As before, we make an implicit assumption that actual demand is equal to the expected demand. Assume that statistical records are used to forecast the number of blood tests required by various medical offices on a monthly basis. Discrepancies that arise between the expected and actual demands can be adjusted at the time of the periodic revisions of plans. The first step in developing alternative production plans is to establish a production level per day for each month. This production level (of completed tests) remains constant within a given month but may change from one month to the next. The calculations for the production level per day and the number of workers required to meet that level are given in Inset 7.

We will now develop a Level plan. It is suggested in the Problem section at the end of this chapter that a Chase plan be developed using the same set of numbers.

4.5.3 Production Plans

The aggregate production plan specifies the number of blood test units to be produced per day. For Example 3, the number remains constant within a given month but may change from one month to the next month. Table 4.13 gives the demand per day for each month in the planning horizon. The calculation of demand per day was explained in Inset 1. The demand per day fluctuates between a minimum of 16.67 in October to a maximum of 29.55 in August. The average demand per day for the 12-month period is 23.33. We now develop our Level plan based on the blood-testing laboratory being currently set up to produce 21 units per day.

4.5.3.1 Level Plan

In the Level plan, the production per day is constant at the average demand level of the 12-month period which is 23.33 units per day. Table 4.15 gives the calculations for this plan. The production in any given month is calculated by multiplying the number of days in that month by the production rate per day (23.33 units). The production numbers are rounded to the nearest integer. For example, in January, 513 test kit units could be produced (23.33 × 22 = 513.26).

However, the demand in January is for only 400 units. Therefore, 113 units go into idle time (513 – 400). Idle technician time cannot be stored in inventory. In February, 467 (23.33 × 20) units could be produced, whereas the expected demand is for 440 test kit units. Therefore, 27 units (467 – 440) are added to the idle time

INSET 7

Monthly production level per day: The production level per day in a given month is a function of the production in that month and the number of working days in that month. The output rate of blood tests accomplished per day can be examined for potential productivity improvement using new technologies.

Production level per day = production per month/working days in that month

Suppose in a given month, say March, we expect to produce 600 test kit units and the number of working days is 22. Then to produce 600 complete test kit units in 22 days,

Production level per day = 600/22 = 27.27: The fractional result can be interpreted in various ways. It can be rounded down and treated as waste. Alternatively, it can be completed using overtime. We treat a production level of 27.27 as producing 27 test kit units on that given day; production will continue to work on unfinished product on the following day.

Number of workers required: The number of workers required on any given day is a function of the production for that day, the number of hours worked on that day, and the production time (time required to produce one unit).

Number of workers = (production × production time per unit)/hours worked

In March, suppose each worker works for 8 h per day and the production time is 4 h per testing kit (unit). To achieve a production output level of 27.27, the required number of workers is: (27.27 × 4)/8 = 13.64. A work force of 13.64 is considered as 13 full-time workers and 0.64 part-time workers. Note: If we rounded demand to 28 blood tests per day we would have (28 × 4)/8 = 14 workers per day in the month of March. The cost of an extra worker per day is (20 × 8 = 160) as compared to the cost of overtime (30 × 8 × 0.64 = 153.60) and the extra worker provides a buffer against unexpected problems, for example, a test gone awry.

category (with zero revenue but labor costs accumulate every hour). The accumulated idle time level at the end of February is 140.

In March, production is 513 units (23.33 × 22) which is 87 units less than the demand of 600 units. This results in 87 units being backordered and they must be carried over into April's demand. Thus, the demand in April is 396 and the 87 backordered units must be added to 396 yielding 483 units. Since production is 490, there are seven units of idle time recorded for April in Table 4.15. In May, there are no idle units. May demand is 480 units and May supply is 467. There is a shortfall of 13 units which will be backordered. No need to continue with the details. Table 4.15 shows all the results.

Table 4.15 Level Plan: Production per Day = 23.33 units for the Blood-Testing Laboratory

Month	*Jan.*	*Feb.*	*Mar.*	*Apr.*	*May*	*Jun.*	*Jul.*	*Aug.*	*Sep.*	*Oct.*	*Nov.*	*Dec.*	*Total*
Expected demand	400	440	600	396	480	610	560	650	450	350	380	540	5856
Working days	22	20	22	21	20	22	19	22	22	21	20	20	251
Production per day	23.33	23.33	23.33	23.33	23.33	23.33	23.33	23.33	23.33	23.33	23.33	23.33	
Total production	513	467	513	490	467	513	443	513	513	490	467	467	5856
Idle time (+) or backorders (–)*	113	27	–87	94	–13	–97	–117	–137	63	140	87	–73	
Idle time (+) or net backorders**	113	27	–87	7	–13	–110	–227	–364	–301	–161	–74	–147	
Idle time	113	27	0	7	0	0	0	0	0	0	0	0	147
Backorders	0	0	87	0	13	110	227	364	301	161	74	147	1484
Idle time cost	2825	675	0	175	0	0	0	0	0	0	0	0	3675
Backorder cost	0	0	8700	0	1300	11,000	22,700	36,400	30,100	16,100	7400	14,700	148,400
Change in production level	2	0	0	0	0	0	0	0	0	0	0	0	
Change in production level—up	2	0	0	0	0	0	0	0	0	0	0	0	2
Change in production level—down	0	0	0	0	0	0	0	0	0	0	0	0	0
Cost of changing production level up	466	0	0	0	0	0	0	0	0	0	0	0	466
Cost of changing production level down	0	0	0	0	0	0	0	0	0	0	0	0	0

* Total production – expected demand.

** Total production – (expected demand + prior period backorders).

Row 6 of Table 4.15 shows the idle time costs as plus values and the backorder costs as minus values. As we have seen in the paragraph above, they are obtained by subtracting row 2 (expected demand) from row 5 (total production). Row 7 shows idle time or net backorders. Idle time values are just carried down from Row 6. Net backorders, on the other hand, must be added to next month's demand. For example, in March there are 87 units backordered. They are added to April's demand of 396 units which yields 483 units of demand for April. Production in April is 490 units which exceeds demand of 483 units so there are seven units of idle time. Idle time accumulates as a simple sum in Row 8—the total of which is 147 units as seen in the rightmost column labeled Totals. Backorders of 13 units in May are added to the June demand level, that is, 610 + 13 = 623 units. June has production capacity of 513 units, so there is a shortfall of 623 – 513 =110 units which is displayed in Row 9 for June. Calculations continue in this way resulting in total backorders of 1484.

We are now ready to determine total cost as follows.

Total idle time costs are the sum of Row 10 or the total of Row 8 (147) multiplied by 25 which is equal to $3675. Total backorder cost is the Row 11's sum which is $148,400. Note that backorders sum to 1484 (Row 9) which is multiplied by $100—the cost of a unit backorder. The total production during regular time is 5855.83. Therefore, the regular time labor cost is $468,480 after rounding total production (5856 × 4 × 20). The total material cost after rounding total production is $585,600 (5856 × 100). There is a production level change. The production level changes from 21 to 23.33 in January. After this, the production level remains constant at 23.33. Therefore, the cost of changing production level is $466 (2.33 × 200). *The total cost of this plan is $1,206,621.*

This is the worst result of all the plans (including those discussed in Example 2) that we have studied. It is 7.5% worse than the worst of all of the prior plans. The result is not surprising because total backorder costs in Example 3 (for service systems) are too large and likely to get out of control. That suggests two things. First, increase the number of technicians and let the idle cost go up; second, opt for a chasing strategy. It might do better. In the Problem section, there is a request for analysis of the chasing strategy for the blood-testing laboratory. Doing the work entailed will provide great benefits. Refer to the chasing model for the manufacturer. Find out how much better the chasing strategy will be for the blood-testing laboratory. We need to recognize that service systems are (generally) more efficient when they follow the chasing demand strategy than when they use the level strategy.

Summary

Everyone leads a job shop life. There are many things to accomplish. Flow-shop type routines are partial, at best, and changing all of the time. Everyone must do AP to cope with such a great variety for the use of their time and money. AP for the commercial job shop is an internal production management function. It constantly matches classes of jobs with the resources needed to produce each of them. Its relationship to business planning and production scheduling is essential.

This chapter explores the methods for AP. Three AP policies are developed and compared. These are (1) level production with problems of making too much or having too little, (2) supply chasing demand, and (3) combinations of (1) and (2). The use of overtime and subcontracting in making aggregate plans has been discussed. There are differences between aggregate plans for manufacturing and service industries and they have been explained. We want to add that aggregation loses information. That may not seem beneficial for the systems approach but it is an important method used by systems analysts. Keep in mind that after aggregation, and problem solution—the opportunity for disaggregation should not be overlooked. With disaggregation, the spotlight is put back onto individuals (people, products, patrons of the arts) and their responsibilities.

Review Questions

1. Why is AP used? What does it do?
2. What is meant by aggregation of units?
3. Explain standard units of work.
4. Explain the use of backordering for goods.
5. Give one or more examples of how backordering can be used for services.
6. Describe the system's nature of AP from the point of view of classes of resources and product-mix families.
7. Explain why AP follows strategic planning.
8. Explain the statement that AP starts a chain reaction in the supply chain of suppliers–producer–customers.
9. Discuss the importance of forecasting for AP.
10. Explain the planning horizon and the updating interval.
11. A job shop manager said, "For the average job shop product, the best planning interval would range from 3 to 6 months." Might this statement provide a reasonable rule of thumb?
12. Why do forecasts for aggregated jobs have an advantage over forecasts for individual (disaggregated) jobs?
13. How can the effects of seasonal demands be taken into account for AP? Explain.
14. In some job shop industries, a smooth production rate is the preferred choice. Explain what this means and when it can be true.
15. In some job shop industries, the workforce size is altered to chase the expected demands. Explain what this means.
16. Compare smooth or level aggregate production policies with chasing policies. Explain when each is likely to be preferred.
17. At one time, the canning industry was totally dependent on harvest dates. As a result, major workforce alterations occurred sporadically. After careful

study, steps were taken to smooth the demand patterns. What measures might have helped?

18. What is a combination aggregate policy?
19. Explain how the trade-off model for workforce adjustment costs and inventory costs yields a better AP policy.

Problems

1. Seven jobs will be in the shop next week. The demand in units for each job and the production rates of the standard operator in pieces per standard operator hour for each type of job are given as follows:

Job	*A*	*B*	*C*	*D*	*E*	*F*	*G*
Demand	600	1000	500	50	2000	20	800
Production rate	60	20	25	10	40	2	40

What production capacity in standard operator hours is required to complete all of these jobs?

2. Six service calls are on hand for next week. The number of steps required for each has been determined and is listed below. The output rate is measured in steps per standard operator hour for each type of job.

Job	*A*	*B*	*C*	*D*	*E*	*F*
Number of steps	500	400	200	150	2000	48
Output rate	100	20	25	75	400	12

What workforce capacity in standard operator hours is needed to complete all service calls?

3. The estimated annual demands for five types of soup made by The Big Soup Company are given below. There are three plants located in the United States. The most productive plant has been chosen as the standard plant. Its output is listed in standard plant output per day.

Soup	*A*	*B*	*C*	*D*	*E*
Demand	900	630	240	1800	1200
Production rate	6	7	2	10	25

The other two plants have indices of 0.9 and 0.7. There are 250 working days in the year, and all numbers are given in thousands of cases. Is it likely that the three plants can handle the annual demand?

4. Using the information in Problem 3, assume that the two other plants have indices of 0.8 and 0.7. Is it likely that the three plants can handle the annual demand?
5. Consider the data for the 6-month period given in the table below and answer the following questions.
   a. What is the total demand in 6 months?
   b. What is the total production in 6 months?
   c. What kind of AP policy is this?
   d. Find the production rate, that is, the number of units that one worker can produce in 1 month?
   e. What is the inventory, if any, at the end of period 6?
   f. What is the shortage, if any, at the end of period 6?

Period	*Demand*	*Production*	*Ending Inventory*	*Ending Shortage*	*Workers*
1	340	380			38
2	420	380			38
3	350	380			38
4	390	380			38
5	360	380			38
6	420	380			38

6. Using the information in Problem 5, what would be the effect of adding another person to the workforce? Specifically, this means increasing the number of workers from 38 to 39. Is this a sensible move?
7. Use the data given in table in Problem 5. In addition, the following information is available. The current number of workers is 37. (Note: This must be changed to the number of workers required in period 1.) Backorders (shortages) cost $100, and carrying inventory costs $25 per unit per time period. Hiring costs $200 per person added, and layoffs cost $300 per person. The cost of an additional person working is $2000 per month. Fractional payroll amounts can be calculated and used as well. Answer the following questions
   a. What is the total inventory carrying cost in 6 months?
   b. What is the total shortage cost in 6 months?
   c. What is the cost of hiring in 6 months?
   d. What is the cost of layoff (firing) in 6 months?
8. Using the information in Problem 7, what would be the effect of adding another person to the workforce? Specifically, this means increasing the number of workers from 38 to 39. Compare various costs calculated in Problems 7 and 8 and present your recommendations concerning adding one person to the workforce. Is this a sensible move?

9. The aggregate demand for a product line for the next 6 months is given below. The firm has regular capacity of 120 units per month, overtime capacity for 40 more units per month, and subcontracting can supply up to 100 units per month.

Month	1	2	3	4	5	6
Demand	220	160	200	210	200	190

Additional Data	
Previous output level	150 units
Beginning inventory	100 units
Shortage cost	$250 per unit per month
Inventory holding cost	$100 per unit per month
Regular time cost	$1200 per unit
Subcontracting cost	$2000 per unit
Overtime cost	$1500 per unit
Hiring workers to increase production level	$200 per unit
Firing workers to decrease production level	$500 per unit

Use the Chase Strategy and answer the following four questions.

a. What is the total cost of shortages?
b. What is the cost of overtime production?
c. What is the cost of subcontracting?
d. What is the cost of increasing production level?

10. Use the data given in Problem 9. If the regular capacity is increased to 210 from the current level of 120 units per month and you meet your demand by only regular time production, then answer the following two questions.

a. What is the cost of changing production level?
b. What is the cost of regular time production?

11. Consider the demand data given in Problem 9. A Level plan that produces 180 units per period in the regular time is being used. The inventory carrying cost per unit per period is $25, and the shortage cost per unit per period is $30.00. The beginning inventory is 100 units. There is no overtime production or subcontracting. Answer the next two questions.

a. What is the total cost of holding inventories?
b. What is the cost of shortages?

12. A company has the following demand forecast for the next 6 months. Assume that an employee contributes eight regular working hours per day. Overtime capacity is limited to a maximum of 20% of regular time capacity. The time to produce one unit is 4 h.

Month	*Forecast*	*Working Days*
January	540	21
February	640	20
March	580	22
April	520	19
May	530	20
June	590	18

a. What is the average production level per day during the 6-month period?
b. If the demand for March (580 units) are produced during regular time in March (22 days), how many workers will you need?
c. Suppose the number of workers is 14 in each month. Each worker produces to the maximum capacity during regular time. How many units can be produced in April during regular time?
d. The management specifies the following policy: Use a level production, that is, the same number of workers is used in each month. No inventories or shortages are allowed. Overtime and/or subcontracting may be used in any month if needed. What level production per day will you use?

13. The aggregate demand for a product line for the next 6 months is given below. The firm has regular capacity of 150 units per month, overtime capacity for 40 more units per month, and subcontracting can supply up to 100 units per month. The chase policy is being used. Find the numbers of units produced during regular time, overtime, and through subcontracting in each month.

Month	1	2	3	4	5	6
Demand	200	130	180	220	210	190

14. In the text there is an example of a system operating at 75% of capacity most of the time. In that section, it is said, "this may be a good thing." Explain why and when it might be beneficial for the system to be run at less than 100% of capacity? Give a numerical illustration.
15. Use the service example of blood-testing to calculate the effect of using the chasing strategy. Use a numerical example.

Readings and References

Alden, J.M., Burns, L.D., Costy, T., and Hutton, R.D., General Motors Increases Its Production Throughput, *Interfaces*, 2006, 36(1), p. 6.

Bowman, E.H., Production Planning by the Transportation Method of Linear Programming, *Journal of Operations Research Society*, February 1956, pp. 100–103.

Bowman, E.H., Consistency and Optimality in Managerial Decision Making, *Management Science*, January 1963, pp. 310–321.

Buxey, G., Production Planning for Seasonal Demand, *International Journal of Operations and Production Management*, 13(7), July 1993, pp. 4–21.

Caixeta-Filho, J.V., van Swaay-Neto, J.M., and de Padua Wagemaker, A., Optimization of the Planning and Trade of Lily Flowers at Jan de Wit Company, *Interfaces*, 2002, 32(1), p. 35.

DeMatta, R., and Miller, T., A Note on the Growth of Production Planning System, *Interfaces*, 23, April 1993, pp. 27–31.

Denton, B., An Integrated Outbound Logistics Model for Frito-Lay: Coordinating Aggregate-Level Production and Distribution Decisions, *Interfaces*, 39(5), 2009, p. 460.

Fisher, M.L., Hammond, W.R., Obenneyer, W.R., and Raman, A., Making Supply Meet Demand in an Uncertain World, *Harvard Business Review*, 72(3), May–June 1994, pp. 83–93.

Fleischmann, B., Ferber, S., and Henrich, P., Strategic Planning of BMW's Global Production Network, *Interfaces*, 36(3), 2006, p. 194.

Gerald, B., Keegan, J., Vigus, B., and Wood, K., The Kellogg Company Optimizes Production, Inventory, and Distribution, *Interfaces*, 2001, 31(6), p. 1.

Kuchta, M., Newman, A., and Topal, E., Implementing a Production Schedule at LKAB's Kiruna Mine, *Interfaces*, 34(2), 2004, p. 124.

Malhotra, M.K., and Ritzman, L.P., Scheduling Flexibility in the Service Sector: A Postal Case Study, *Production and Operations Management*, 3, Spring 1994, pp. 100–117.

Mentzer, J.T., and Moon, M.A., Understanding Demand, *Supply Chain Management*, May–June 2004, pp. 38–45.

Metters, R., and Vargas, V., A Comparison of Production Scheduling Policies on Costs, Service Levels, and Schedule Changes, *Production and Operations Management*, 17(3), 1999, pp. 76–91.

Plambeck, E.L., and Taylor, T.A., Sell the Plant? The Impact of Contract Manufacturing on Innovation, Capacity, and Profitability, *Management Science*, 51(1), January 2005, pp. 133–150.

Radas, S., and Shugan, S.M., Managing Service Demand: Shifting and Bundling, *Journal of Service Research*, 1, August 1998, pp. 47–64.

Sheu, C., and Wacker, J.G., A Planning and Control Framework for Nonprofit Humanitarian Organizations, *International Journal of Operations and Production Management*, 14(4), 1994, pp. 64–77.

Shostack, G., Lynn, Designing Services that Deliver, *Harvard Business Review*, 62, January–February 1984, pp. 133–139.

Smith, B.C., Leimkuhler, J.F., and Darrow, R.M., Yield Management at American Airlines, *Interfaces*, 22, January–February 1992, pp. 8–31.

Smith-Daniels, V., Scheweikhar, S., and Smith-Daniels, D., Capacity Management in Health Care Services: Review and Future Research Directions, *Decision Sciences*, 91, 1988, pp. 889–919.

Sridharan, V., and LaForge, R.L., Freezing the Master Production Schedule: Implications for Customer Service, *Decision Sciences*, 25, May–June 1994, pp. 461–469.

Taubert, W., A Search Decision Rule for the Aggregate Scheduling Problem, *Management Science*, February 1968, pp. B343–B359.

Trottman, M., Choices in Stormy Weather: How Airline Employees Make Hundreds of Decisions to Cancel or Reroute Flights, *Wall Street Journal*, February 14, 2006, pp. Bl–B3.

Chapter 5

Inventory Management

Reader's Choice—"We want to turn our inventory faster than our people."—James Sinegal, former founder and CEO of Costco

Bell, P.C., and Noori, H., Managing Inventories Through Difficult Economic Times: A Simple Model, *Interfaces*, 15(5), 1985, p. 39. This paper presents an inventory model for unusual demand volatility, price, and financial conditions in difficult economic times. The model results when applied to a company's inventory planning for the 1981–1983 recession period paralleled the company's actual inventory situations.

Canen, A.G., and Galvao, R.D., An Application of ABC Analysis to Control Imported Material, *Interfaces*, 10(4), 1980, p. 22. An ABC analysis of a Brazilian company improved firm's performance. The inventory-related costs were reduced by almost 30%.

Cohen, M., Kamesam, P.V., Kleindorfer, P., Lee, H., and Tekerian, A., Optimizer: IBM's Multi-Echelon Inventory System for Managing Service Logistics, *Interfaces*, 20(1), 1990, p. 65. This paper describes IBM's "Optimizer" system for managing service levels and inventory of spare parts. The use of Optimizer achieved inventory reductions, better service, improved flexibility to meet service requirements, and better planning and control.

Farasyn, I., Perkoz, K., and Van de Velde, W., Spreadsheet Models for Inventory Target Setting at Procter & Gamble, *Interfaces*, 38(4), 2008, p. 241. This paper describes the application of spreadsheet models for inventory planning in Procter & Gamble (P&G). The models identify the best inventory levels that provide the required customer service levels under the supply chain constraints. The models are being used worldwide by supply chain planners and have helped in reducing inventory levels by $350 million.

Karmarkar, U.S., Kekre, S., Kekre, S., and Freeman, S., Lot-Sizing and Lead-Time Performance in a Manufacturing Cell, *Interfaces*, 15(2), 1985, p. 1. This paper examines how lot-sizing policies affect manufacturing lead times. The paper compares the simulation model developed by Eastman Kodak Co. (Rochester, New York) and a mathematical model (Q-LOTS) available in literature. Both models gave similar results.

Kleutghen, P.P., and McGee, J.C., Development and Implementation of an Integrated Inventory Management Program at Pfizer Pharmaceuticals, *Interfaces*, 15(1), 1985, p. 69. This paper describes Pfizer's inventory management system that includes demand forecasting, production planning, materials requirements planning, purchasing, and inventory management. Pfizer reduced inventories by $23.9 million and back orders by 95% over a 3-year time period.

Lee, H.L., Billington, C., and Carter, B., Hewlett-Packard Gains Control of Inventory and Service Through Design for Localization, *Interfaces*, 23(4), 1993, p. 1. This paper presents a model to study the impact of different design alternatives on inventory costs and delivery service. The focus is on design for localization in the manufacture of DeskJet Plus printers at Hewlett-Packard.

Pasternack, B.A., Filling Out the Doughnuts: The Single Period Inventory Model in Corporate Pricing Policy, *Interfaces*, 10(5), 1980, p. 96. This paper addresses the single period inventory problem for donuts, a product with short shelf life, to help managers of individual donut shops that are part of a decentralized chain to make pricing decisions that align closely with the firm's objective. The modifications led to increased profitability.

This chapter discusses inventory control policies for effective material management in an organization. Whether making things with materials or using them to provide services, materials are the major circulation system of supply chains. Materials are used in making goods and providing services. Materials flow through the supply chains, which must always begin with raw materials. P/OM transforms the raw materials into materials that are work in process and finally into finished goods.

After reading this chapter, you should be able to:

- Explain what inventory management entails.
- Describe the difference between static and dynamic inventory models.
- Discuss demand distribution effects on inventory situations.
- Discuss lead-time effects on inventory situations.
- Describe all costs relevant to inventory models.
- Differentiate inventory costs by process types.
- Explain order point policies (OPPs) and when they are used.
- Discuss the use of economic order quantity (EOQ) models for determining the optimal order size for batch delivery.
- Discuss the use of economic production quantity (EPQ) models for determining the optimal production run for continuous delivery.
- Explain the operation of the perpetual inventory model and explain why it is the most widely used inventory control system.
- Explain the operation of the periodic inventory model and describe the special circumstances that make its use desirable.
- Describe the quantity (price) discount model and explain how it indicates when a discount should be taken.
- Perform ABC classification of materials.

5.1 Introduction

Materials management is a system of broad-based planning and control over one of the most important components of the cost of goods sold (COGS). Two trends have been pervasive. The first is the marked decline of the direct labor component of the COGS. The second is the marked rise in the direct and indirect (overhead) cost of materials. Because material costs are now critical to profitability, most organizations have created positions of high responsibility to oversee the many parts of the system that have to be integrated for materials management.

Materials management involves organizing and coordinating all management functions that are responsible for every aspect of materials movements and transformations—called the materials management system. This system is triggered by demands (including those forecasted) that deplete stocks, causing inventory management to request replenishment through purchasing agents or direct contact with suppliers or vendors.

There are three main classes of materials that have to be purchased and managed.

First, there are raw materials. These are generally extracted from the ground and then refined, but they are still the basic ingredients. Examples include mined metals such as copper, gold, and platinum; chemicals such as sodium and potassium salts, manganese, and phosphates; grains such as wheat and rye; beans such as coffee; natural gas, and petroleum.

Raw materials have value-added by operations. Value-adding occurs when purchased components are further transformed by the company's production process. Thus, refining, processing, packaging, and shipping when done by the organization are value-adding and profit-making processes. All buyers of raw materials specify their required quality standards. Grains can be too dirty. All soy is not alike. Coffee prices vary with the perceived quality of the taste of the beans. Raw materials can be bulky and require special spacious storage bins. Companies prefer to locate their refining operations near the source of raw materials, so they do not have to transport tons of materials from which pounds or even ounces are eventually derived for use.

Organizations that are in the business of supplying raw materials at the very start of the upstream acquisition process are themselves dependent on purchasing. A quick summary includes the equipment to dig in the mines or harvest the crop. The deposits in which the mines are located must be acquired. The land to be planted and farmed must be procured. Mining requires tools and lubricants, and farming demands seeds and fertilizer. This leads to the somewhat trite statement that "every organization has a supplier."

Second, components and subassemblies are purchased materials that have greater value-added than the raw materials. They are, in fact, composed of raw materials that already have experienced value-adding. Components and subassemblies are characterized by some degree of fabrication, assembly, and manufacture.

They are assembled into higher-order products by combining them with each other and with other parts made by the producer. This produces the *third* class of materials that need to be managed. Work in process can be stored and eventually shipped as finished goods. Work in process has more value-added than the purchased subassemblies. There is a progression of value-adding that starts at raw materials and moves up the supply chain to finished goods—sold and shipped.

Inventory management encompasses the widest spectrum of activities related to materials. Let us start with the primary activities which include when and how much to make or purchase. In addition, timing of replenishments and decisions about storage are important decisions as well.

Inventories serve several functions in an organization. The main function of inventories is to reduce the interdependency of various stages of the production and delivery system. Consider three subsystems of an organization representing the supplier, production, and the market. These three subsystems are rigidly connected with each other, without any inventories, as shown in Figure 5.1.

Raw materials and purchased parts are flowing from various suppliers to manufacturing, and finished products are going from manufacturing to the market. Within the manufacturing department, semifinished goods move from one

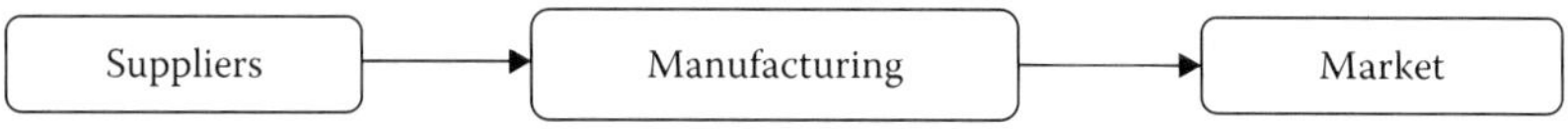

Figure 5.1 Flow of materials.

machine to another. The three subsystems are linked in such a way that there is no inventory stocked in-between them. Therefore, the impact of any disruptions in any one of the subsystems will be felt in the other two. For example, if production output stops in manufacturing because of some reason such as power failure, machine breakdown, a strike, etc., then it will not be possible to supply the goods in time to the customers, hence affecting the market subsystem. Similarly, if suppliers fail to supply raw materials and purchased parts in time, then the manufacturing subsystem, and in turn, the marketing subsystem will be affected. The manufacturing subsystem itself may consist of several subsystems representing several productive facilities. These manufacturing subsystems are also interlinked and the performance of any one of them is affected by the performance of the others.

On the other hand, if we have an inventory bank in-between each pair of the subsystems, then inventory can dampen the effect of disruptions in one system on the other. In this case, if the supplier is unable to supply the raw materials and purchased parts in time, then production will not suffer for the want of raw material because the stored inventory of raw material can be utilized. Similarly, if the demand is fluctuating in the market, then the inventory of finished goods can help to absorb these fluctuations and their influence is not felt by the production department. In the same way, production subsystems operate independent of each other because of the inventory of semifinished goods.

Thus, inventories help in reducing the dependence of one subsystem over the other. Stock on hand acts as a shock absorber and gives a cushioning effect to the turbulence in one subsystem so that the same is not transmitted to other subsystems. It should be pointed out that inventory only helps in diminishing the intensity of the impact of disrupted operations in one subsystem over other subsystems and does not eliminate it completely because the amount of inventories is limited.

FUNCTIONS OF INVENTORIES

- Reduces dependency of one subsystem *on* the others.
- Helps in developing smooth production plans during aggregate production planning process.
- Inventories are built in anticipation of shortages and price increases.
- Helps in talking advantage of price discounts.

Inventories also help in developing level production plans when production does not fluctuate. Inventories are also built in anticipation of shortages and price increases. Sometime the vendors give price discounts for buying-in large quantities. That leads to purchases in large quantities which increases inventories.

Since it is essential to keep inventories for the smooth and efficient operations of an organization, the question is not, "whether to keep inventories or not to keep inventories" but the question is, "how much inventory should be kept?" We propose to answer this question in this chapter. Inventory is a necessary evil which cannot be eliminated completely and at the same time large levels of inventories may prove disastrous for a company. Surplus inventory may become the graveyard of business and a major factor contributing to the business failure. Inventory plays a crucial role in the efficient and smooth operation of an organization. It decouples the various stages of a business system.

This chapter explains how to manage inventories of materials that have continuous demand. This class of inventory models is known by the acronym OPP (order point policies). These materials are independent, which means they are not components of a high variety of finished goods. The two basic models include the economic order quantity (EOQ) model that is designed for buying items to satisfy replenishment requirements of continuous demand items and the economic production quantity (EPQ) model that is designed for determining the batch (also called lot) sizes in which an item will be produced. This model is also known as economic batch size or economic lot size (ELS) model. We will use the terms EPQ and ELS interchangeably in this chapter.

These concepts are fundamental to understanding how perpetual and periodic inventory systems work and when each is preferred. Quantity discount models (not discussed in this chapter) are a byproduct of the EOQ model, providing further insights for inventory managers in situations where discounts are offered to buy-in large quantities.

Without the systems perspective, everyone, everywhere will be ordering what they need as they need it. Lack of coordination diminishes buying power and loses the knowledge-based benefits of the centralized system. Strategic thinking is required to optimize the production plans of suppliers and producers to best meet the needs of their customers throughout the supply chain. In that way, the lowest costs and best deliveries can be achieved for mutual benefit. The information system of a multinational company with a centralized inventory management system has to keep track globally of where everything is, where and when it will be needed, when to re-order it, and where to store it. It is an enormous system to manage. The strategic thinking that went into designing any such system must be robust and not casual. The need for re-planning is always there when things do not work out as planned.

5.2 Types of Inventory Situations

Inventory is those stocks or items used to support production (raw materials and work in process items), supporting activities (maintenance, repair, and operating supplies), and customer service (finished goods and spare parts). [This definition is taken from the APICS Dictionary. The American Production and Inventory Control Society (APICS) is a professional society that has played an influential role in the inventory management area.] We should note that APICS changed its name to The Association of Operations Management to emphasize that operations management encompasses inventory control as well as many other (equally critical) aspects of all production systems.

Who manages inventory? Management of inventory is a major P/OM responsibility. How to manage inventory is dependent on the type of inventory that is involved. Most types of inventory situations are best handled by well-designed computer systems that utilize as much centralization of record keeping and order placement as is feasible. The manifold advantages of the systems perspective with centralized buying includes the fact that larger quantities provide stronger supplier relationships, bigger discounts, a more informed choice of suppliers, and less chance for mistakes (with respect to crucial factors such as price, quality, and reliability).

Each of the six classes of inventory situations described below requires its own type of management even though they can all be centralized or decentralized according to the dispersion of use for producers, suppliers, and customers:

1. Order repetition—static versus dynamic situations.
2. Demand distribution—certainty, risk, and uncertainty.
3. Stability of demand distribution—fixed or varying.
4. Demand continuity—smoothly continuous or sporadic and occurring as lumpy demand; independent.
5. Lead-time (LT) distributions—fixed or varying. LT is the interval between order placement and order receipt (including recognized need to create an order and place it).
6. Dependent or independent demand—when components are dependent on one or more end items, the information system must be able to calculate linked demands.

5.2.1 Static versus Dynamic

To explain order repetition, static inventory models have no repetition. They portray "one-shot" ordering situations, whereas dynamic inventory models place orders repetitively over long periods of time. A few examples underscore the practical nature of this distinction between only one order and a repetitive stream of orders for the same item placed over time. Dynamic models are the primary focus of this chapter.

The pure static case also is called a one-period model even though under some circumstances a corrective "second shot" may be allowed. The "Christmas tree problem" is a good illustration of the static situation. The owner of a tree nursery that sells Christmas trees locally said that she placed her orders with a Canadian tree farm north of Montreal back in July. Reasoning that it would be a good year—because people were feeling more prosperous than the prior year—she had bought the maximum number of trees that her organization could truck and accommodate. Unfortunately, the 2 weeks before Christmas were unusually rainy. This discouraged people from buying trees for the holiday. In the last week before Christmas, the owner had posted sale signs slashing prices. She felt that helped but nevertheless, 25% of the trees remained unsold on Christmas eve. Five percent of them were live trees that could be saved, but 20% would have to be scrapped. The seller's problem: how many trees to order in July for next December? Most of the sales take place a few days before Christmas. There is no time to take corrective action. What is to be done if too few trees are stocked? Driving up to Canada to replenish supplies is impractical. Also, there may not be any trees left there to sell. If sales are unexpectedly strong, buying locally will cost too much. If too many trees are stocked, the best alternative is to advertise discounts, hoping to get anyone who was going to buy a tree to buy it from the overstocked dealer. The dealer never really knows whether the order size was over or under, or just right, until Christmas day. See Pasternack (1980) that addresses the single-period inventory problem for donuts.

Other static examples include the storekeeper who has to decide how many *Wall Street Journal* newspapers to buy for each day. Only one decision can be made to buy n newspapers. In its purest form, there is no opportunity to correct that decision based on later information. Another example is the problem of the hot dog vendor at the ballpark. Consider the department store buyer who places an order in July for toys to be sold at Christmas time. If the toy is a dud, there is severe overstock. If the toy is hot, there will not be enough stock to meet demand. Both types of situations occur regularly. Another example is the spare-parts order for a complex machine. When placed with the original order, the parts are relatively inexpensive. When required later, because of unanticipated failure, the costs are exorbitant. This spare-parts model is a static decision problem. In the case of overestimated demand, salvage value is sometimes available. For example, a department store that overbuys on toys, shipped from abroad in time for the holiday season, may be able to sell those toys at a discount after the selling season is finished.

Dynamic situations require different considerations because the demand for such items is constant. Orders are placed repetitively over time. The problem becomes one of adjusting inventory levels to balance the various costs so that total variable costs are minimized. Variable, in this case, means these costs change with order size. Service systems use many kinds of supplies with dynamic demand patterns (as in hospitals, hotels, restaurants, theme parks, airlines, and educational institutions).

5.2.2 Type of Demand Distribution—Certainty, Risk, and Uncertainty

There are decision problems that have outcomes known with *certainty*, for example, signing a contract to supply a given number of units converts demand from a *risk* to a *certainty*. Locating a supplier down the block reduces delivery time uncertainty.

Generally, there is a cost for certainty. Because contracts reduce the producer's risk, the buyer expects the price will reflect this fact. The supplier that locates across the street from the buyer expects compensation for being at that location.

Sometimes *certainty* is a reasonable assumption. This only works when the degree of variability will not affect the solution. Then, certainty is assumed for convenience and it does not violate the spirit of the model. LT for delivery is often treated in this way. Suppose that 2 weeks is the average LT for a specific case. The variability around the average will determine if 2 weeks can be used as if it were certain. It is important that users of inventory models know when the certainty assumptions are allowable. When LT variability could cause a stock-out, it cannot be treated as constant. Buffer stock is stock carried to prevent outages when demand exceeds expectations. Basic inventory methods dealing with OPP make assumptions about the demand distribution that get translated into buffer-stock levels. The OPP models on which this chapter focuses are based on demand and not lead time variability. Order points are stock levels at which new orders are triggered and placed.

Uncertainty means that there is no good forecast for the probabilities of demand and/or LT distributions. When uncertainty exists, the probabilities for various levels of demand occurring are speculative. However, when there is a known risk, some planning can be done. Delivery of critical materials from the port city of Kobe in Japan always was totally reliable until the January 18, 1995, earthquake caused serious delays. Some companies factor such possibilities into their planning and can react quickly. Another example is the 2003 strike in Venezuela, which led to an increase in the price of oil from $23 to $33 per barrel. The August 29, 2005, hurricane Katrina devastated New Orleans and closed its busy port for months. Serious damage had been predicted years in advance. A number of companies including BMW, Hyundai, P&G, Home Depot, and Wal-Mart had plans in place to deal with the situation. The earthquake and tsunami in Fukushima Daiichi (March 2011) and the flooding in Thailand (October 2011) disrupted production for Toyota and Honda in the United States for at least 6 months. Finite probabilities always exist for various catastrophes. Having a team on hand, ready to assess the situation, and find solutions can provide significant advantages.

Order cancellations may be a less severe disaster, but they are usually costly enough to make planning for the event worthwhile. Many production departments have exposure to the risk that an order in process will be cancelled. Contingency planning for cancellations shows better production management abilities than those who react to cancellations after the fact.

When forecasting is difficult because there is no history, systems are developed to search for advance warning. They are tuned in to whomever might have information about orders. Every effort is made to find the key players who originate the orders and to keep them in the communication loop at all times. Other factors that might trigger orders also are tracked. Efforts are made to gain some control over the uncertain occurrences.

5.2.3 Stability of Demand Distribution—Fixed or Varying

Uncertainty about setting the right capacity for supply includes the possibility that the demand distribution is changing over time. This could be caused by a stable shift factor such as seasonality. On the other hand, the drivers of change could also be unknown. In the latter case, the instability of causal factors is generally acknowledged to reflect high risk. If it is known how demand is changing, then the risk levels may be controllable because forecasts of some merit are possible. OPP methods are applicable to predictable demand distributions. The best of these is stable demand but forecasts can be modified if it is known how the distributions are changing. Otherwise, the system is searching in the dark. Other methods than OPP should be used to manage these situations. Material requirements planning (MRP) is the most favored alternative when the demand distribution is unknown or unstable. MRP is not discussed in this book.

5.2.4 Demand Continuity—Smoothly Continuous or Lumpy

Much of the prior discussion applies to demand continuity. OPP needs demand continuity that can be described as persistence of the stable demand pattern over a relatively long time period.

MRP as an alternative inventory methodology can deal with the lack of smoothly continuous demand. It should be noted that assuming smooth and continuous demand is akin to converting a known risk situation into one of certainty. The assumption is usually valid and can be tested by simulating different patterns that are more or less smooth and continuous and measuring the extra costs incurred for assuming perfect smoothness when, in fact, it is good but not perfect. Such testing (called sensitivity analysis) is most readily done with computer simulation programs.

5.2.5 Lead-Time Distributions—Fixed or Varying

LT is the interval between order placement and receipt (order need recognition and placement times are included). Variability of the LT will be a factor in setting the size of the buffer (safety) stock. As LTs get longer, inventory systems become more sensitive to problems that can arise in the supply chain. One of the main reasons is that with long LTs, correction of errors takes longer. The more critical the materials are for production, the worse this situation becomes. For materials that are not critical, the assumption of fixed LTs does not do too much damage. When materials are critical, inventory planning had best take into account forecasts for the LT distribution. In that event, LT distributions will assign additional units to buffer stocks. Another name for buffer stock is safety stock (SS) because these units are held to provide protection against variability in demand and (where applicable) LTs.

5.2.6 Dependent or Independent Demand

The focus is now on independent-demand systems. This means that orders can be placed for nondependent items without considering what demand is forecast for the end products of which they are part. For example, labels for a specific kind of jam are dependent on the demand for that particular kind of jam. Beach plum jam is highly seasonal because of the nature of that crop and does not meet the criteria for stable, continuous demand that characterizes OPP models. Alternatives to OPP should be used when dependent demand systems are involved. This is applicable for components and subassemblies that are used as parts of one or more finished products that have sporadic demand. The dependency is greatest when both the timing and the quantity for end-product demands are not predictable.

5.3 Inventory-Related Costs

The core of inventory analysis lies in finding and measuring relevant costs. Six main types of costs are discussed and then, in a seventh category, a few others are mentioned.

The six costs that will be discussed are

1. Costs of ordering
2. Costs of setups and changeovers
3. Costs of carrying inventory
4. Costs of discounts
5. Out-of-stock costs
6. Costs of running the inventory system

5.3.1 Costs of Ordering

The cost of ordering includes those cost items that go into making up and placing an order, following up of the order and receiving the order. It consists of writing up the purchase requisition form, making phone calls in connection with specifications, and ordering, e-mailing, faxing, or mailing purchase orders to the supplier. All costs that increase as a function of the number of purchase requisitions qualify for inclusion.

5.3.2 Costs of Setups and Changeovers

In a manufacturing situation, the ordering cost is replaced by the cost of *setting up* the machines to do a production run. This requires cleaning up from the prior job, which also is called *taking down*. The process is known as changeover, and the costs can be significant.

A number of parts may have to be made before the setup is complete. The cost of learning is involved and defectives play a role. When acquiring equipment, changeover times and costs can be as important to consider as output rates.

5.3.3 Costs of Carrying Inventory

Inventory is a form of investment. Capital is tied up in materials and goods. The alternative uses of inventory include spending for R&D, new product and/or process development, advertising, promotion, and going global. Some firms even put the money into financial instruments, the stock market, or the savings bank.

Expanded capacity and diversification are typical opportunities that, when ignored, incur the cost of not doing that much better with the investment funds. By holding inventory, the company foregoes investing its capital in these alternative ways. Such opportunity costs account for a large part of the costs of carrying inventory.

These are the costs over which P/OM can exercise control using inventory policies. Thus, if a company has shelf space for 1000 units, but can get a discount if it purchases at least 2000 units, then to get this discount it must expand storage capacity. It can buy or rent additional space.

There are options for ordering large quantities to obtain discounts without requiring expanded storage capacity. One of these is vendor releasing, whereby the supplier agrees to deliver small increments of the larger order over time.

Another method for reducing storage space requirements is the use of cooperative storage. Commonly used supplies are ordered at discount in large quantities, stored in cooperative warehouses, and dispensed to participating hospitals, in the same metropolitan area, on an as-needed basis. Airlines share the storage costs and investment-based carrying costs of commonly used parts such as jet engines. Cooperative sharing reduces storage costs and increases the availability of expensive components that require large dollar expenditures.

Items carried in stock are subject to costs of pilferage, obsolescence, deterioration, and damage. These costs represent real losses in the value of inventory. Pilferage, which is petty theft, is characteristic of small items such as tools. Department stores suffer extensively from stolen merchandize. Hotels lose ashtrays and towels. Pencils and stamps disappear in offices.

Obsolescence may be the most important component of carrying cost because it happens so often and so fast. Obsolescence occurs quite suddenly because a competitor introduces technological change. Also, it can be the kind of loss that is associated with style goods, toys, and Christmas trees. Out-of-season and out-of-style items can lose value and must be sold at a special reduced rate. The determination of how much inventory to carry will be affected by the nature of the inventories and the way in which units lose value over time.

Deterioration affects the carrying cost of a broad range of products. Industrial products that deteriorate with advancing age include adhesives, chemicals, textiles, and rubber. Weather deteriorates iron and wood. Rubber gaskets in pumps can fail.

Food is date stamped but even without that documentation the owners of restaurants are keenly aware of the effects of sour milk and stale bread.

Companies put spoilage retardants in foods, but increasingly customers avoid additives. Adding ingredients to prevent spoilage increases material and production costs while decreasing the carrying cost for product deterioration. The net market effect must be factored in as well. Spoiled milk and stale bread are quickly perceived, but customers can avoid buying products before they spoil by noting freshness dating information as required for milk and bread. Whatever cannot be sold because of real or dated deterioration is added to the carrying cost. Some drugs like aspirin are dated. Although drugs are vulnerable to spoilage, many people use them anyway because though ineffective they do not taste bad.

Some products that are not required to use freshness dating are testing its effect on their market share. Consumer advocacy groups are suggesting regulations that automobile tires be dated because there is age deterioration in tire materials. Freshness dating not only increases the carrying cost component, but it also increases demands on operations to make and deliver product as much before the freshness date as possible. Delays in getting dated products to market will decrease revenues, productivity, and increase carrying costs. It is evident that consumers want many product categories to show elapsed time. Freshness dating increases production and inventory costs.

An added component of carrying cost is both taxes and insurance. If insurance rates and taxes are determined on a per unit basis, then the amount of inventory that is stocked will determine directly the insurance and tax components of the carrying costs.

5.3.4 Costs of Discounts

Accepting discounts by buying at least a certain amount of material involves extra costs that may make taking the discount unprofitable. An appropriate inventory cost analysis must be used to determine whether a discount that is offered should be taken. The extra costs for taking the discount are compared to the savings obtained from the discount. Extra expenses include additional carrying costs. Part of these carrying costs is incurred for additional storage space.

5.3.5 Out-of-Stock Costs

When the firm does not have stock to fill an order, there is a penalty to be paid. Perhaps the customer goes elsewhere, but will return for the next purchase. Then the penalty is only the value of the order that is lost. If the customer is irritated by the out-of-stock situation and finds a new supplier, the customer may be lost forever. The loss of goodwill must be translated into a cost that is equivalent to the termination of the revenue generation of that customer. The specific cost is how much lower the lifetime value of that customer becomes, or the revenue from that customer may terminate entirely.

If the buyer is willing to wait to have the order filled, the company creates a backorder. This calls for filling the order as soon as capacity is available or materials arrive. Backorder costs include the penalties of alienated customers. To avoid this penalty, some mail-order companies prefer to fill customers' orders with a more expensive substitute instead of creating a backorder. The outage cost is related to the decrease in profit margin. However, the goodwill generated by the gesture is an intangible addition to long-term profit. Depending on the system that is used (i.e., backordering, substitutions, fill or kill, and so on), various costs of being out of stock will occur. The lost-goodwill cost is most difficult to evaluate.

Organizations that are not close to their customers frequently ignore lost goodwill. This may occur because the firm does not know how to measure or recognize goodwill. Many bureaucracies are identified as neglecting customer satisfaction inadvertently by overlooking the job satisfaction of their employees who in turn deal with the customers. Disgruntled employees alienate customers which means there are fewer happy customers.

5.3.6 Costs of Running the Inventory System

Processing costs that are associated with running the inventory system are referred to here as systemic costs. These costs are often associated with information system costs called IT costs. This category of costs is usually a function of the size of the inventory that is carried and the importance of knowing exact stock levels on-line immediately.

A big part of systemic expense is keeping up with incessant supplier cost changes, customer demand alterations, carrying cost modifications, labor costs for keeping records, and technology alteration costs. Costs are related to the number of people operating the inventory system. Operating costs also include systems assistance, programming, and training costs. Training costs are high in well-run IT systems, especially when labor turnover is significant and technology improvements accelerate. The amount of time that the system operates (many times round-the-clock) and the number of locations that are networked into the centralized data system (round-the-world) have to be factored into this cost.

Systems involving many stock-keeping units (SKUs) are dependent upon having an organized information system. There are different SKU part numbers for each model type, as well as for each size and color. Part numbers for SKUs identify specific suppliers and indicate where inventories are stored. When there are frequent online transactions with many SKUs, the amount of detail is great. Such systems are labor-intensive and expensive to operate.

It makes sense to focus on the SKUs for materials and items that are critical for production. Also, it makes sense to pay special attention to items that have high dollar volume because these items waste the majority of the money allocated if handled badly.

The status of work in process (WIP) can be monitored by means of bar codes, radio frequency identification (RFID), and optical readers. Often, these technologies are combined. There still is no sign of an entirely paperless factory. Systemic costs continue to be formidable.

5.3.7 Additional Inventory Policy Costs

The six costs previously discussed are generally the most relevant in determining inventory policy. However, other costs can play a part in specific cases. The costs of delay in processing orders can take on great significance when a heart transplant is involved. Not quite as dramatic, every manufacturer recognizes the costs of delay when lengthy setups are required. Service organizations are sensitive to the costs of delay. Organizations set goals for the maximum time that can be allowed to elapse before answering a ringing phone. Airline travel delay is regularly published and the airlines are keenly sensitive to the measure.

The costs of production interruptions have previously been related to the critical nature of inventory items. Lost inventory in the warehouse and errors in bookkeeping are causes of interruptions. Lost luggage in airline travel is an example of a cost to both the traveler and the airline. In 2010, the Air Transport Industry (ATI) Baggage Report stated that more than 29 million pieces of luggage were delayed, damaged, or lost at the world's airports.

Salvage costs can play important roles, as can the costs required for expediting orders. The costs of spoilage for food might be better handled as separate costs instead of as a part of the carrying cost related to obsolescence. The pro-and-con costs of central warehousing when compared to dispersed warehousing can be crucial. In various circumstances, one or more of these costs can dominate the inventory policy evaluation. A strong systems study will look outside the typical boundaries of inventory costs to spot those factors that are influencing the cost/benefit system.

We will study the various analytical models to solve inventory problem under various situations.

5.4 EOQ Model

The EOQ model, also called the square root model, forms the basis of inventory control techniques. The EOQ model requires determination of the order quantity which is the amount of material (Q) to be purchased each time an order is placed, given the annual demand (D), the ordering cost per order (S), the inventory holding (carrying) cost per unit per year (H), and the unit cost per item (C). Q is the decision variable. The best value of Q, which minimizes the total annual cost, is called the EOQ. The total annual cost includes the annual ordering cost, annual inventory holding cost, and annual cost of the item. The annual total cost is given by

$$\text{Annual total cost, } TC = \text{annual ordering cost} + \text{annual inventory holding cost} + \text{annual item cost.}$$

5.4.1 Annual Ordering Cost

The annual ordering cost is the number of orders per year multiplied by the ordering cost per order. The number of orders in one year is annual demand (*D*) divided by the order quantity (*Q*), that is, D/Q. Therefore, annual ordering cost = $(D/Q) * S$. An increase in the value of *Q* will decrease the annual ordering cost and vice versa.

5.4.2 Annual Inventory Holding Cost

The annual cost of holding inventory is the average inventory multiplied by the inventory holding cost per unit per year. The average inventory in a year is $Q/2$ as explained below.

Suppose the annual demand $D = 1200$ units. Figures 5.2 through 5.4 show changes in inventory levels for different values of *Q*. Figure 5.2, for $Q = 1200$, shows that 1200 units are purchased at the beginning of the year. So the inventory level at the beginning is 1200 units. The inventory level at the end of the year (12th month) is zero. The average inventory is, therefore, $(1200 + 0)/2 = 600$ which is $Q/2$. Figure 5.3 shows the inventory level variations for $Q = 600$. The

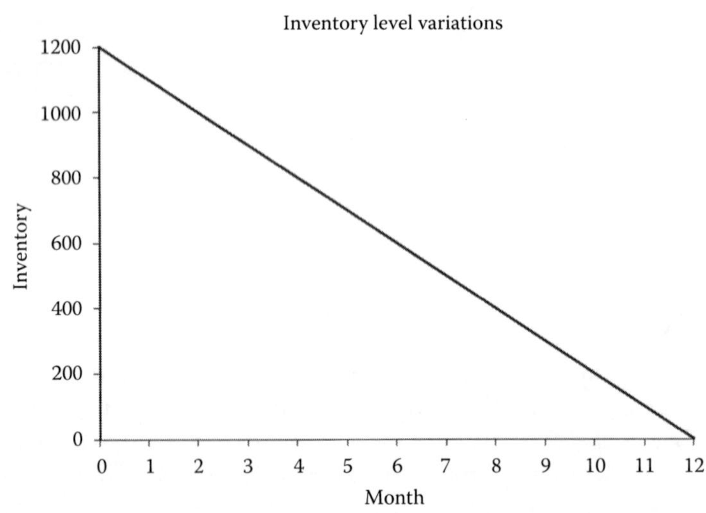

Figure 5.2 Order size, $Q = 1200$.

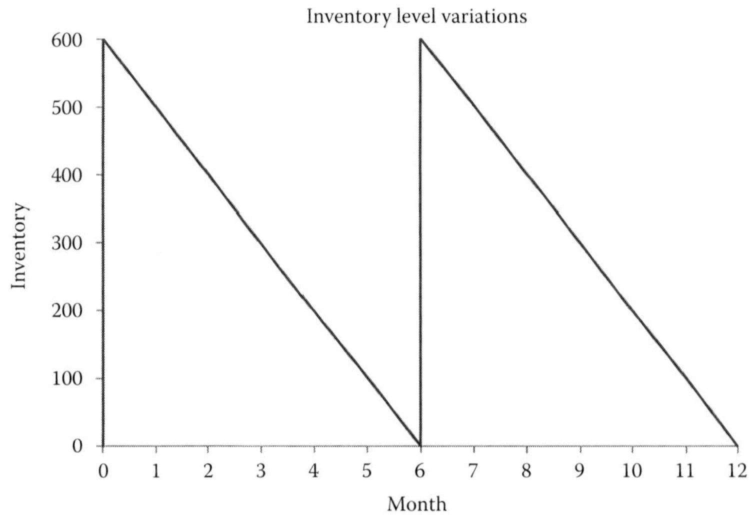

Figure 5.3 Order size, $Q = 600$.

year starts with an inventory level of 600 units that becomes zero at the end of sixth month. The average inventory during the first 6 months is, therefore, $(600 + 0)/2 = 300$. At the end of the sixth month, an additional 600 units are purchased that raises the inventory level to 600 again. The inventory level is zero again at the end of the year. So the average inventory during the last 6 months is also $(600 + 0)/2 = 300$. In other words, the average inventory throughout the year is $300 = 600/2 = Q/2$. Figure 5.4 shows the inventory level variations if

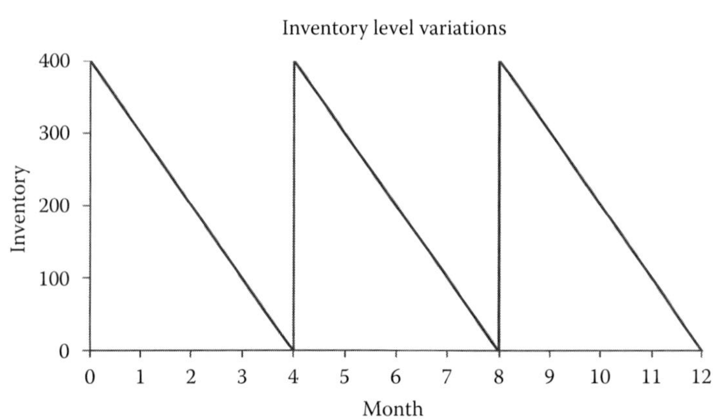

Figure 5.4 Order size, $Q = 400$.

$Q = 400$. In this case, the material will be purchased three times during the year and the average inventory will be $Q/2 = 400/2 = 200$.

The annual inventory holding cost is, therefore, given as $(Q/2) * H$. An increase in the value of Q will increase the annual inventory holding cost and vice versa.

5.4.3 Annual Item Cost

The annual item cost is calculated by multiplying the annual demand D by the item cost C, that is, annual item cost $= D * C$. A change in the value of Q does not change the annual item cost. It may, however, be pointed out that the annual item cost will change with a change in the value of Q if quantity discounts are involved. The quantity discount model is discussed later in this chapter.

5.4.4 Annual Total Cost

The annual total cost is given by

$$TC = \left(\frac{D}{Q}\right) * S + \left(\frac{Q}{2}\right) * H + D * C.$$

5.4.5 Annual Total Variable Cost

In the equation for TC, the values of D, H, and S are known (or can be estimated). The only unknown *variable* is Q which is the decision variable. Therefore, the only costs that are affected by changing the value of Q are the annual ordering cost and the annual holding cost. The sum of these two costs is called the annual total variable cost (TVC). TVC is given by

$$TVC = \left(\frac{D}{Q}\right) * S + \left(\frac{Q}{2}\right) * H.$$

5.4.6 Example for Cost Calculations

Suppose $D = 1200$ units, $S = \$5.00$, $H = \$1.20$, and $C = \$12.00$. Table 5.1 gives the values of annual ordering cost, annual inventory carrying cost, and annual TVC for several values of Q ranging from 50 to 300. The annual item cost will remain constant at \$14,400 (= 1200 * 12) irrespective of the value of Q. The range 50–300 is arbitrarily chosen to explain the changes in the values of these costs with changes in Q. It can be seen in Table 5.1 that as Q increases from 50 to 300, the annual ordering cost continues to decrease and the annual inventory holding cost continues to increase. Both TVC and TC first decrease, go down to a minimum

Table 5.1 Various Costs for Different Values of *Q*

Order Size (Q)	*Ordering Cost ($)*	*Inventory Cost ($)*	*TVC ($)*
50	120	30	150
75	80	45	125
100	60	60	120
125	48	75	123
150	40	90	130
175	34	105	139
200	30	120	150
225	27	135	162
250	24	150	174
275	22	165	187
300	20	180	200

at $Q = 100$ and then start increasing again. The changes in annual ordering cost, annual inventory holding cost, and *TVC* are shown graphically for various values of *Q* in Figure 5.5. From this table, it appears that $Q = 100$ is the best quantity to buy. But is it the absolute minimum? The EOQ formula in the following section will answer this question.

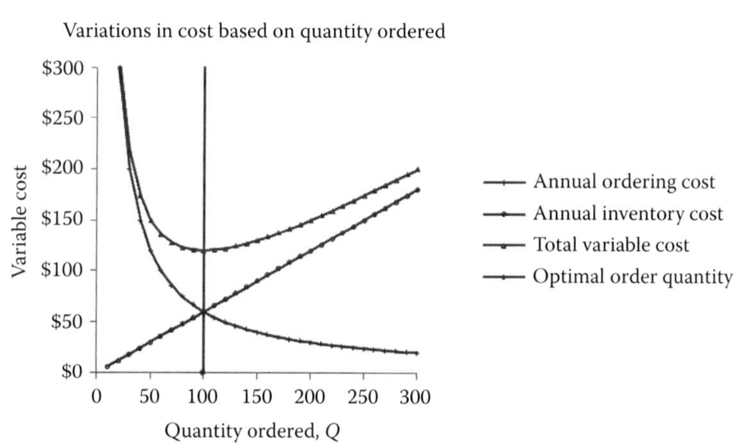

Figure 5.5 Cost variations based on quantity ordered.

5.4.7 EOQ Formula

As discussed above, if Q increases, then the total orders per year will decrease and therefore the annual order cost will also decrease; a higher value of Q will increase the average inventory and hence the total annual inventory cost will also increase. Following the same reasoning, if Q decreases, then annual order cost will increase and annual inventory cost will decrease. The change in the value of Q does not change the total annual item cost.

The total annual cost (TC) is given by

$$TC = \left(\frac{D}{Q}\right) * S + \left(\frac{Q}{2}\right) * H + D * C.$$

The optimal value of Q that minimizes TC is obtained when the annual ordering cost and the annual inventory holding cost are equal, that is, $(D/Q) * S = (Q/2) * H$. Solving this equation gives the formula for EOQ (optimal quantity), Q_O.

EOQ formula:

$$Q_O = \sqrt{\frac{2DS}{H}}.$$

The Q_O for this problem is 100.

$$Q_O = \sqrt{\frac{2 * 1200 * 5}{1.2}} = 100.$$

It may be noted in Table 5.1 that annual ordering cost and annual inventory holding cost are equal ($60.00) for $Q = 100$. The following inset lists various symbols and equations that have been used in deriving the EOQ formula.

- D: annual demand
- S: order cost/order
- H: inventory holding (or carrying) cost per unit per year
- Q: order quantity
- Number of orders per year = D/Q
- Annual ordering cost = $(D/Q) * S$
- Average inventory throughout the year = $Q/2$
- Annual inventory holding cost = $(Q/2) * H$
- Total variable cost, $TVC = (D/Q) * S + (Q/2) * H$
- EOQ is obtained when, $(D/Q) * S = (Q/2) * H$

5.5 EPQ Model

The economic production quantity (EPQ) model (also known as the economic lot size (ELS) model) is used in manufacturing situations where inventory increases at a finite rate and depends on the production rate and the usage rate of the item under consideration. In addition to the variables (D, S, H, Q, and C) defined earlier, we define two more variables: p = production rate per day (daily production rate) and d = demand rate per day (daily demand rate). The values of p and d must be in the same time unit. For example, these values could be weekly rates instead of daily rates. However, daily rates are most common. Q in this case is the production quantity (rather than order quantity) to be made in one lot and S is the cost of setting up the machine to produce that one lot. Therefore, S is called the setup cost per set up (rather than order cost per order).

Let us first study a numerical problem before deriving the equation for EPQ. Suppose, $p = 50$ units/day, $d = 10$ units/day, and $Q = 500$ (production quantity). The optimal value of Q is called the EPQ. The time to produce 500 units, t_p is 10 days = 500/50, that is, $t_p = Q/p$.

During these 10 days, 50 units are produced per day and at the same time 10 units (d) are used per day. Therefore, the inventory level is increasing at the finite rate of $40 = 50 - 10$ ($= p - d$) units per day. At the end of 10 days, the total number of units in inventory will be $400 = 10 * 40$. This is the maximum inventory level, I_{max}.

At the end of the 10th day, we stop producing this item and then continue to meet the demand from the inventory of 400 units. The inventory will last for 40 days (400/10) because we have 400 units in stock and the demand rate is 10 units/day.

The production cycle thus consists of 50 days. For the first 10 days, we produce and use the item. For the next 40 days, there is no production and there is only the usage of the item.

After 50 days, the next batch consisting of EPQ units is scheduled for production. This is how the cycles continue. Figure 5.6 shows the inventory level variations.

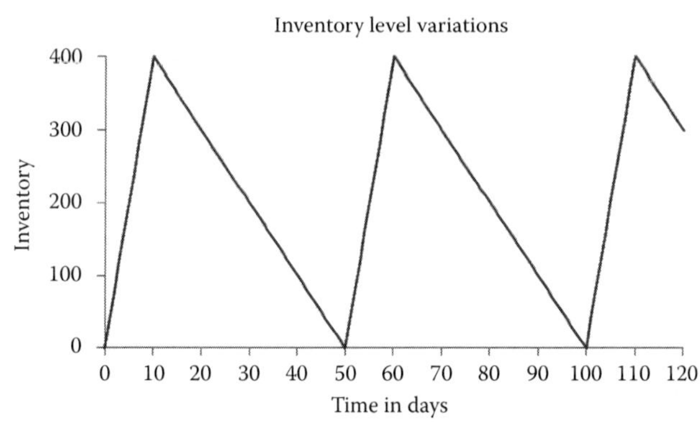

Figure 5.6 EPQ model.

The maximum inventory level as explained earlier, $I_{max} = 400$.

$$I_{max} = Q * (1 - d/p)$$

Average inventory = $I_{max}/2$
Annual setup cost = $(D/Q)^*S$
Annual holding cost = Average inventory * $H = (I_{max}/2) * H = H * (Q/2) * (1 - d/p)$
TVC is given by the following equation:

$$TVC = \frac{D}{Q}S + H\frac{Q}{2}\left(1 - \frac{d}{p}\right).$$

Q_{EPQ}, is obtained by equating the annual setup cost with annual holding cost and then solving for Q.

$$Q_{EPQ} = \sqrt{\frac{2DS}{H(1 - d/p)}}.$$

It is worth repeating that the term EPQ is being used instead of EOQ in this model.

Example: Find the EPQ for the following problem:

Also identify the storage capacity required.

Annual demand = 50,000 units; setup cost = \$25.00; inventory holding cost = \$5.00 per unit per year.

Production rate = 500 units per day; number of working days = 250.

Demand per day for this question will be 50,000/250 = 200.

It is assumed that demand occurs only during the working days, that is, 250 days in this case.

Using the EPQ formula: $Q_{EPQ} = 912.87$.

I_{max} for this problem will be 548. Therefore, the storage capacity is 548.

$$Q_{EPQ} = \sqrt{\frac{2 * 50{,}000 * 25.00}{5.00 * (1 - 200/250)}} = 912.87.$$

$$I_{max} = 912.87 * (1 - 200/500) = 547.72.$$

5.6 ABC Classification

Materials management—from the inception of the purchasing process all the way through production and the shipping of finished goods—can be improved by utilizing a powerful systems concept called the ABC classification. This concept alerts everyone: Some materials are more important than others—it is prudent to do what is important before doing what is less important. This concept applies to P/OM in a number of different. Two are discussed here.

5.6.1 Material Criticality

It is necessary for materials management to categorize the critical nature of parts, components, and other materials. There are various definitions of "critical" that fit different situations. When a part failure causes product or process failure, it is a critical part. Many parts of an airline engine may be critical, in this sense.

The following scenario is relevant. This airline considers parts to be most critical that have greater than a 10% probability of failing within 1 year. Membership in this top-ranking group is called A-critical. Perhaps the top 25% of all critical parts can be classified as being A-critical.

A second set, called B-critical, is identified by the airline as any part that has greater than a 10% probability of failing within 5 years. A-critical items are excluded. Perhaps the next 25% of all critical parts can be considered to be B-critical.

The third set, called C-critical, might be associated with greater than 10% probabilities of failing after 5 years. Because 50% of the parts are in the A and B groups, the remaining 50% of the parts are in the C group. The numbers can be adjusted so that A, B, and C are 25%, 25%, and 50%, respectively, of the total number of parts.

As an alternative definition, part failure can have a probability (not a certainty) of stopping the process or product. Thus, a possible description of critical parts relates to the probability of total process or product failure when the part fails.

The parts are rank ordered by the probability of causing process or product failure. Say that if the first part fails, there is a 30% probability of product failure. The next part might have a 20% probability. Perhaps the top 25% of critical parts has an 80% probability of causing (each independently) product failure. This kind of situation is depicted in Figure 5.7.

Some processes do not fail as a result of part failure but instead the production output is reduced by a significant amount. Curves similar to Figure 5.7 can be created for such situations.

Criticality is a coined term that can mean crucial to performance or dangerous to use. Thus, an alternative definition of criticality could apply to the danger involved in using materials. Flammability, explosiveness, and toxicity of fumes are crucial safety factors for materials management. Whichever definition of criticality is used, the procedure is to list first the most critical parts. Next, systematically rank-order parts according to their relative criticality. The concept of criticality should reflect the costs of failures, including safety dangers, loss of life, and losses in production output.

Spare parts and other backup materials should be provided to conform to remedial failure strategies. Also to be considered are replacement parts for preventive

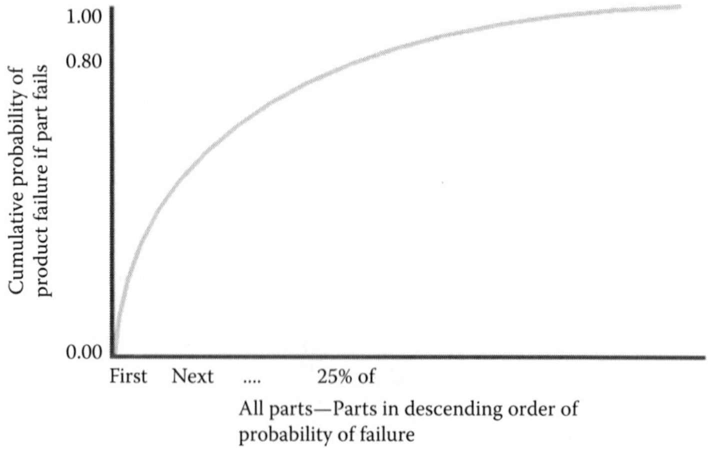

Figure 5.7 ABC analysis—based on material criticality.

failure strategies. Further, spare parts must be inspected to ensure their continued integrity over time.

A large South American refinery decided to double its rubber spare parts inventory. Within 2 years, many of the parts had begun to deteriorate and could not be called upon to replace critical failed parts. The high penalty resulted in a return to the original policy. Organizations have found that poor spare parts policies can be an "Achilles' heel."

5.6.2 Annual Dollar Volume of Materials

More widely used is a second set of ABC categories based on sorting materials by their annual dollar volume. Dollar volume is the surrogate for potential savings that can be made by improving the inventory management of specific materials. Accordingly, all parts, components, and other materials used by a company should be listed and then rank-ordered by their annual dollar volume. Start with those items that have the highest levels of dollar volume and rank order them from the highest to the lowest levels. The top 25% of these materials are called A-type items. The next 25% are called B-type items. The bottom 50% are called C-type items.

The ABC method requires individual study of each A-item in depth to improve its performance. Details include how much to order at one time, which determines how often to order; who to order from; what quality standards to set; delivery LTs; the consistency of LTs; as well as all special agreements with suppliers.

B-type items are studied in groups and with less attention to detail. Policies for C-type items are set to be as simple as possible to administer. C-type items have low-dollar volume, which means that they have low price per unit or they may have low volume, or both. Most are C-type because of the low price. Small penalties are paid for overstocking such items, so they do not have to be ordered as often. Still, it may turn out that a C-type item can lead to a critical situation, for example, the CEO's bathroom runs out of toilet paper. Toilet paper is a typical C-type item.

If materials managers do not use this ABC systems approach, which sorts all materials into those that have more potential savings than others, then a major strategic capability for systematic and continuous improvement is being overlooked.

Companies differ with respect to what percent of all items stocked account for 75% of their total annual dollar volume. Usually, a small percentage of all items (such as 25%) accounts for a very large percentage of total annual dollar volume (such as 75%) as shown in Figure 5.8. The A-type items are labeled in Figure 5.8. Figure 5.8 portrays a typical case where 20–30% of all items carried account for as much as 70–80% of the company's total dollar volume.

There is no fixed convention that A, B, and C class breaks must occur at 25% and 50%. Some companies use only A and B classes. It is historically interesting that Del Harder of the Ford Motor Company is said to have developed the ABC concept in the 1940s. Still, almost 70 years later, it is not well known among those who have not studied P/OM that inventories of materials conform to a curve that reflects disproportionately large effects (of criticality and dollar demand) derived from a small percent of all items in the supply chain. Because the costs of inventory studies tend to be proportional to the number of items under consideration, materials management always chooses to study and update A-class items.

We will use the following example to illustrate the process to classify items in A, B, and C categories.

Example: Consider the data given in Table 5.2. There are 10 items labeled P though Y in the first column. The second column gives the annual volume (number

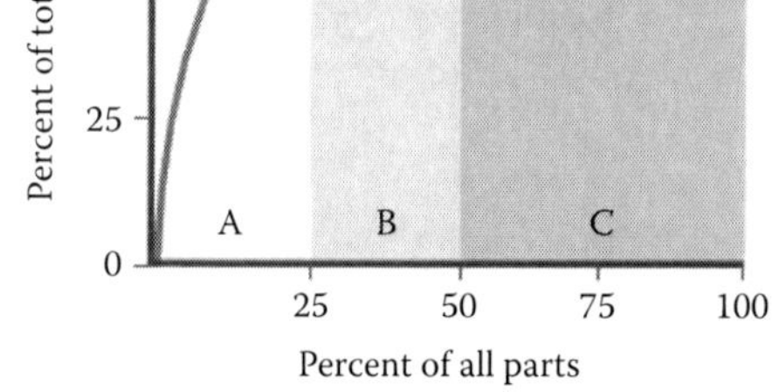

Figure 5.8 ABC analysis—based on dollar volume. Parts are rank ordered by dollar volume.

Table 5.2 ABC Analysis

Item Stock Number	*Annual Volume (units)*	*Unit Cost ($)*	*Annual Dollar Volume*	*Percentage of Annual Dollar Volume*	*Cumulative % of Annual Dollar Volume*	*% of Number of Items Stocked*	*Cumulative % of Number of Items Stocked*	*Category*
P	1250	92.00	115,000.00	40.17	40.17	10.00	10.00	A
Q	530	168.00	89,040.00	31.10	71.26	10.00	20.00	A
R	1970	18.75	36,937.50	12.90	84.17	10.00	30.00	B
S	430	42.20	18,146.00	6.34	90.50	10.00	40.00	B
T	990	13.80	13,662.00	4.77	95.27	10.00	50.00	B
U	680	12.50	8,500.00	2.97	98.24	10.00	60.00	C
V	2150	0.98	2,107.00	0.74	98.98	10.00	70.00	C
W	210	9.80	2,058.00	0.72	99.70	10.00	80.00	C
X	1250	0.52	650.00	0.23	99.93	10.00	90.00	C
Y	335	0.64	214.40	0.07	100.00	10.00	100.00	C
		Total	286,314.90					

of units) and the third column gives the cost per unit of the item. The fourth column gives the annual dollar volume which is obtained by multiplying the annual volume (units) by the cost per unit. For example, the annual dollar volume for item *P* is $115,000 (= 125.0 * $92.00). All the 10 items are rank-ordered by their annual dollar volume. That is, the item that has the highest annual dollar volume is on the top. The next step is to calculate the percentage of annual dollar volume. This is obtained by dividing the annual dollar volume of an item by the total annual dollar volume of all items which is $286,314.90. For example, the percentage of annual dollar volume of item *P* is 40.17% (= 115,000/286,314.90) and that of *Q* is 31.10% (= 89,040.00/286,314.90). The next column in the table gives the cumulative percentage of annual dollar volume. For example, items *P* and *Q* together account for 71.26%. Similarly, the first three items *P*, *Q*, and *R* account for 84.17% (= 40.17% + 31.10% + 12.90%). The next column gives the percentage of the number of items stocked. Since we are analyzing 10 items, each item accounts for 10%. The next column gives the cumulative percentage of the number of items stocked. *P* and *Q* together are equal to 20% of all items. The first three items *P*, *Q*, and *R* account for 30% of all items and so on.

The ABC analysis puts these items in three categories. The first two items *P* and *Q* (20%) account for 71.26% of the annual dollar volume. We can classify them in the A category. The last five items *U* through *Y* (50% of all items) account for 4.80% (2.97% + 0.74% + 0.72% + 0.23% + 0.07%) of annual dollar volume. These items can be conveniently classified as C items. The remaining three items *R*, *S*, and *T* (30%) account for 24.01% (= 12.90% + 6.34% + 4.77%) of annual dollar volume and are classified as B items. The items at the boundary lie in a gray area and need executive judgment for appropriate classification. For example, item *R* could have been classified in the A category, pushing category A to three items (30%) with an annual dollar volume of 84.17%.

5.7 Quantity Discount Model

In certain ordering situations, the supplier may give a discount on the price if the item is purchased in large quantities. This scenario is called quantity discount or price discount. However, the price discount on large quantities may not always lead to total lower costs because large quantities will increase the inventory carrying costs. Consider the following example to understand how the quantity discount model works.

Example: The annual demand (D) for an item is 240,000 units. The ordering cost per order (S) is $30.00. The inventory carrying cost per unit per year (H) is 30% of the cost (price) of the item, that is, H = 30% of C. The vendor has quoted the following costs (prices) per unit.

- Price 1: $2.80 for order quantity less than or equal to 29,999
- Price 2: $2.77 for order quantity 30,000 and above

To solve this problem, we will compare the total costs for both prices. As in the EOQ model, the EOQ is given by

$$Q_O = \sqrt{\frac{2DS}{H}},$$

and, the total cost (*TC*) is given by

$$TC = \left(\frac{D}{Q}\right) * S + \left(\frac{Q}{2}\right) * H + D * C.$$

We start the calculations by assuming that we plan to buy at the lower price ($2.77). The inventory carrying cost for this price is $0.83 (=30% of $2.77) per unit per year. The EOQ for this price is 4163 (using the EOQ formula). However, we cannot buy 4163 units at the price of $2.77 because the minimum quantity specified by the vendor is 30,000. Therefore, we have to buy at least 30,000 units to avail of this price discount. We calculate the total cost at this quantity *TC* (30,000). Using the *TC* equation:

$$TC(30{,}000) = \left(\frac{240{,}000}{30{,}000}\right) * 30 + \left(\frac{30{,}000}{2}\right) * 0.83 + 240{,}000 * 2.77 = \$677{,}505.00.$$

We now calculate the EOQ for the higher price $2.80. The value of *H* for this price is $0.84 (30% of $2.80). The EOQ is 4140. This quantity is feasible because we can by up to 29,999 units at $2.80 per unit. The total cost at this quantity, *TC* (4140) will be

$$TC(4140) = \left(\frac{240{,}000}{4{,}140}\right) * 30 + \left(\frac{4140}{2}\right) * 0.84 + 240{,}000 * 2.80 = \$675{,}477.93.$$

The order quantity for this example is 4140 since *TC*(4140) < *TC*(30,000). In case *TC*(30,000) was lower than *TC*(4140), 30,000 will be the answer. This example illustrates that a quantity discount is accepted only when it lowers the total costs.

There are other qualitative factors that need to be considered while making decisions about the order quantity. A materials manager may say "no" to the idea of buying in a larger quantity even if this larger quantity minimizes the total cost. Several reasons could explain this position. More cash will be tied up and more storage space required while buying in large quantities. On the other hand, if there is a possibility of a shortage of materials from a delivery disruption, then the larger order will be preferred. Choosing a smaller order size may indicate an intuition that in the future prices will fall.

Some factors that make a larger purchase attractive could include

- Is there a possibility of a strike?
- Can the purchase of a large quantity make it difficult for competitors to buy it?
- Can the purchase of a large quantity promote further price reductions?

Some factors that make a smaller purchase attractive could include

- Should there be a change in the material being used?
- Does this item spoil easily?
- Does it have to be protected against pilferage?
- Does it experience obsolescence?

The reasoning process that has been applied can be extended to any number of price breaks for quantity discounts. Problems on multiple price breaks are also included in the set of practice questions at the end of this chapter.

5.8 Lead Times

LT is the interval that elapses between the recognition that an order should be placed and the delivery of that order. The required LTs for both EOQ and EPQ models can be identified by studying their respective graphs as shown in Figure 5.9. Note how the diminishing stock level reaches a threshold (or limen) called Q_{RP} on both diagrams. Q_{RP} stands for the stock level of the reorder point (RP). That threshold triggers the order for replenishment. The level of stock that is at the RP is based on having enough stock on hand to meet orders until the replenishment supply arrives and is ready to be used. The interval between reordering and receiving that order, and readying the units for use, is called the LT. Figure 5.9 shows how

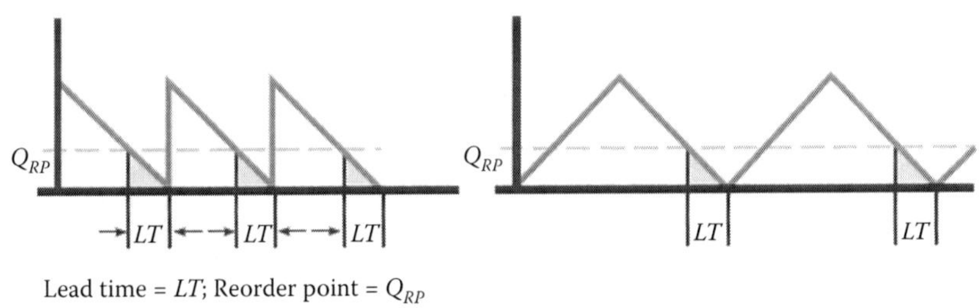

Figure 5.9 Inventory reorder points and lead times.

the horizontal Q_{RP} line intersects the *place order and initiate LT line* on the down-sloping side of both the EOQ and the EPQ models.

Eight LT considerations that apply to EOQ or EPQ or both:

1. The amount of time required to recognize the need to reorder. If the RP is monitored continuously, then this period is as close to zero time as the system permits. If the stock level is read at intervals, then the average interval for noting that the RP has been reached is part of the LT interval.
2. The interval for doing whatever clerical work is needed to prepare the order. This includes determining how much to order and from whom to order. It might even include preparing for bidding. If multiple suppliers are to be used, this also includes determining how to divide the order. Self-supply with EPQ pertains.
3. Mail, e-mail, EDI, or telephone intervals to communicate with the supplier (or suppliers) and to place the order(s).
4. How long does the supplier's organization take to react to the placement of an order? Is a system in place for communicating with everyone who participates in filling the order? How long does it take to find out if the requested items are in stock? If the items are not in stock, how long will it take for the supplier to set up the process to make them? How long will it take to produce, pack, and ship?
5. Delivery time includes loading, transit, and unloading transporters. If transshipment is needed, the LT period must be extended to take that into account. This period begins when the product leaves the supplier's control and ends when the buyer takes control of the items.
6. Processing of delivered items by the receiving department includes quantity and quality checks as well determining storage locations and moving items.
7. Inspection to be sure items match specifications either by sampling or 100% inspection is generally required. Time may be required to deal with problems uncovered by the inspection.
8. Computers used for updating records are sometimes unavailable until after normal office hours. The effect of such delays on the production schedule must be considered.

These eight LT components, when added together, form the total LT. Determination of good LT estimates requires awareness of all relevant systems factors. Proper LT determination requires a systems study that traces out all contributing factors including the averages and variances of these factors. Each situation requires familiarity with what is actually being done and what needs to be done.

LTs are usually variable. There are delays of varying length in each of the eight LT components listed earlier. When variance is significant, estimates must be made of the degree of variability. Although LTs are usually variable, they are often treated as if they were fixed, single values. There are two reasons for this:

1. The amount of variability is considered to be small and relatively negligible.
2. It is complex to assume variable demand and variable LT simultaneously.

Then, to take variable LT into account in a quick and easy way, the idea of increasing protection against stock-outs is used. This is accomplished by assuming a larger (worse case) average than the actual average for the LT. To illustrate, if LT is known to average 10 days, and it is also known to range ±2 days almost all of the time, then an estimate of 12 days for LT might be a self-protective move. Extra inventory would be carried to buffer the 2 days that LT might slip. Airlines add extra minutes to their flight schedules to reduce the regularly reported frequency of being late. The uses of safety (buffer) stock is a just-in-case, rather than a just-in-time, tactic. The much-used perpetual inventory system always employs some amount of extra stock as protection against surges in demand and delays in delivery. The question is "how much extra stock?"

5.9 Order Point Policies

Order point policies (OPPs) define the stock level at which an order will be placed. In other words, as withdrawals decrease the number of units on hand, a particular stock level is reached. That number of units, called the RP, triggers an order for more stock to be placed with the appropriate supplier. OPP models specify the number of units to order. The time between orders varies depending on the order quantity and the rate at which demand depletes the stock on-hand. So, order size is fixed and the interval between orders varies. This discussion is about one type of order point system called the *perpetual* inventory system. It continuously records inventory withdrawals. Most often it is used online and in real time.

Next, another type of OPP will be described. It is called a *periodic* inventory system. On consecutive specific dates separated by the same amount of time, the item's record is called up and an order is placed for a variable number of units. The order size is determined by the amount of stock on hand when the record is read. It is the date that triggers the review and the order being placed. Therefore, in the periodic case, the interval between orders is fixed while the ordered amount varies. The order amount is a function of the rate at which demand occurred in the period between reviews of the item's records.

Periodic OPP has become outmoded in some companies because computer systems enable the perpetual inventory system to be used inexpensively. The perpetual system has advantages that will be explained. Nevertheless, the periodic system will be described for two reasons. First, it is used to order certain items together from the same supplier.

This may be done by small- and medium-sized organizations to obtain carload freight rates because they do not have sufficient volume to fill a freight car with one item. They may also want to deliver a number of different items to retailers on a (say) once-a-week basis. Having a truck run to fulfill requirements of the perpetual inventory system could be far too costly. Second, some advanced inventory models mix the use of the perpetual and the periodic systems with the result that costs are further reduced. Therefore, it is useful to understand the mechanism of the periodic model.

Inventory models (such as the two mentioned here) are based on the concepts and mathematical structure of an EOQ—the fixed amount to order, or an EPQ—the fixed amount to produce. EOQ and EPQ have been discussed earlier.

There is another set of problems that are not resolved by OPP inventory models. Those "other" situations are dealt with by MRP—not discussed in this book. At this moment, the point to make is that MRP is used when the assumptions about order continuity and consistency of demand that are required by EOQ and EPQ are not met.

The most important requirements for EOQ and EPQ (which are OPP models) are consistency of independent demand over time with relatively smooth and regular withdrawal patterns of units from stock. MRP works for sporadic dependent demand by using information systems control of messy demand patterns. MRP requires detailed information-processing capabilities, which have grown enormously since MRP was first developed in the 1960's. When OPP is an online system, dealing with the perpetual inventory management of large numbers of SKUs, significant information capabilities are also required.

5.10 Perpetual (Fixed Quantity) Inventory Systems

Perpetual, also known as fixed quantity, inventory systems continuously record inventory received from suppliers and withdrawn by employees. Most perpetual inventory systems are used online, in real time with some demand variability. The EOQ and EPQ models discussed so far do not include demand variability. In many practical situations, this assumption is unrealistic. With modifications, the EOQ and EPQ models can deal with demand variability.

Variability of demand arises from variations in order sizes. Variability is passed along the supply chain, affecting everyone. Customers are the primary cause of demand variability. There are other causes of demand variability as well. These include stock on-hand replacement to make up for items lost as a result of warehouse fires, employee pilferage, and the discovery that items in stock cannot be shipped because they are defective.

Consider what occurs when a machine setting changes, creating a large number of rejects that cannot be reworked. Production must run larger numbers of items to compensate for the scrap. More materials must be withdrawn from inventory to be used by the process. The supplier furnished the usual number of units per order that was based on the estimate of the expected demand (D). Because demand is larger than expected, the RP, Q_{RP}, will be reached sooner. The perspective that will be used to explain how the EOQ model can be modified to protect the buyer will assume batch delivery of the quantity:

$$Q_O = \sqrt{\frac{2DS}{H}}.$$

The same conditions apply. The processes must be ongoing with item demand being relatively constant over time. It is evident that such modeling does not apply to job shops with small orders because they lack continuity of demand. The focus now is on items that are stocked regularly, where the demand can vary. The buyer orders the EOQ, but has on hand some buffer stock (which is sometimes called reserve stock or SS). The buffer stock is extra stock so that when demand is heavier than expected, orders can still be filled.

An option is to design a perpetual inventory system. It is perpetual because it is an online system, tracking the stock on-hand (SOH) at each transaction of withdrawal or stock entry. To design the system, it will be useful to answer the following questions:

1. How many buffer stock units should be carried?
   The answer involves balancing the added carrying costs for extra stock against the costs of running out of stock. As one cost goes up, the other comes down.
2. When should an order be placed for the buffer stock?
   The buffer stock should be created as soon as it is determined that it is beneficial to have it. It does not have to be replenished when it has been drawn on to avoid stock-outs because the use of buffer stock is expected to average out at zero. The only times that this will not apply will be if the demand system is changing over time or is unstable.
3. How often should an order be placed for buffer stock?
   Once, and then whenever the buffer-stock level needs to be updated.
4. How many units should be ordered to meet expected demand?
   Use the same economic order quantity.
5. How often and when should the orders be placed for units to meet the demand?
   An order is placed whenever the reorder point, Q_{RP}, is reached.
6. How does this perpetual system work?
   Withdrawal quantities are entered in the computer each time one or more units are taken out of stock. These quantities are subtracted from the previous stock-level balance to determine the new balance quantity of stock on-hand.

5.10.1 Reorder Point and Safety (Buffer) Stock

The RP quantity is designated for each item and is entered in the computer program. When the RP has been reached, the program recognizes this fact and an order is placed for the EOQ. The stock level of the RP (Q_{RP}) is equal to the expected (average) demand during the LT period ($\bar{D}_{LT}$) plus the safety stock (SS) quantity. Thus, $Q_{RP} = \bar{D}_{LT} + SS$.

The expected demand during LT is a function of average demand per day (d) and the magnitude of LT and is determined as, $\bar{D}_{LT} = d \times LT$. It may be noted that calculation of $\bar{D}_{LT}$ becomes complex if LT also varies.

The value of SS depends on the variability of demand and the service level that a company wants to provide. The service level is a measure of the stock-out situations allowed. For example, a 95% service level means that there will be no out-of-stock situation 95% of the time during LT.

Assuming that the demand follows a normal distribution, the value of SS can be determined as $SS = z\sigma_{LT}$, where, σ_{LT} is the standard deviation of demand during LT and z is a measure of the service level that we want to provide. z is called standard normal random variable and can be found from its statistical table. For the 95% service level, the value of $z = 1.65$. The z table is included in Appendix B.

Determining SS level requires an economic balancing situation between the cost of going out of stock versus the cost of carrying more inventory. Outages happen whenever actual demand in the LT period exceeds Q_{RP}. The likelihood of an outage occurring will be decreased by increasing the value of buffer stock. The large buffer stock means that the carrying cost of stock is high to make sure that the actual cost of stock-outages is small.

5.10.2 Operating the Perpetual Inventory System

The first step to understanding how the perpetual inventory system works is by studying its graphic representation. Figure 5.10 illustrates the two parts that have been put together. The EOQ model sits on top of the buffer stock (which is the shaded rectangle).

Follow the stock on-hand line as it forms an irregular saw-tooth pattern. When demand becomes greater, the line moves down faster, and when the demand rate decreases, the line moves down more slowly. In other words, when the slope of the line becomes steeper, the RP (Q_{RP}) is reached more quickly. The interval between orders is variable depending on whether the demand has been faster or slower than the average rate.

Note in Figure 5.10, the first of the replenishment cycles stops right at the buffer-stock line. At that point, SOH is zero and no buffer stock is withdrawn. Demand dips into the buffer-stock region only for the second replenishment cycle. In other words, buffer stock is called upon once in three replenishment cycles. The buffer stock withdrawn will be replaced when the next order is received and the units are put in stock.

The third of the three replenishment cycles shows an order arriving when the stock level is above the buffer-stock level. If the demand distribution is stable over time, then the buffer-stock level will average out to the planned number of units for buffer stock.

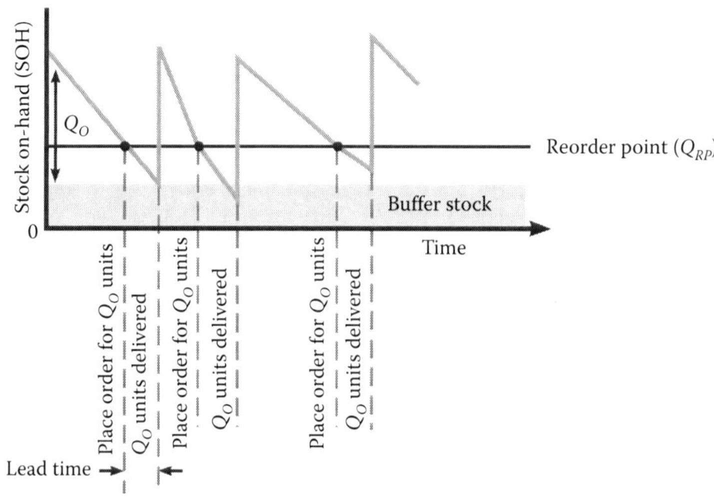

Figure 5.10 Perpetual (fixed quantity) inventory system.

5.10.3 Two-Bin Perpetual (Fixed Quantity) Inventory Control System

The two-bin system is a smart way of continuously monitoring the order point. It is a simple self-operating perpetual inventory system. Figure 5.11 shows the two bins with familiar labels. Two bins are marked 1 and 2. The optimal order size of Q_o units has been delivered. First, Bin 1 is filled to the RP level. Bin 1 now contains units to cover expected demand in the LT period plus SS. The remaining units from the delivery of EOQ = Q_o units are put into Bin 2. The bottom of Bin 2 is labeled the RP. When Bin 2 is emptied, the number of units remaining are all in Bin 1 and they are the RP in number. Spending a few minutes with Figure 5.11 will make clear the delineation of the RPs in Bins 1 and 2. Withdrawals are made from Bin 2 first. When Bin 2 is

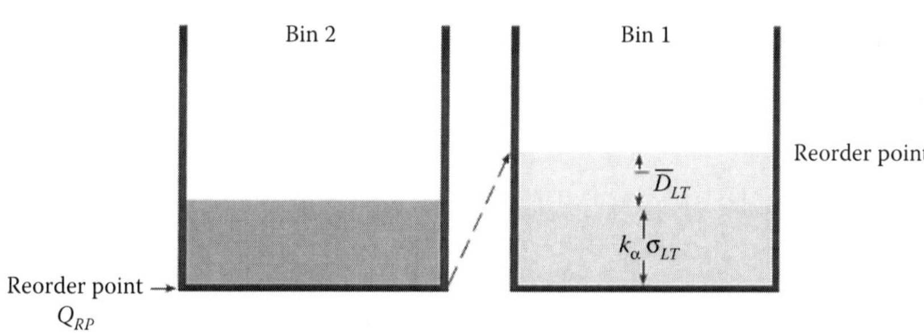

Figure 5.11 Two-bin perpetual inventory control system.

depleted, an order is placed for the EOQ, and further units are taken from Bin 1. Each time Bin 2 is emptied, a new order is placed—it is equivalent to reaching the RP. The two-bin system is not feasible for many kinds of items. When applicable, much clerical work is eliminated. This two-bin system is well suited to small items such as nuts, bolts, and fasteners. These are items too small and too numerous to make withdrawal entries for each transaction. The same reasoning applies to recording withdrawals of liquids, for which the two-bin system approach is also ideal. See, for example, the application of two-bin system concept in effective management of a nursing ward for replenishment of supplies method (http://www.hec.ca/pages/sylvain.landry/en).

5.11 Periodic Review (Fixed Time) Inventory Systems

Periodic inventory systems are based on review of inventory levels at regular fixed review periods. These systems were more popular than perpetual inventory systems before inventory information was digitized and put online. These were ideally suited for manual entries and when actions should be taken periodically rather than randomly.

Computers outmoded periodic manual systems primarily designed to save money on the clerical aspects of tracking inventory. However, periodic inventory systems continue to be used for other reasons. These include requirements of suppliers concerning the timing for accepting new orders, requirements of shippers about timing deliveries, meeting the schedules of customers, and fulfilling the need to combine orders to obtain volumes sufficient for shipment discounts.

Some organizations have central warehouses that will only accept orders from their regional distributors once in a week. Each region expects deliveries on a different day of the week. Further, some industries prefer the regularity of the periodic method, which can be linked to changeover intervals for production processes as well as the phases of projects. For example, the stages of buildings must be synchronized with what suppliers deliver.

Periodic inventory systems also play a part in an advanced class of inventory models (called Ss policies) that combine the ordering rules of perpetual and periodic order systems to obtain lower total costs. These blended methods can be encountered in big inventory systems installations such as the Armed Forces use.

The optimal interval for periodic review, t_0, is based on the square root relationship given in the following equation:

$$t_0 = \sqrt{\frac{2S}{DH}}.$$

The equation for t_0 can be derived as follows:

$$t_0 = \frac{Q_0}{D} = \frac{1}{D} * \sqrt{\frac{2DS}{H}} = \sqrt{\frac{2S}{DH}}.$$

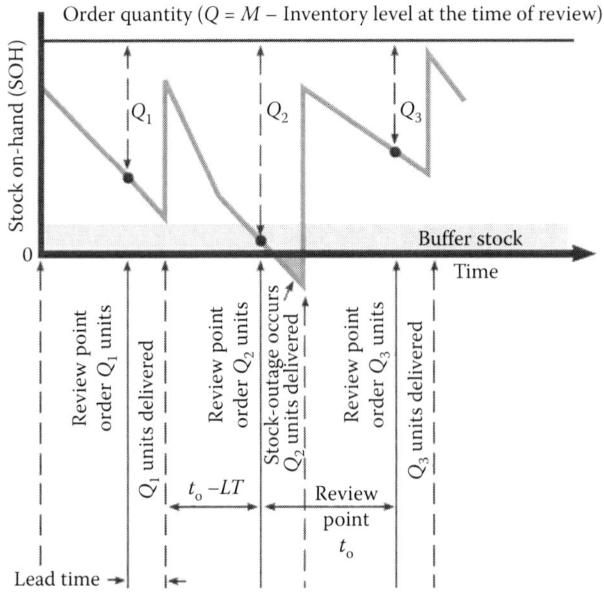

Figure 5.12 Periodic review (fixed time) inventory systems.

High-cost items and large carrying costs increase the desirability of short review periods. High ordering costs and/or large setup costs call for long review periods.

To use the periodic model, first find and set the review interval. Then set a maximum inventory level, say *M*. *M* is equal to the demand during the review period and LT and SS. That is, $M = d(t_0 + \text{LT}) + \text{SS}$, where *d* is the average demand per unit time (day, week, month, etc.).

Then, at each review, SOH is determined and an order is placed for quantity *Q*, where $Q = M - \text{SOH}$. The value of *Q* will be different each time an order is placed because in the equation for *Q*, *M* is fixed and the value of SOH depends on the demand during the review period. This is in contrast to the perpetual inventory model where the order was a fixed quantity at variable intervals. Figure 5.12 illustrates the way that the periodic order system functions.

5.11.1 Safety (Buffer) Stock in Periodic (Fixed Time) Inventory Systems

The size of the buffer stock for the periodic review system is calculated in the same manner as for the perpetual inventory system. However, the SS for the periodic inventory system must include protection against unexpectedly great demand—in a review period plus one LT interval.

SS for the periodic inventory system is calculated by using the distribution of demand during (t_0 + LT) because after an order is placed, variations in demand can be experienced during the period t_0. Then a review occurs and an order is placed. Demand variations can continue during the LT. Therefore, exposure to variable demand before correction can be made occurs over a review period and an LT. This requires extra stock for protection, which is a major drawback of the periodic model. The standard deviation of demand during (t_0 + LT) is calculated as follows: $\sigma_{(t_0+\mathrm{LT})} = \sigma\sqrt{(\mathrm{LT} + t_0)}$, where σ is the standard deviation of demand within one unit time period (day, week, month, etc.).

SS is then calculated as $\mathrm{SS} = z\sigma_{(t_0+\mathrm{LT})}$, where $\sigma_{(t_0+\mathrm{LT})}$ is the standard deviation of demand for the interval (t_0 + LT) and z is the standard normal random number and is obtained from statistical tables (see Appendix B) based on the desired service level.

The periodic model appeals to managers who like bookkeeping and regularity of actions. On the review date, the books are opened. The ledger is added up and the order is placed. The market drives actions in the perpetual, whereas the calendar drives actions in the periodic. The difference is both economic and psychological.

Summary

This chapter addresses "independent demand" inventory management. It explains OPP models in terms of demand continuity. The costs of inventory are treated. These include costs of ordering, setups, and changeovers, carrying inventory, discounts, stock-outages, and running the inventory system. Process types differentiate inventory costs.

Then, EOQ models are developed for batch delivery. The TVC equations are set up and solved. EPQ models for continuous delivery of product are contrasted with EOQ models. The EPQ are applied to intermittent flow shops. There is discussion of LT structure, uncertainty, and variability.

Next, the perpetual inventory system with its RP and buffer-stock calculations is detailed. This includes operation of the two-bin perpetual inventory control system. The periodic inventory system is explained including why this system is being outmoded by computer capabilities—yet it still is needed in certain situations. This chapter also explains how quantity discount models work. ABC classification of materials is another powerful concept included in this chapter.

Review Questions

1. It has been stated that, as a rule of thumb, "the best inventory is no inventory." Discuss this heuristic.
2. How does this rule of thumb apply to gasoline? Is gasoline an OPP commodity?

3. For a toothpaste manufacturer, how is the decision made concerning how many caps should be ordered? Could it be a different number than the number of tubes that are ordered at one time?
4. How often should an order size be updated?
5. What method should be used to determine the order quantity for a raw material that is used continuously within a flow shop?
6. What method should be used to determine the order quantity for a raw material that is used continuously within an intermittent flow shop?
7. Who are the people that are responsible for placing orders?
8. Benetton is a well-known manufacturer and retailer of clothing all over the world. The Benetton factories are tied in with retailers so that demand information is relatively immediate and complete concerning what is selling and what is not. How does this information affect lot-size planning?
9. Salespeople use hand-held telecommunication devices to communicate inventory status to the warehouse in a major toy company. Why is this system needed and what does it affect?
10. What is carrying cost composed of and what is the range of values that will be found for this cost under varying conditions?
11. What is the logic for a buyer to accept a discount offer?
12. What is the logic for a seller to offer a discount?
13. Why is there an ordering cost and of what is it composed?
14. What is the difference between an ordering cost and a setup cost?
15. When is it likely that an ordering cost will be larger than a setup cost?
16. When is it likely that a setup cost will be larger than an ordering cost?
17. Relate the number of orders placed and the order quantity.
18. Why are OPPs called by that name?
19. Why are *TVC* equations written for OPP models instead of total cost equations?
20. When discounts are being considered, total cost equations must be used instead of *TVC* equations. Why is this so?
21. Differentiate between EOQ and EPQ models.
22. When is lead-time variability a problem and what can be done about it?
23. How can lead-time variability be modeled?
24. What is a two-bin system? When is it applicable?
25. Describe a perpetual inventory system.
26. Under what circumstances is a perpetual inventory system preferred?
27. Describe a periodic inventory system.
28. When is a periodic inventory system preferred?
29. How can a quantity discount model be used by a buyer and supplier to negotiate a price break point schedule that benefits both of them?
30. Why is it that multiple price break points for many discount levels can be examined in the same fashion as one price break point for a single discount?
31. Is the two-bin inventory system perpetual or periodic?
32. Distinguish between inventory problems under certainty, risk, and uncertainty.

Problems

1. Water testing at the Central Park reservoir requires a chemical reagent that costs $500 per gallon. Use is constant at 1/3 gallon per week. Carrying cost rate is considered to be 12% per year, and the cost of an order is $125. What is the optimal order quantity for the reagent?
2. Continuing with the information about the Central Park reservoir given in Problem 1, the city could make this reagent at the rate of 1/8 gallon per day, at a cost of $300 per gallon. The setup cost is $150. Use a 7-day week. Compare using the EOQ and the EPQ systems. What course of action do you recommend?
3. Water testing at the Delaware reservoir requires a chemical reagent that costs $400 per gallon. Use is constant at 1/3 gallon per week. Carrying cost rate is considered to be 10% per year, and the cost of an order is $100. What is the optimal order quantity for the reagent?
4. Continuing with the information about the Delaware reservoir given in Problem 3, the town could make this reagent at the rate of 1/4 gallon per day, at a cost of $500 per gallon. The setup cost is $125. Use a 7-day week. Compare using the EOQ and the EPQ systems. What course of action do you recommend?
5. The Drug Store carries Deodorant R, which has an expected demand of 15,000 jars per year (or 60 jars per day with 250 days per year). Lead time from the distributor is 3 days. It has been determined that demand in any 3-day period exceeds 200 jars only once out of every 100 3-day periods. This outage level (of 1 in 100 LT periods) is considered acceptable by P/OM and their marketing colleagues. The EOQ has been derived as 2820 jars. Set up the perpetual inventory system. (EOQ with stock-outs.)
6. Use the information given in Problem 5, plus the fact that it has been determined that demand in any 50-day period exceeds 4000 jars only once out of every 100 50-day periods. This outage level (of 1 in 100 LT periods) is considered acceptable by P/OM and their marketing colleagues. Set up the periodic inventory system. (EOQ with stock-outs.)
7. Compare the results derived in Problems 5 and 6. What do you recommend doing? Explain how you have taken into account all of the important differentiating characteristics of perpetual and periodic inventory systems. (EOQ with stock-outs.)
8. Consider the recommendation made in Problem 5, taking into account the fact that The Drug Store must combine orders for Deodorant R with other items in order to have sufficient volume to qualify for the distributor's shipping without charge. With this constraint, what are your recommendations?
9. The information required to solve an EOQ inventory problem is as follows: demand per year (D) = 5000, ordering cost per order (S) = $10.00, cost of the item (C) = $10 per unit, and inventory carrying cost per unit per year (H) = 16% of the cost of the item (C). What is the optimal order quantity? (EOQ model).

10. Using the information in Problem 9, instead of buying from a supplier the decision is to make the item in the company's factory. The new equipment is able to produce 30 units per day. Cost per item is now \$6.00. The set up cost is \$150.00 per set up. What is the optimal run size (EPQ)?
11. Using the information in Problems 9 and 10, which is better: make or buy?
12. Using the information in Problems 9–11, what factors that are not part of the mathematical model might shift the decision?
13. The following quantity discount schedule has been offered for the situation described in Problem 9. Should either of these discounts be accepted?
    \$10.00 for $Q \leq 299$
    \$9.00 for $300 \leq Q \leq 499$
    \$8.00 for $Q \geq 500$
14. The quantity discount schedule offered in Problem 13 prompted a competitor to offer the following discount schedule. Should any of these discounts be accepted
    \$10.00 for $Q \leq 250$
    \$9.00 for $251 \leq Q \leq 599$
    \$7.00 for $Q \geq 600$
15. Compare the answers to Problems 13 and 14 and discuss these results, making appropriate recommendations.
16. Murphy's is famous for their coffee blend. The company buys and roasts the beans and then packs the coffee in foil bags. It buys the beans periodically in quantities of 120,000 pounds and assumes this to be the optimal order quantity. This year it has been paying \$2.40 per pound on a fairly constant basis. The company ships 1,200,000 1-pound bags of its blended coffee per year to its distributors. This is equivalent to shipping 24,000 1-pound foil bags in each of 50 weeks of the year. This can be considered to be constant and continuous demand over time. What carrying cost in percent per year is implied (or imputed) by this policy if an order costs \$100 on the average? Discuss the results.
17. Using the information in Problem 16, suggest a better ordering policy.
18. The manager of the greeting card production department has been buying two rolls of acetate at a time. They cost \$200 each. Card production requires 10 rolls per year. Ordering cost is estimated to be \$4 per order. What carrying cost rate is imputed? Is it reasonable?
19. Using the information in Problem 18, if the cost of rolls of acetate increases to \$250 each, what happens to the imputed carrying cost rate? Is this reasonable?
20. In an inventory control system, the annual demand is 12,000 units; the ordering cost is \$30 per order and the inventory holding cost is \$3.00 per unit per year. The order quantity is 1000 units and the cost per unit of the item is \$150? What is the total cost (*TC*) per year? *TC* includes the inventory holding cost, ordering cost, and the cost of the item.
21. A company is planning for its financing needs. What is the total cost (*TC*) per year given an annual demand of 12,000 units, setup cost of \$32 per order, a

holding cost per unit per year of \$4, an order quantity of 400 units, and a cost per unit of inventory of \$150? *TC* includes the yearly set up cost, inventory holding cost, and the item cost.

22. If annual demand is 50,000 units, the ordering cost is \$25 per order, and the holding cost is \$5 per unit per year, what is the optimal order quantity?
23. A company is using the EOQ model to manage its inventories. Suppose its annual demand doubles, while the ordering cost per order and inventory holding cost per unit per year do not change. What will happen to the EOQ?
24. Find the EPQ for the following problem. Also identify the storage capacity required.
    Annual demand = 50,000 units; setup cost = 25; inventory holding cost = 5 per unit per year. Production rate = 500 units per day; number of working days = 250.
25. Consider the following data and answer the next four questions.
    A plant operates 5 days a week, 52 weeks a year, and can produce at 60 units per day.
    The setup cost for production run is \$450.00.
    The cost of holding inventory is \$5.00 per unit per year.
    The annual demand for this product is 5000 units.
    a. What is the EPQ?
    b. What is the total annual cost of inventory and set up if the batch size (*Q*) is 800 units?
    c. What is the number of set ups per year if the lot size (*Q*) is 1500 units?
    d. What is the maximum inventory level if the batch size (*Q*) is 600?

Readings and References

Abernathy, F.H. et al., Control Your Inventory in a World of Lean Retailing, *Harvard Businesss Review*, 78(6), November–December 2000, pp. 169–176.

Aggarwal, S., MRP, JIT, OPT, FMS? *Harvard Business Review*, 66(5), September–October 1985, pp. 8–16.

Ansari, A. and B. Modarress, The Potential Benefits of Just-in-Time Purchasing for U.S. Manufacturing, *Production and Inventory Management Journal*, 28, Second Quarter, 1987, pp. 30–35.

Ansberry, C., Hurry-Up Inventory Method Hurts Where It Once Helped, *Wall Street Journal Online*, June 25, 2002.

Barba-Gutierrez, Y., B. Adenso-Diaz, and S.M. Gupta, Lot Sizing in Reverse MRP for Scheduling Disassembly, *International Journal of Production Economics*, 111(2), February 2008, p. 741.

Bell, P.C. and N. Hamid, Managing Inventories Through Difficult Economic Times: A Simple Model, *Interfaces*, 15(5), 1985, p. 39.

Berry, W.L., Lot Sizing Procedures for Requirements Planning Systems: A Framework for Analysis, *Production and Inventory Management*, 13, Second Quarter, 1972, pp. 19–34.

Billesback, T.J., A Study of Implementation of Just-in-Time in the United States, *Production and Inventory Management Journal*, 32, Third Quarter, 1991, pp. 1–4.

Bird, L., High-Tech Inventory System Coordinates Retailer's Clothes with Customers' Taste, *Wall Street Journal*, June 12, 1996, pp. B1, B5.

Bowers, M.R. and A. Agarwal, Lower In-Process Inventories and Better On-Time Performance at Tanner Companies Inc, *Interfaces*, 25, July–August 1995, pp. 30–43.

Bradley, J.R. and R.W. Conway, Managing Cyclic Inventories, *Production and Operations Management*, 12(4), Winter 2003, pp. 464–479.

Callioni, G., X. de Montgros, R. Slagmulder, L.N. Van Wassenhove, and L. Wright, Inventory Driven Costs, *Harvard Business Review*, March 2005, pp. 135–141.

Canen, A.G. and R.D. Galvao, An Application of ABC Analysis to Control Imported Material, *Interfaces*, 10(4), 1980, p. 22.

Chen, F., Salesforce Initiative, Market Information, and Production/Inventory Planning, *Management Science*, 51(1), January 2005, pp. 60–75.

Chopra, S., G. Reinhardt, and D. Maqbool, The Effect of Lead Time Uncertainty on Safety Stocks, *Decision Sciences*, 35(1), Winter 2004, pp. 1–24.

Cohen, M., K.V. Pasumarti, K. Paul, L. Hau, and T. Armen, Optimizer: IBM's Multi-Echelon Inventory System for Managing Service Logistics, *Interfaces*, 20(1), 1990, p. 65.

Farasyn, I., K. Perkoz, and W. Van de Velde, Spreadsheet Models for Inventory Target Setting at Procter & Gamble, *Interfaces*, 38(4), 2008, p. 241.

Flores, B.E. and D.C. Whybark, Implementing Multiple Criteria ABC Analysis, *Journal of Operations Management*, 7(1–2), October 1987, pp. 79–85.

Flowers, A.D. and J.B. O'Neill II, An Application of Classical Inventory Analysis to a Spare Parts Inventory, *Interfaces*, February 1978, pp. 76–79.

Gossard, G., A New Focus on Inventory Performance and Bottom-Line Profits, *Hospital Material Management Quarterly*, 18(2), November 1996, pp. 69–76.

Inman, R.A. and S. Mehra, JIT Applications for Service Environments, *Production and Inventory Management Journal*, 32, Third Quarter, 1991, pp. 16–21.

Kanet, J. and V. Sridharan, The Value of Using Scheduling Information in Planning Material Requirements, *Decision Sciences*, 29(2), Spring 1998, pp. 479–498.

Karmarkar, U., Getting Control of Just-in-Time, *Harvard Business Review*, 67(5), September–October 1989, pp. 122–131.

Karmarkar, U.S., Kekre, S., K. Sunder, and F. Susan, Lot-Sizing and Lead-Time Performance in a Manufacturing Cell, *Interfaces*, 15(2), 1985, p. 1.

Keren, B., The Single Period Inventory Model, *Omega*, 37(4), August 2009, p. 801.

Kleutghen, P.P. and J.C. McGee, Development and Implementation of an Integrated Inventory Management Program at Pfizer Pharmaceuticals, *Interfaces*, 15(1), 1985, p. 69.

Krupp, J.A., Are ABC Codes an Obsolete Technology? *APICS—The Performance Advantage*, April 1994, pp. 34–35.

Krupp, J.A., Managing Demand Variations with Safety Stock, *Journal of Business Forecasting*, 16(2), Summer 1997, pp. 8–13.

Lee, H.L. and C. Billington, Managing Supply Chain Inventory: Pitfalls and Opportunities, *Sloan Management Review*, Spring 1992, pp. 65–73.

Lee, H.L., Billington, C., and Carter, B. Hewlett-Packard Gains Control of Inventory and Service Through Design for Localization, *Interfaces*, 23(4), 1993, p. 1.

Leschke, J.P., The Setup-Reduction Process: Part 1, *Production and Inventory Management Journal*, 38(1), First Quarter, 1997a, pp. 32–37.

Leschke, J.P., The Setup-Reduction Process: Part 2-Setting Reduction Priorities, *Production and Inventory Management Journal*, 38(1), First Quarter 1997b, pp. 38–42.

Lieberman, M.B., S. Helper, and L. Dmeester, The Empirical Determinants of Inventory Levels in High-Volume Manufacturing, *Production and Operations Management*, 8, Spring 1999, pp. 44–55.

Liu, X. and Z. Lian, Cost-Effective Inventory Control in a Value-added Manufacturing System, *European Journal of Operational Research*, 196(2), July 2009, p. 534.

Noblitt, J.M., The Economic Order Quantity Model: Panacea or Plague? *APICS—The Performance Advantage*, February 2001, pp. 53–57.

Pasternack, B.A., Filling Out the Doughnuts: The Single Period Inventory Model in Corporate Pricing Policy, *Interfaces*, 10(5), 1980, p. 96.

Reid, R.A., The ABC Method in Hospital Inventory Management: A Practical Approach, *Production and Inventory Management Journal*, Fourth Quarter, 1987, pp. 67–70.

Rubin, P. and W.C. Benton, A Generalized Framework for Quantity Discount Pricing Schedules, *Decision Sciences*, 34(1), Winter 2003, pp. 173–188.

Starr, M.K. and D.W. Miller, *Inventory Control—Theory and Practice*, Englewood Cliffs, NJ: Prentice-Hall, 1962.

Timme, S.G. and C. Williams-Timme, The Real Cost of Holding, *Supply Chain Management Review*, July–August 2003, pp. 30–37.

Vujosevic, M., D. Petrovic, and R. Petrovic, EOQ Formula When Inventory Cost is Fuzzy, *International Journal of Production Economics*, 45(1–3), August 1, 1996, pp. 499–504.

Wagner, H.M. and T.M. Whitin, Dynamic Version of the Economic Lot Size Model, *Management Science*, 5(1), 1958, pp. 89–96.

White, R.E., J.N. Pearson, and J.R. Wilson, JIT Manufacturing: A Survey of Implementations in Small and Large U.S. Manufacturers, *Management Science*, 45, January 1999, pp. 1–15.

Whitson, D., Applying Just-In-Time Systems in Health Care, *IIE Solutions*, 29(8), August 1997, pp. 32–37.

Chapter 6

Scheduling

Readers' choice: A schedule defends from chaos and whim—Annie Dillard

Cayirli, T., and Veral, E., Outpatient Scheduling in Health Care: A review of Literature, *Production and Operations Management*, 12(4), Winter, 2003, pp. 519–549. This paper gives a comprehensive review of various techniques available for scheduling of outpatients.

Deutsch, H., and Mabert, V.A., Queuing Theory and Teller Staffing: A Successful Application, *Interfaces*, 10(5), 1980, pp. 63. Bankers Trust Company of New York cuts down about $1 million from its annual wage expenses by using the staffing plan proposed in this paper.

Glover, F., Jones, G., Karney, D., Klingman, D., and Mote, J., An Integrated Production, Distribution, and Inventory Planning System, *Interfaces*, 9(5), 1979, p. 21. This paper describes the application of an integrated production, distribution and inventory system that saved Agrico Chemical Co. $18 million in first three years.

Gupta, D., and Denton, B., Appointment Scheduling in Health Care: Challenges and Opportunities, *IIE Transactions*, 40, 2008, pp. 800–819. This paper uses tardiness measure among others and discusses specific literature in the body of the paper. It has an excellence reference list.

Hodgson, T.J., On-Line Scheduling in the Production Environment: Part II. A "Successful" Case History, *Interfaces*,

10(2), 1980, p. 60. The paper develops a scheduling strategy to schedule operations in a bottleneck department.

Jain, S.K., A Simulation-based Scheduling and Management-information System for a Machine Shop, *Interfaces*, 6(1), 1975, p. 81.This paper describes a production scheduling model for processing about 1000 rolls daily on 40 machines in Bethlehem Steel Corporation.

Krajewski, L.J., Ritzman, L.P., and McKenzie, P., Shift Scheduling in Banking Operations: A Case Application, *Interfaces*, 10(2), 1980, p. 1. The shift scheduling system outlined in this paper has reduced cost for Ohio National Bank. It is now possible for management to change mix of full-time and part-time employees to account for changes in the work force productivity.

Lesaint, D., Voudouris, C., and Azarmi, N., Dynamic Worforce Scheduling for British Telecomunications plc, *Interfaces*, January–February 2000, pp. 45–56. The scheduling system implemented in 1997–1998 by British Telecommunications plc reached close to 20,000 engineers and saved about $150 millions yearly in operational costs.

Love, R.R., Jr., and Hoey, J.M., Management Science Improves Fast-Food Operations, *Interfaces*, 20(2), 1990, p. 21. The owner of several fast food restaurants was able to standardize the scheduling process across restaurants resulting in higher quality schedules.

Moss, S., Dale, C., and Brame, G., Sequence-dependent Scheduling at Baxter International, Interfaces; 30(2), 2000, p. 70. Baxter International was able reduce set-up time by using cluster analysis resulting in an annual savings of $165,000.

Smith, R., Find the Best Checkout Line, Wall Street Journal, December 8, 2011. The author describes the mindset of a customer while selecting a checkout line. Retailers try to make the system more efficient by speeding up. This article describes what works and what doesn't.

This chapter discusses scheduling of resources. The resources can take many forms: facilities and people (departments, machines, work centers, employees) for manufacturing and service organizations as well as government and not-for-profit operations. In essence, production scheduling is planning when jobs are to be started and finished, who is going to do them, where, and in what order.

After reading this chapter, you should be able to:

- Explain why production scheduling must be done by every organization whether it manufactures or provides services.
- Discuss the application of the loading function.
- Draw a Gantt chart and explain its information display.
- Describe the role of sequencing and how to apply sequencing rules for one facility and for more than one facility.
- Classify scheduling problems according to various criteria that are used in practice.
- Explain the purpose of priority sequencing rules and the means for using them.
- Describe various priority rules for sequencing—that are used in practice.
- Apply Johnson's rule to the 2-machine flow shop problem.
- Analyze dynamic scheduling problems.

6.1 Introduction

The production-scheduling plans are accomplished by the *shop loading* function and by the *sequencing function* in the production department.

Shop loading consists of assigning jobs that are on hand to departments or work centers or machines. *Shop loading* is done for hospitals, restaurants, schools, police, and firemen assignments; in other words, services, manufacturing, and all other organizational systems. *Sequencing* for both services and manufacturing is the next logical procedure. It completes assignments by specifying the order in which jobs are to be done. In this chapter, we have described various sequencing policies.

Scheduling decisions rarely achieve optimal assignments for every job at every facility. Best assignments can seldom be made on a one-by-one basis. Instead, the problem must be looked at as a whole. As an analogy, NASA found that if every component of a space vehicle is optimized with respect to its function and only its function, the "bird will not fly." Instead, the vehicle must be designed together as a coordinated system of components. The same applies to the scheduling function. The assignment of work and its timing need to be orchestrated. Bottlenecks must be taken into account, and the goal of synchronized manufacturing requires coordination of the entire production department with respect to the mix of jobs that is being orchestrated.

The systems approach is called for so that the total set of assignments is optimized. The system's total costs are minimized or the total profits are maximized. Total quality (the sum of individual item qualities) and total productivity (the sum of individual item productivities) are maximized. This means that each and every item is not at maximum quality and productivity. The nature of the goal is to select assignments that—although less than best individually—result in the overall system's best.

Production scheduling is always a system's problem because jobs, people, and teams compete with each other for best facilities. Jobs (as surrogates for customers) also compete with each other concerning which gets done first. Facilities compete with each other for jobs. Departments that do similar work compete with each other for preferred work. Orders placed with suppliers are tied to priorities with respect to jobs and customers, so suppliers' orders also rival each other in terms of importance and how they are treated.

Strategic planning is required to design the facilities, staff the departments, select suppliers, and design the product mix—for the expected blend of job shop orders. In contrast, the flow shop has a specific set of items to be made in volume on dedicated facilities. The flow shop system is predesigned to optimize the production schedule through synchronization. The difference between a job shop and flow shop is explained later in this chapter.

With the job shop, the company must achieve systems optimization to rationalize the various preferences in a way that is not self-defeating. When the systems point of view prevails, the solutions are company-wide optimizations even though people and facilities experience suboptimal assignments. Over time, the facilities need to be altered in such a way as to minimize the degree of suboptimization. That requires strategic thinking.

The final step of production scheduling assigns actual jobs to designated facilities with unambiguous stipulations that they be completed at specific times. The steps in scheduling are reviewed below, moving from generic resource planning to actual assignments at workstations.

1. Aggregate planning (scheduling) developed resource plans based on forecasts of orders in generic units such as standard hours. This was studied in Chapter 4.
2. Later, with actual orders on hand or with reasonable predictions about orders, the master production schedule (MPS) assigns jobs to time slots to permit orders to be placed for required materials using material requirements planning (MRP). These time period assignments are so well defined that they are referred to as time buckets. MRP is not the subject matter of this book.
3. The next step in production scheduling is to load facilities, which means taking the actual orders and assigning them to designated facilities. The loading function answers the question: Which department is going to do what work? Sequencing answers the question: What is the order in which the work will be done?

It is this third step in production scheduling that does both loading and sequencing which is the subject matter of this chapter.

6.1.1 Loading

Loading, also called shop loading, is required to assign specific jobs or teams to specific facilities. Loading is needed for machine shops, hospitals, and offices. Specifically, loading assigns the work to various facilities like divisions, departments, work centers, load centers, stations, machines, and people. We will often use the term "machines" in this chapter when we refer to a facility. Loading releases jobs to facilities. Although loading assigns work to facilities, it does not specify the order in which jobs should be done at the facility. Sequencing methods (described later) determine the order of work at the facility.

Aggregate planning (or aggregate scheduling) used standard hours based on forecasts to determine what resources should be assembled over the planning horizon. However, loading takes place in the production department when the real orders are on hand. The loading function loads the real jobs and not the forecast. If the aggregate scheduling job was done well, then the appropriate kinds and amounts of resources are available for loading. Loading assumes that the material requirement analysis has been done and that orders have been properly placed for required materials and for needed parts and subassemblies. It further assumes that the parts will be on hand, as planned. A problem in supplier shipments can delay scheduling production of the affected order. The MPS also made resource assignments that could be modified if capacity was not adequate. Planning actual shop assignments is a regular, repetitive managerial responsibility. Releasing the jobs, as per assignment, is another.

Each facility carries a backlog of work, which is its "load"—hardly a case of perfect just-in-time in which no waiting occurs. The backlog is generally much larger than the work in process, which can be seen on the shop floor. This is because work not assigned yet is waiting. It may appear to be clutter or it can be hidden in storage spaces. Even when it is visible, it is hard to know the backlog which translates into an inventory investment which is idle and receiving no value-adding attention. A major objective of loading is to spread the load so that waiting is minimized, flow is smooth and rapid, and congestion is avoided.

These objectives are constrained by the fact that not all workstations can do all kinds of jobs. Some workstations and people are better suited for specific jobs than are others. Some stations cannot do jobs that others can do. Some are faster than others and tend to be overloaded. The scheduling objective is to smooth the load with balanced work assignments at stations.

Methods for loading include assignment method, the transportation method, and various heuristics. However, these methods are not discussed in this book.

6.1.2 Sequencing Operations

Sequencing models and methods follow the discussion of loading models and methods. Sequencing involves shop floor control, which consists of communicating the status of orders and the productivity of workstations. Sequencing establishes the order for doing the jobs at each facility. Sequencing reflects job priorities according to the way that jobs are arranged in the queues. Say that Jobs *x*, *y*, and *z* have been assigned to workstation 1 (loading). Jobs *x*, *y*, and *z* are in a queue (waiting line). Sequencing rules determine which job should be first in line, which second, etc. There are different costs associated with the various orderings of jobs. The objective function can be to minimize system's costs, or to minimize total system's time, or (if margin data are available) to maximize total system's profit. We discuss several objective functions later in the chapter.

Good sequencing provides less waiting time, decreased delivery delays, and better due date performance. There are costs associated with waiting and delays. Total savings from regularly sequencing the right way the first time can accumulate to substantial sums. Re-sequencing can be significantly more costly. When there are many jobs and facilities, sequencing rules have considerable economic importance.

6.1.3 Scheduling Example

As discussed above, scheduling determines the order in which the jobs assigned to a machine or a group of machines will be processed. Each job is characterized by its routing that specifies the information about the operations to be performed, the sequence of these operations, the machines required for processing these operations, and the times required for processing these operations. Consider the example given in Table 6.1 after the loading decisions have been made.

There are three jobs A, B, and C that have to be produced. The loading function has assigned these jobs to be processed on four machines that are designated as M1, M2, M3, and M4. Each machine performs a specific function and the machines are not interchangeable. The three jobs consist of 4, 3, and 4 operations, respectively. The operations of job A are designated as A1, A2, A3, and A4. The operations of job B are designated as B1, B2, and B3. Similarly, the four operations of job C are designated as C1, C2, C3, and C4. Each operation is processed on a specific machine. Table 6.1 gives the machines required for each operation of each job and its processing time. For example, it shows that the first operation of job A, A1, is processed on machine M1 and requires 5 days; second operation, A2, is processed on machine M3 and requires 3 days, and so on. A machine may be required more than once for a job. For example, operation A4 might have needed M3 again instead of M2. In that case, the sequence of machines required for job A will be M1–M3–M4–M3.

The operations of all jobs have to follow their processing sequences. For example, operation A3 of job A cannot be processed before operation A2. Also, each machine can process only one job at a time. We also assume that once a job is

Table 6.1 Example of a Job Shop with Three Jobs and Four Machines

Job	*Operation Number*	*Machine Number*	*Processing Time (Days)*
A	A1	M1	5
	A2	M3	3
	A3	M4	7
	A4	M2	4
B	B1	M2	2
	B2	M3	6
	B3	M4	8
C	C1	M1	4
	C2	M2	6
	C3	M3	8
	C4	M4	2

started on a machine, its processing cannot be interrupted, that is, preemption is not allowed. The machines are continuously available and will not break down during the planning horizon. This assumption is rather unrealistic but we allow it to avoid complexity at this stage.

The scheduling function entails the determination of the sequence (order) in which these jobs will be processed. A production schedule gives the timetable that specifies the times at which the jobs in a production department will be processed on various machines. The schedule gives the starting and ending times of each job assigned to the machines on which the job has to be processed. Figure 6.1 graphically represents a schedule for the problem given in Table 6.1. As will be seen, there are various possible schedules for this problem. This figure depicts a Gantt chart.

A Gantt chart represents machines or facilities on horizontal rows. The Gantt chart in Figure 6.1 has four horizontal rows—one for each machine. A processing sequence has to be specified to draw a Gantt chart. Suppose the sequencing decision is to process the jobs in the order A–B–C. Let us discuss how the starting and ending times for each job are determined.

Schedule of Job A: The first operation of job A, A1, is to be done on Machine M1 and requires 5 days. Therefore, the Gantt charts shows that A1 is scheduled on days

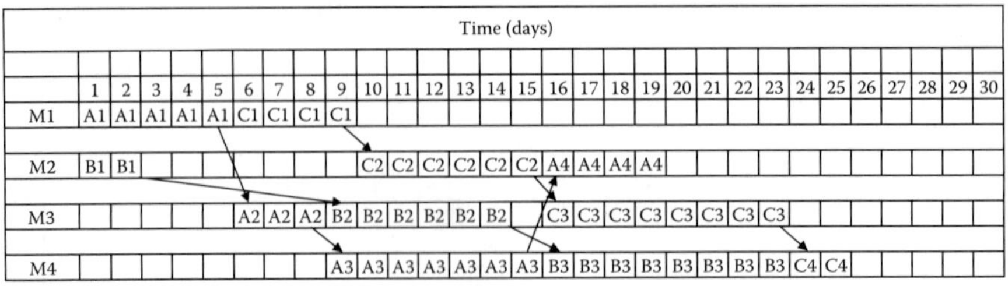

Figure 6.1 Gantt chart for sequence A–B–C (make-span = 25 days).

1, 2, 3, 4, and 5. It is common practice to say that starting time of A1 is zero (0) and the ending time is 5. The next operation A2 is to be done on machine M3 and can start as soon as A1 is completed. It is shown to be scheduled on days 6, 7, and 8. Again, common practice dictates that we say that starting time of A2 is 5 (which is the ending time of A1) and the ending time is 8. In other words, the interval 0 to 1 is day 1, the interval 1 to 2 is day 2, the interval 2 to 3 is day 3, etc. Continuing in this way, A3 is scheduled on M4 from day 8 to day 15; and A4 is scheduled on M2 from day 15 to day 19. The job A is finished in 19 days which is called the completion time of job A or the flow time of job A. It may be noted that the total time of all operations of job A, time of A1 + time of A2 + time of A3 + time of A4 = 5 + 3 + 7 + 4 = 19. In this case, the flow time of job A is 19 days. However, in general, the flow time of a job is not equal to the sum of processing times of its operations because a job might have to wait for its turn to be processed. See the flow times of jobs B and C below.

Schedule of Job B: The first operation of job B, B1, starts on machine M2 at time zero and is finished on time 2 (days 1 and 2). The machine required for operation B2 is M3. The operation B2 can start on M3 as soon as B1 is completed. However, B2 takes 6 days. If we schedule B2 starting at time 2, the operation A2, scheduled to start on day 6 will have to be pushed forward. However, we have made an assumption that preemption is not allowed which means that A2 cannot be rescheduled—pushed forward. So B2 will have to wait and start after A2 has been completed on day 8. Proceeding in this way, B3 is completed on day 23. The flow time of job B is 23. The total time for job B is time for B1 + time for B2 + time for B3 = 2 + 6 + 8 = 16. However, the flow time of job B is 23 because job B had to wait for its turn to be processed. It waited on days 3, 4, 5, 6, 7, and 8 (6 days) on M3 when A2 was being processed and on day 15 on M4 when A3 was processed. The total wait-in-time of job B is 7 days. The total processing time is 16 days. Therefore, the flow time is 23 days (total job time + total wait-in-time).

Schedule of job C: Using the process described above for jobs A and B, the start and end times of all operations of job C can be determined. Job C will be finished at time 25. The flow time of job C is 25, whereas the sum of the process-

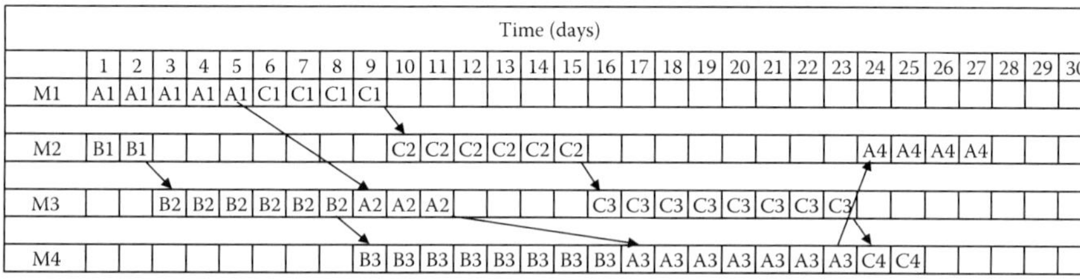

Figure 6.2 Gantt chart for sequence B–A–C (make-span = 27 days).

ing times of its operations is 20 days. Job C had to wait for 5 days on M1 (see Gantt chart).

The maximum of these flow times is called the *make-span* or the required schedule time. So the make-span for this problem is 25 which is the largest of the flow time of job A (19), flow time of job B (23), and flow time of job C (25).

Figures 6.2 and 6.3 give two different schedules. In Figure 6.2, the sequence is B–A–C (make-span 27) and in Figure 6.3 the sequence is C–A–B (make-span 30). None of these sequences is better than the sequence A–B–C (make-span 25).

> One of the objectives in the scheduling problems is to minimize make-span. This is also known as schedule time or maximum flow time or maximum job completion time.

One of the *most frequently used* objectives in the scheduling problems is to minimize make-span. Of the three solutions presented here, the schedule that orders jobs in the order A–B–C given in Figure 6.1 is the best. However, can we say that this is the best solution to this problem? There are many other solutions to this problem. For a 3-job problem, there are six possible sequences that include ABC, ACB, BAC, BCA, CAB, and CBA. For a 4-job problem, there will be 24 sequences; and 120 sequences for a 5-job problem. In general, there are *n*! (*n*-factorial) sequences for

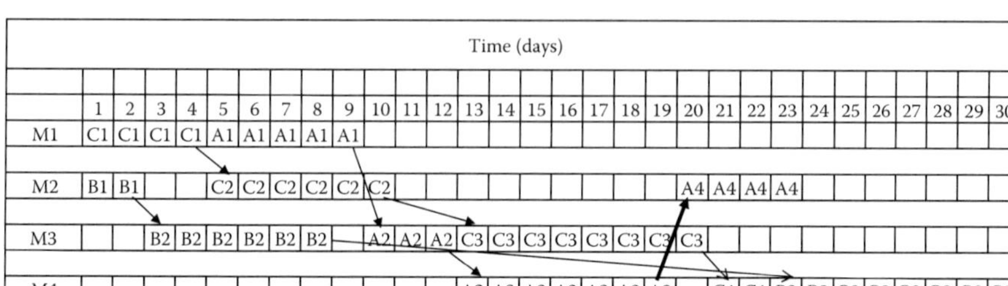

Figure 6.3 Gantt chart for sequence C–A–B (make-span = 30 days).

an n-job problem. We are assuming that the same sequence is followed on each machine.

We will study an optimization technique (Johnson's rule) to find the sequence that minimizes make span. We also discuss various other objective functions and techniques to find an optimal solution for them.

6.2 Classification of Scheduling Problems

The scheduling problems in the literature have been classified according to many criteria. Some of the important criteria are based on

- Sequence of machines
- Number of machines
- Processing times
- Job arrival time
- Objective functions

6.2.1 Sequence of Machines

The scheduling problems can be classified as flow shops and job shops based on the sequence of machines required by the jobs to be processed. A flow shop consists of several jobs and several machines. All jobs require machines in the same order for being processed. Table 6.2 gives an example of a flow shop in which three jobs A, B, and C are processed on four machines M1, M2, M3, and M4. The sequences of machines to process these jobs are same (M1–M3–M4–M2).

In a job shop, the sequence of machines will be mixed, that is, the jobs may require machines in different sequences. The example given in Table 6.1 is that of a job shop.

6.2.2 Number of Machines

Scheduling problems are classified as single-machine problems, two-machine problems, and multiple (3 or more) machine problems.

Table 6.2 Example of a Flow Shop

Job	*Operation #1*	*Operation #2*	*Operation #3*	*Operation #4*	*Machine for Operation #1*	*Machine for Operation #2*	*Machine for Operation #3*	*Machine for Operation #4*
A	A1	A2	A3	A4	M1	M3	M4	M2
B	B1	B2	B3	B4	M1	M3	M4	M2
C	C1	C2	C3	C4	M1	M3	M4	M2

6.2.3 *Processing Times*

If processing times of all jobs are known and constant, the scheduling problem is called a deterministic problem. The scheduling problem is called probabilistic (or stochastic) if the processing times are not fixed, that is, the processing times must be represented by a probability distribution.

6.2.4 *Job Arrival Times*

Based on this criterion, scheduling problems are classified as static and dynamic problems. In the case of static problems, the number of jobs is fixed and will not change until the current set of jobs has been processed. In the case of dynamic problems, new jobs enter the system and become part of the current set of unprocessed jobs. The arrival rate of jobs is given in the case of dynamic problems.

6.2.5 *Objective Functions*

Scheduling researchers have studied a large variety of objective functions. In this chapter, we will study the following objectives:

- Minimize make-span
- Minimize average flow time (or job completion time)
- Average number of jobs in the system
- Minimize average tardiness
- Minimize maximum tardiness
- Minimize number of tardy jobs

Minimizing make-span has been discussed above and is relevant for two or more machines. In this chapter, we will discuss the scheduling rules for static and deterministic flow shop problems consisting of two machines. The other five objectives can be used for any number of machines, both deterministic and probabilistic processing times, and for static as well as dynamic problems. However, we will study these objective functions for a single machine, deterministic, and static problems. The scheduling rule for job shops and for more than three machines are complex and beyond the scope of this chapter.

6.3 Two Machines Flow-Shop Problem

Consider a problem with five jobs (A, B, C, D, and E) and two machines M1 and M2. All five jobs consist of two operations each. The first operation of each job is processed on machine M1; and the second operation is processed on machine M2. Table 6.3 gives the machines required for each job and the processing times for each operation of each job.

Table 6.3 Data for a 5-Job 2-Machine Flow Shop Problem

Job	*Operation #1*	*Operation #2*	*Machine for Operation #1*	*Machine for Operation #2*	*Time for Operation #1 (Days)*	*Time for Operation #2 (Days)*
A	A1	A2	M1	M2	8	3
B	B1	B2	M1	M2	5	7
C	C1	C2	M1	M2	6	9
D	D1	D2	M1	M2	7	1
E	E1	E2	M1	M2	4	6

The scheduling objective is to find an optimal sequence that gives the order in which the five jobs will be processed on the two machines to minimize make-span. Let us find the make-span for one of the sequences, say, A–B–C–D–E, before attempting to find the optimal answer. We will draw a Gantt chart to find make-span. The sequence A–B–C–D–E tells us that A is the first job to be processed, B is the second job, and so on. E is the last job to be processed. The Gantt chart for the sequence A–B–C–D–E is given in Figure 6.4. We must assume that the sequence is the same on both machines. This is also called "no passing" in the scheduling literature. The value of make-span (time to complete all jobs) is 36 days. Our objective is to identify the sequence that minimizes the value of make-span. We will study Johnson's rule to sequence jobs to achieve this objective (see Johnson, 1954).

6.3.1 *Johnson's Rule*

Johnson's rule is a proven method to give an optimal solution. There are five sequence positions 1–5. Johnson's rule assigns each job to one of these positions in an optimal manner. This rule also requires that the same optimal sequence is used on both machines. The rule also assumes that no preemption (no passing) is allowed which means that once a job is started it cannot be interrupted.

Position 1	Position 2	Position 3	Position 4	Position 5

Gantt chart for sequence A–B–C–D–E

Time (days)

	1	2	3	4	5	6	7	8	9	10	11	12	13	14	15	16	17	18	19	20	21	22	23	24	25	26	27	28	29	30	31	32	33	34	35	36
M1	A1	A1	A1	A1	A1	A1	A1	A1	B1	B1	B1	B1	B1	C1	C1	C1	C1	C1	C1	D1	D1	D1	D1	D1	D1	D1	E1	E1	E1	E1						
M2									A2	A2	A2			B2	B2	B2	B2	B2	B2	B2	C2	C2	C2	C2	C2	C2	C2	C2	C2	D2	E2	E2	E2	E2	E2	E2

Figure 6.4 Gantt chart for sequence A–B–C–D–E.

We use the following four-step process to find the optimal sequence.

Step 1: Find the minimum processing time considering times on both machines.
Step 2: Identify the corresponding job and the corresponding machine for the minimum time identified at Step 1.
Step 3: Scheduling rule
  a. If the machine identified in Step 2 is machine M1, then the job identified in Step 2 will be scheduled in the first available schedule position.
  b. If the machine identified in Step 2 is machine M2, then the job identified in Step 2 will be scheduled in the last available schedule position.
Step 4: Remove the job from consideration whose position has been fixed in Step 3 and go to Step 1.

Continue this process until all jobs have been scheduled.

Iteration 1 (see Figure 6.5)

Step 1: The minimum time is 1.
Step 2: The job is D and the machine is M2.
Step 3: Since the machine identified at Step 2 is machine M2, the job D will be assigned to the last available sequence position which is position 5; and the resulting partial sequence is given below.
Step 4: Delete job D from consideration.

Position 1	Position 2	Position 3	Position 4	D

Job	*Operation #1*	*Operation #2*	*Machine for Operation #1*	*Machine for Operation #2*	*Time for Operation #1 (Days)*	*Time for Operation #2 (Days)*
A	A1	A2	M1	M2	8	3
B	B1	B2	M1	M2	5	7
C	C1	C2	M1	M2	6	9
D	D1	D2	M1	M2	7	1
E	E1	E2	M1	M2	4	6

Figure 6.5 Iteration 1.

Job	*Operation #1*	*Operation #2*	*Machine for Operation #1*	*Machine for Operation #2*	*Time for Operation #1 (Days)*	*Time for Operation #2 (Days)*	
A	A1	A2	M1	M2	8	3	
B	B1	B2	M1	M2	5	7	
C	C1	C2	M1	M2	6	9	
~~D~~	~~D1~~	~~D2~~	~~M1~~	~~M2~~	~~7~~	~~1~~	Scheduled
E	E1	E2	M1	M2	4	6	

Figure 6.6 Iteration 2.

Iteration 2 (see Figure 6.6)

Step 1: The next minimum time is 3.
Step 2: The job is A and the machine is M2.
Step 3: The job A will be assigned to the last available schedule position, which is position 4. After assigning job A to position 4, the partial sequence is given below.
Step 4: Delete job A from consideration.

Position 1	Position 2	Position 3	A	D

Iteration 3 (see Figure 6.7)

Step 1: The minimum time is 4.
Step 2: The job is E and machine is M1.
Step 3: The job E will be assigned to the first available schedule position, which is position 1. The partial sequence after assigning job E to position 1 is given below.
Step 4: Delete job E from consideration.

E	Position 2	Position 3	A	D

Iteration 4 (see Figure 6.8)

Step 1: The minimum time is 5.
Step 2: The job is B and machine is M1.

Job	*Operation #1*	*Operation #2*	*Machine for Operation #1*	*Machine for Operation #2*	*Time for Operation #1 (Days)*	*Time for Operation #2 (Days)*	
~~A~~	~~A1~~	~~A2~~	~~M1~~	~~M2~~	~~8~~	~~3~~	Scheduled
B	B1	B2	M1	M2	5	7	
C	C1	C2	M1	M2	6	9	
~~D~~	~~D1~~	~~D2~~	~~M1~~	~~M2~~	~~7~~	~~1~~	Scheduled
E	E1	E2	M1	M2	4	6	

Figure 6.7 Iteration 3.

Job	*Operation #1*	*Operation #2*	*Machine for Operation #1*	*Machine for Operation #2*	*Time for Operation #1 (Days)*	*Time for Operation #2 (Days)*	
~~A~~	~~A1~~	~~A2~~	~~M1~~	~~M2~~	~~8~~	~~3~~	Scheduled
B	B1	B2	M1	M2	5	7	
C	C1	C2	M1	M2	6	9	
~~D~~	~~D1~~	~~D2~~	~~M1~~	~~M2~~	~~7~~	~~1~~	Scheduled
~~E~~	~~E1~~	~~E2~~	~~M1~~	~~M2~~	~~4~~	~~6~~	Scheduled

Figure 6.8 Iteration 4.

Step 3: The job B will be assigned to the first available schedule position, which is position 2. The partial sequence after assigning job B to position 2 is given below.

Step 4: Delete job B from consideration

E	B	Position 3	A	D

Iteration 5 (see Figure 6.9)

The only unscheduled job at this stage is C and it will be assigned to the remaining unassigned position 3.

Job	*Operation #1*	*Operation #2*	*Machine for Operation #1*	*Machine for Operation #2*	*Time for Operation #1 (Days)*	*Time for Operation #2 (Days)*	
~~A~~	~~A1~~	~~A2~~	~~M1~~	~~M2~~	~~8~~	~~3~~	Scheduled
~~B~~	~~B1~~	~~B2~~	~~M1~~	~~M2~~	~~5~~	~~7~~	Scheduled
C	C1	C2	M1	M2	6	9	
~~D~~	~~D1~~	~~D2~~	~~M1~~	~~M2~~	~~7~~	~~1~~	Scheduled
~~E~~	~~E1~~	~~E2~~	~~M1~~	~~M2~~	~~4~~	~~6~~	Scheduled

Figure 6.9 Iteration 5.

The final sequence is given below.

E	B	C	A	D

The value of make-span for this sequence will be determined by drawing the Gantt chart as described below.

6.3.2 *Finding Make-Span*

The sequence E–B–C–A–D identified by Johnson's rule guarantees the minimum value of make-span. However, Johnson's rule does not give the value of make-span. It only identifies the best sequence. The value of make-span is obtained either by drawing the Gantt chart or a computerized algorithm can be used. The Gantt chart for this optimal sequence is given in Figure 6.10. The value of make-span is 31 days. This is the optimal answer.

6.3.3 *Multiple Sequences*

It must be noted that multiple optimal sequences are possible for a given problem. It means that several sequences can have the same minimum value of make-span.

Gantt chart for sequence E–B–C–A–D																																				
Time (days)																																				
	1	2	3	4	5	6	7	8	9	10	11	12	13	14	15	16	17	18	19	20	21	22	23	24	25	26	27	28	29	30	31	32	33	34	35	36
M1	E1	E1	E1	E1	B1	B1	B1	B1	B1	C1	C1	C1	C1	C1	C1	A1	A1	A1	A1	A1	A1	A1	A1	D1	D1	D1	D1	D1	D1	D1						
M2					E2	E2	E2	E2	E2	E2	B2	B2	B2	B2	B2	B2	B2	C2	C2	C2	C2	C2	C2	C2	C2	C2	A2	A2	A2		D2					

Figure 6.10 Gantt chart for sequence E–B–C–A–D.

For example, for the problem studied above, the sequence E–C–B–A–D also gives a make-span of 31 days. However, Johnson's rule identifies only one of these sequences.

6.3.4 Breaking Ties

It might happen at Step 1 that there are more than one minimum times. In such a situation, which job should be picked for assigning a position in the sequence? We will discuss three different cases of these ties.

Case 1: Minimum time is on both machines but for different jobs.
Case 2: Minimum time is on the same machine but for a different job.
Case 3: Minimum time is on both machines but for the same job.

Ties—Case 1: Minimum time is on both machines but for different jobs.

Consider the problem given in Table 6.4. The minimum time is 1 but it occurs at two places—job A on M1 and job D on M2. In this situation, the ties are broken at random. Job A may be selected before job D or vice versa.

The sequence if A is selected first and then D, or if D is selected first and then A will be the same and is given in Figure 6.11. In either case, the resulting partial sequence is the same. The scheduling algorithm continues until all jobs are scheduled. The final sequence is also shown in Figure 6.11.

Ties—Case 2: Minimum time is on the same machine but for a different job.

Consider the problem given in Table 6.5. The minimum time is 1 but occurs at two places—job B on M2 and job D on M2. The ties are broken at random. B may be selected before D or D may be selected before B. In this case, two different partial sequences will result based on which job is selected first.

These partial sequences are shown in Figure 6.12. The scheduling algorithm continues until all jobs are scheduled. The two final sequences are also shown

Table 6.4 Example Data for Case 1 of Ties

Job	*Operation #1*	*Operation #2*	*Machine for Operation #1*	*Machine for Operation #2*	*Time for Operation #1 (Days)*	*Time for Operation #2 (Days)*
A	A1	A2	M1	M2	1	3
B	B1	B2	M1	M2	5	7
C	C1	C2	M1	M2	6	9
D	D1	D2	M1	M2	7	1
E	E1	E2	M1	M2	4	6

Partial scequence after scheduling A and D				
A	Position 2	Position 3	Position 4	D
Final sequence				
A	E	B	C	D

Figure 6.11 Breaking ties for Case 1.

Table 6.5 Example Data for Case 2 of Ties

Job	*Operation #1*	*Operation #2*	*Machine for Operation #1*	*Machine for Operation #2*	*Time for Oper-ation #1 (Days)*	*Time for Oper-ation #2 (Days)*
A	A1	A2	M1	M2	8	3
B	B1	B2	M1	M2	5	1
C	C1	C2	M1	M2	6	9
D	D1	D2	M1	M2	7	1
E	E1	E2	M1	M2	4	6

Partial Sequence if B is selected first and then D				
Position 1	Position 2	Position 3	D	B
Final sequence if B is selected first and then D				
E	C	A	D	B
Partial Sequence D is selected first and then B				
Position 1	Position 2	Position 3	B	D
Final sequence if D is selected first and then B				
E	C	A	B	D

Figure 6.12 Breaking ties for Case 2.

Table 6.6 Example Data for Case 3 of Ties

Job	*Operation #1*	*Operation #2*	*Machine for Operation #1*	*Machine for Operation #2*	*Time for Operation #1 (Days)*	*Time for Operation #2 (Days)*
A	A1	A2	M1	M2	8	3
B	B1	B2	M1	M2	5	7
C	C1	C2	M1	M2	2	2
D	D1	D2	M1	M2	7	9
E	E1	E2	M1	M2	4	6

in Figure 6.12. In this case, the two final sequences are the same except for the sequence positions of D and B.

Ties—Case 3: Minimum time is on both machines but for the same job.

Consider the problem given in Table 6.6. The minimum time is 2 but occurs at two places—job C on M1 and also job C on M2. The ties are broken at random. Job C on M1 may be selected before job C on M2 or vice versa. In this case, two different partial sequences will result based on which combination is selected first.

These partial sequences are shown in Figure 6.13. The scheduling algorithm continues until all jobs are scheduled. The two final sequences are also shown in Figures 6.13. Note that C can be the first job or the last job in the optimal sequence.

Partial Sequence if C on M1 is selected first				
C	Position 2	Position 3	Position 4	Position 5
Final sequence if C on M1 is selected first				
C	E	B	D	A
Partial Sequence if C on M2 is selected first				
Position 1	Position 2	Position 3	Position 4	C
Final sequence if C on M2 is selected first				
E	B	D	A	C

Figure 6.13 Breaking ties for Case 3.

Table 6.7 Data for Multiple Ties

Job	*Operation #1*	*Operation #2*	*Machine for Operation #1*	*Machine for Operation #2*	*Time for Operation #1 (Days)*	*Time for Operation #2 (Days)*
A	A1	A2	M1	M2	2	3
B	B1	B2	M1	M2	5	7
C	C1	C2	M1	M2	2	2
D	D1	D2	M1	M2	8	9
E	E1	E2	M1	M2	4	2

Comments about ties

In general, all three cases of ties may exist in a given problem. See data in Table 6.7 for multiple ties. The problem is not solved here but is included as Problem 10. The examples that we considered show ties in the first iteration. However, the ties may occur during any iteration. The general rule is to break the ties at random. However, in this chapter, we will break the ties in the alphabetic order, that is A before B, etc. If the tie is for the same job on the two machines (Case 3), the ties will be broken by the rule: machine M1 before machine M2. All resulting sequences, irrespective of the tie-breaking rule, will give the same minimum value of the make-span.

6.4 Single-Machine Scheduling

There is a single machine on which several jobs have to be processed. The order in which these jobs will be processed needs to be specified. This schedule will not be changed until all jobs have been processed. This is the "static" version of the problem. In the "dynamic" version, the schedule can be altered. Dynamic version is studied in Section 6.5.

6.4.1 Objective Functions

Many criteria exist for evaluating production schedules. We considered minimizing make-span while studying the 2-machine flow-shop problems. In the case of a single-machine problem, minimizing make-span is irrelevant because all sequences will give the same make-span which is the completion time of the last job in the sequence. For a single-machine problem, minimizing average flow time which is the average amount of time required to complete each job in the group is a more appropriate criterion. Another measure is the average number of jobs in the system. Both of these measures are functions of the start time and end time of each job.

At this point, we introduce another term "due date" for each job. The due date for a job gives the time by which a job needs to be completed. If a job is completed after the due date, the job will be tardy. We will study three additional criteria to evaluate a production schedule that is based on the tardiness of the jobs.

Thus, we study a total of five objective functions for the single-machine problem. These include the minimization of

- Average flow time
- Average number of jobs in the system
- Average tardiness
- Maximum tardiness
- Number of tardy jobs

6.4.2 Scheduling Rules

There are several rules that can be used to find the order of processing the jobs. However, we will study the following three scheduling rules in this section.

- *First come first served (FCFS).* The jobs are processed in the order in which they arrived at the machine.
- *Shortest processing time (SPT).* This is also called as shortest operation time. Among jobs on hand, the job that requires the minimum processing time is processed first and then the other jobs are processed in the ascending order of their processing times.
- *Earliest (shortest) due date (EDD).* The job that has the smallest due date is processed first and then the other jobs are processed in the ascending order of their due dates.

Other scheduling rules include *slack time* and *critical ratio*. Slack time is defined as the due date–processing time, and critical ratio is defined as processing time divided by due date.

6.4.3 Example

Consider the example given in Table 6.8. There are five jobs A, B, C, D, and E. A is the first job that arrived in the production department. B, C, D, and E followed A in this order. The processing times and due dates of all jobs are also given. The order in which these jobs have to be processed needs to be specified.

Table 6.8 Data for a Single Machine

	Days	
Job	*Time*	*Due Date*
A	17	45
B	12	35
C	22	27
D	18	54
E	26	47

6.4.3.1 FCFS Rule

Table 6.9 gives answers by using the FCFS rule. The order of processing is A, B, C, D, and E. In addition to the data given in Table 6.8, we have added the completion time and the tardiness of each job in Table 6.9. A is the first job to be processed. It will start at time zero and will be completed at time 17 because its processing time

Table 6.9 FCFS Rule

Job	*Time*	*Due Date*	*Completion Time*	*Tardiness*
A	17	45	17	0
B	12	35	29	0
C	22	27	51	24
D	18	54	69	15
E	26	47	95	48
		Total	261	87
Average completion time				52.2
Average number of jobs in system				2.75
Average tardiness				17.4
Maximum tardiness				48
Number of tardy jobs				3

is 17. Its due date is 45. Therefore, the job is not tardy and its tardiness is zero. As soon as job A is completed, job B starts at time 17. The processing time of job B is 12. So job B will be completed at time 29 (= 12 + 17). The due date is 35. The job is not tardy and therefore its tardiness is zero. In this way, the completion time and tardiness of all jobs are completed.

In general, the completion time and tardiness are calculated using the following equations:

$$\text{Completion time} = \text{Start time} + \text{Processing time.}$$
$$\text{Tardiness} = \text{Completion time} - \text{Due date.}$$

However, the tardiness equation may give a negative number. For example, using this equation, the tardiness of job A = 17 – 45 = –28. When there is a negative tardiness, we make it zero. We can also write the tardiness equation as

$$\text{Tardiness} = \text{Larger of } [0 \text{ or } (\text{Completion time} - \text{Due date})].$$

It may be of interest to note that in the scheduling literature lateness is defined as Lateness = Completion time – Due date, and this number can be positive or negative. A negative value of lateness means the job is early. A zero value of lateness means that the job is on time and a positive value for lateness means the job is tardy.

6.4.3.2 Calculation of Objective Functions

Average completion time: Add completion times of all jobs and divide it by the number of jobs. For the data in Table 6.9, the average completion time of all jobs is = 52.2 = 261/5.

Average number of jobs in the system: This is obtained by dividing the total of completion times of all jobs by the completion time of the last job. For the data in Table 6.9, the average number of jobs in the system is = 2.75 = 261/95.

Average tardiness: Add tardiness of all jobs (including zero tardiness) and divide it by the number of jobs (including jobs with zero tardiness). For the data given in Table 6.9, the average tardiness = 17.4 = 87/5.

Maximum tardiness: This is the maximum of all numbers in the tardiness column. The maximum tardiness is 48 (job E) for the data given in Table 6.9.

Number of tardy jobs: Count the number of jobs that are tardy. Three jobs (C, D, and E) are tardy for the data given in Table 6.9.

6.4.3.3 SPT Rule

The jobs are processed in the increasing order of their processing times when using the SPT rule. The job with the minimum processing time B (12) is processed first.

Table 6.10 SPT Rule

Job	*Time*	*Due Date*	*Completion Time*	*Tardiness*
B	12	35	12	0
A	17	45	29	0
D	18	54	47	0
C	22	27	69	42
E	26	47	95	48
		Total	252	90
Average completion time				50.4
Average number of jobs in system				2.65
Average tardiness				18
Maximum tardiness				48
Number of tardy jobs				2

B (12) is followed by A (17), D (18), C (22), and E (26). Table 6.10 shows the ordering of jobs (under the column Job). The calculations of the objective functions follow the same procedure as described for the FCFS rule.

6.4.3.4 EDD Rule

The jobs are processed in the increasing order of their due dates. The job with the minimum due date C (27) is processed first, and is followed by B (35), A (45), E (47), and D (54). Table 6.11 shows the ordering of jobs (under the column Job). The calculations of the objective functions follow the same procedure as described for the FCFS rule.

Note: It should be noted that make-span is the same (95) for all scheduling rules discussed above. Therefore, minimizing make-span is not used as a criterion in single-machine problems as mentioned earlier.

6.4.3.5 More on FCFS or First-In, First-Out Sequence Rule

The most natural ordering for doing work is in the order that the jobs are received. This means that the first jobs into the shop get worked on first. This is sometimes

Table 6.11 EDD Rule

Job	*Time*	*Due Date*	*Completion Time*	*Tardiness*
C	22	27	22	0
B	12	35	34	0
A	17	45	51	6
E	26	47	77	30
D	18	54	95	41
		Total	279	77
Average completion time				55.8
Average number of jobs in system				2.94
Average tardiness				15.4
Maximum tardiness				41
Number of tardy jobs				3

called FIFO for "first-in, first-out." Supermarkets like to use FIFO for their expiration-dated products (do not use after 4 April 2011). There is a cost advantage in getting older products on the shelves to be purchased first. LIFO, which is "last-in, first-out," frequently causes spoiled milk problems. That is because the first-in with the earlier date is waiting until all of the later-date items are purchased or shipped. If there is no age-spoilage problem then, as its advocates point out, LIFO can save warehouse-handling costs. LIFO items are more readily accessible. So, where the product date does not matter, and where you have to move a lot of things away to gain access to the first ones in, LIFO may save money.

FIFO is an appealing sequencing policy because it seems to be the fairest rule to follow. Sometimes—to emphasize the fair treatment sense—as has been stated before, FIFO is called "first-come, first-serve" (FCFS). Customers can get angry when someone seems to jump to the head of the line (last-come, first-serve). The cost of angry customers is not to be trivialized.

In another sense, FIFO seems to be unfair because it penalizes the average customer. The penalty is extra waiting time for processing time of the average order. This means that on-average regular customers will wait longer—even though FIFO satisfies first-come, first-serve. We should note that the SPT rule provides the best situation for the average customer since it offers the minimum average waiting time.

Customers who regularly submit orders with short processing times will benefit if the job shop does not employ the FIFO rule, but uses instead an SPT priority rule. Customers who regularly submit long orders may be discriminated against by the shortest processing rule. If so, compensatory steps can be taken at the discretion of those doing the sequencing.

6.5 Dynamic Scheduling Problems

A scheduling problem is classified as a dynamic problem if the number of jobs is not fixed. The examples include new production orders, customers in a bank, shoppers in a store, cars at a gas station, etc. The new jobs (production orders, customers, cars, etc.) keep on coming into the system, and the schedule needs to integrate new arrivals every time that a new schedule is prepared which is usually whenever a job is completed.

6.5.1 Example

Consider a single-machine problem for which the data are given in Table 6.12. There are five jobs A, B, C, D, and E that are waiting to be processed. Suppose the SPT rule is being used. Table 6.13 gives the order in which these jobs will be processed using the SPT rule. B is the first job to be processed followed by A, D, C, and E.

Job B starts at the current time (zero) and will finish at time 12. When job B is finished, the next job to be processed will be job A if no other jobs have arrived in the system. Suppose two new jobs F and G arrive in the system when B is being processed. F arrives on the fifth day and G arrives on the 10th day. Also assume that the processing time of job F is 8 days and that of G is 20 days.

After job B has been processed, there are six jobs (A, C, D, E, F, and G) that are waiting to be processed. Since the scheduling rule is SPT, the job with the minimum processing time from among the jobs that are waiting to be processed will be

Table 6.12 Data for Dynamic Scheduling Problem

Job	*Time (Days)*
A	17
B	12
C	22
D	18
E	26

Table 6.13 SPT Sequence

Job	*Time (Days)*
B	12
A	17
D	18
C	22
E	26

Table 6.14 SPT Sequence after Job B is Processed

Job	*Time (Days)*
F	8
A	17
D	18
G	20
C	22
E	26

scheduled next. Table 6.14 gives the schedule at this time using the SPT rule. The next job to be processed is F which is followed by A, D, G, C, and E. Job F will be completed at time 20 (12 + 8) where 12 is the completion time of job B. If more jobs arrive in the system while F is being processed, they will be integrated with the current jobs and a new schedule will be developed.

These are called dynamic problems since the schedule is continuously updated. We considered the example of a single-machine problem. However, the dynamic situation is faced in multiple-machine problems also. This is a tradeoff example of where the systems approach must be used to consider the costs of interrupting the prior schedule in order to obtain the advantages of continuous updating.

6.5.2 Objective Functions for Dynamic Problems

Objective functions for dynamic problems are defined in the same way as for single-machine static problems. The values of completion time and tardiness of each job are recorded, and the values of the objective functions can be calculated at any time based on the number of jobs completed at that time.

Summary

Production scheduling is the culminating series of steps that determines when orders are to be worked on, where, and by whom. The function goes back to the earliest days of systematic production and assignment of service jobs to work crews.

The quote with which we began this chapter from the writer Annie Dillard captures the point that schedules prevent operating under chaotic conditions and mistaken whimsy. Scheduling methods are designed to bring order and efficiency to the work process. They are also a sensible way to deal properly with what counts. Stephen Covey (www.forbes.com/sites/kevinkruse/2012/07/16/the-7-habits/) stated, "The key is not to prioritize what's on your schedule, but to schedule your priorities."

The Gantt load chart was developed in early 1900's, and it is still used. Loading decisions concern which jobs are to be assigned to which teams or facilities. Jobs on-hand are relatively risk-free to schedule. Jobs that are on the books as forecasts are problematic. Depending on the real situation, they may not be scheduled because there is too high a risk of cancellation.

Loading is nonspecific to job order, so it is always followed by sequencing and scheduling. The classification scheme for scheduling based on various criteria is discussed. There is strong methodology to determine the best order for job processing. Which job goes first, second, and so on? Service system managers often require first-come, first-serve so that their customers do not get upset. The same applies to production deliveries. Customer systems are criticized when the "fairness criterion" [first-in, first-out (FIFO)] is violated. Nevertheless, the alternative of processing orders so that SPTs go first provides benefits to everyone except those who regularly have orders that take a long time to process. Choosing the criteria is a systems problem that should reflect the basic values of the organization.

Gantt layout charts are used to organize sequencing assignments. The sequencing situation depends on the number of jobs (n) and the number of machines (m) that are available to work on the jobs. Solution methods differ according to the number of facilities (machines, m). Various scheduling rules are developed for single-machine problems. Johnson's rule is described for the two-machine flow-ship problems. The problems of dynamic scheduling are also discussed.

Review Questions

1. Describe the loading function.
2. Explain a Gantt chart.
3. What is a flow shop?
4. Distinguish between flow and job shops.
5. When does FIFO make sense as a sequencing rule?
6. When does LIFO make sense as a sequencing rule?
7. When should SPT be used?

8. What is a dynamic scheduling problem?
9. Describe the conditions when customers receiving services in a bank constitute a dynamic problem. Differentiate the conditions for a static problem.
10. What factors will you consider in choosing a checkout line in a grocery store?
11. Why may it not be desirable to complete a job before its due date?
12. The schedule has been set when suddenly a "very important customer" sends in a new order and demands it go to the head of the line. What considerations will you discuss with your colleagues?

Problems

1. What is the order of processing the jobs using Johnson's rule for the data given in the following table?

Processing Times (Days)		
Job	*Machine 1*	*Machine 2*
A	8	12
B	4	9
C	11	7
D	2	6
E	10	5

2. What happens if you add job F to the above table in Problem 1 which takes 2 days on Machine 1 and 1 day on Machine 2?
3. Draw a Gantt chart and find the value of make-span (time to complete all jobs) using the sequence A–B–C–D–E for the following problem:

Processing Time (Days)		
Job	*M1*	*M2*
A	8	6
B	9	4
C	3	7
D	2	6
E	6	5

Use the data given in the following table to answer the next four problems. The table gives the order in which five jobs arrived in the production department; their processing times and due dates.

Job	*Processing Time (Days)*	*Due Date (Days)*
A	18	46
B	13	33
C	21	25
D	19	52
E	24	48

4. What is the average number of jobs in the system using the EDD rule?
5. What is the average tardiness using the FCFS rule?
6. What is the number of tardy jobs using SPT rule?
7. What is the average completion time using FCFS rule?
8. Suppose SPT rule is being used in a "dynamic" scheduling problem. There are four jobs A, B, C, and D ready to be processed at the present time. The processing times for the four jobs are 3, 7, 6, and 2 days, respectively. A new job E will arrive on the 4th day. The processing time for job E is 2 days. On which day (from present time) job E will be completed? The present time is zero (0).
9. The due date for a job is day 20. It is finished on day 18.
   a. What is the lateness of the job?
   b. What is the tardiness of the job?
10. Find all optimal sequences for the data given in the following table. (Hint: use Johnson's rule. There are multiple ties.)

Job	*Operation #1*	*Operation #2*	*Machine for Operation #1*	*Machine for Operation #2*	*Time for Operation #1 (Days)*	*Time for Operation #2 (Days)*
A	A1	A2	M1	M2	2	3
B	B1	B2	M1	M2	5	7
C	C1	C2	M1	M2	2	2
D	D1	D2	M1	M2	8	9
E	E1	E2	M1	M2	4	2

11. The Door Knob Company has four orders on hand, and each must be processed in the sequential order:

First: Department A—press shop

Second: Department B—plating and finishing The following table lists the number of days required for each job in each department. For example, job IV requires one day in the press shop and one day in the finishing department. Assume that no other work is being done by the departments and that "no passing" of jobs is allowed. Use a Gantt chart to show the best-work schedule. (Best-work schedule means minimum time to finish all four jobs.)

	Job I	*Job II*	*Job III*	*Job IV*
Department A	8	8	5	1
Department B	8	3	4	1

12. The Market Research Store has four orders on hand, and each must be processed in the sequential order:

    First: Department A—computer analysis
    Second: Department B—report writing and printing

    The following table lists the number of days required by each job in each department. For example, job IV requires two days during computer analysis and one day in report writing and printing.

	Job I	*Job II*	*Job III*	*Job IV*
Department A	3	6	5	2
Department B	8	3	4	1

Assume that no other work is being done by the departments and that "no passing" of jobs is allowed. Use a Gantt sequencing chart to show the best-work schedule. (Best-work schedule means minimum time to finish all four jobs.)

13. There are six jobs that were named in alphabetical order (A, B, C, D, E, and F) as they arrived in the production department. Processing times for these jobs are given in the following table.
    a. What is the SPT sequence? What are the total and mean flow times for that sequence?
    b. What is the first-come, first-serve (or FIFO) sequence? What are the total and mean flow times for that sequence?
    c. Compare SPT and FIFO.

Job	*Time (Days)*
A	5
B	4

Job	*Time (Days)*
C	6
D	9
E	12
F	8

14. The problem faced by Information Search, Inc. (ISI) has the processing times (in hours) shown below: what is the optimal sequence and what are its total and mean flow times?

	A	*B*	*C*	*D*	*E*
Look up references	4	8	9	4	6
Write report	3	2	5	7	4

15. Consider the data given in the following table. There are six jobs. As these jobs arrived, they were named in alphabetical order (A, B, C, D, E, and F). The processing times and due dates are also given the following table.

Job	*Processing Time (Days)*	*Due Date (Days)*
A	5	10
B	4	16
C	6	18
D	9	20
E	12	36
F	8	25

In which order, the jobs will be processed using the following sequencing rules.

a. First-come first-served
b. Shortest processing time
c. Earliest due date

For each of the sequences obtained above, calculate the following values:

d. Average flow time
e. Average number of jobs in the system
f. Average tardiness
g. Maximum tardiness
h. Number of tardy jobs

Note: there will be three (one for each sequencing rule) answers for each of the questions d through h.

Readings and References

Baker, K., M. Magazine, Workforce Scheduling with Cyclic Demands and Days Off Constraints, *Management Science*, 24(2), October 1977, pp. 161–167.

Bard, J.F., Staff Scheduling in High Volume Service Facilities with Downgrading, *IIE Transactions*, 36, 2004, pp. 985–997.

Bechtold, S.E., M.J. Brusco, Microcomputer Based Working Set Generation Methods for Personnel Scheduling, *International Journal of Operations & Production Management*, 15(10), 1995, pp. 63–74.

Bowman, E.H., Production Planning by the Transportation Method of Linear Programming, *Journal of Operations Research Society*, February 1956, pp. 100–103.

Browne, J.J., Simplified Scheduling of Routine Work Hours and Days Off, *Industrial Engineering*, December 1979, pp. 27–29.

Buxey, G., Production Planning for Seasonal Demand, *International Journal of Operations and Production Management*, 13(7), July 1993, pp. 4–21.

Cayirli, T., and E. Veral, Outpatient Scheduling in Health Care: A Review of Literature, *Production and Operations Management*, 12(4), Winter 2003, pp. 519–549.

Clark, W., *The Gantt Chart: A Working Tool of Management*, New York: Ronald Press, 1922.

Davidow, W.H., B. Uttal, Service Companies: Focus or Falter, *Harvard Business Review*, 67, July–August 1989, pp. 77–85.

Davis, S.G., Reutzel, E.T., Joint Determination of Machine Requirements and Shift Scheduling in Banking Operations, *Interfaces*, 11(1), 1981, p. 41.

DeMatta, R., Miller, T., A Note on the Growth of Production Planning System, *Interfaces*, 23, April 1993, pp. 27–31.

Deutsch, H., Mabert, V.A., Queuing Theory and Teller Staffing: A Successful Application, *Interfaces*, 10(5), 1980, p. 63.

Easton, F.F., Rossin, D.R., Borders, W.S., Analysis of Alternative Scheduling Policies for Hospital Nurses, *Production and Operations Management*, 1, Spring 1992, pp. 159–174.

Farmer, A., Smith, J.S., Miller, L.T., Scheduling Umpire Crews for Professional Tennis Tournaments, *Interfaces*, 37(2), March–April 2007, pp. 187–196.

Fisher, M.L., Hammond, W.R., Obenneyer, W.R., Raman, A., Making Supply Meet Demand in an Uncertain World, *Harvard Business Review*, 72(3), May–June, 1994, pp. 83–93.

Foote, B.L., Queuing Case Study of Drive-In Banking, *Interfaces*, 6(4), August 1976, p. 31.

Forbes magazine, www.forbes.com/sites/kevinkruse/2012/07/16/the-7-habits/

Friedman, H.H., Friedman, L.W., Reducing the 'Wait' in Waiting-Line Systems: Waiting Line Segmentation, *Business Horizons*, 40(4), July–August 1997, pp. 54–58.

Glover, F., Jones, G., Karney, D., Klingman, D., Mote, J., An Integrated Production, Distribution, and Inventory Planning System, *Interfaces*, 9(5), 1979, p. 21.

Gupta, D. and Denton, B., Appointment Scheduling in Health Care: Challenges and Opportunities, *IIE Transactions*, 40, 2008, pp. 800–819.

Hodgson, T.J., On-Line Scheduling in the Production Environment: Part II. A "Successful" Case History, *Interfaces*, 10(2), 1980, p. 60.

Jain, S.K., A Simulation-based Scheduling and Management-information System for a Machine Shop, *Interfaces*, 6(1), 1975, p. 81.

Johnson, S.M., Optimal Two-and Three-Stage Production with Setup Times Included, *Naval Research Quarterly*, March 1954.

Kimes, S.E., Yield Management: A Tool for Capacity Constrained Service Firms, *Journal of Operations Management*, 8(4), October 1989, pp. 348–363.

Kimes, S.E., Chase, R.B., The Strategic Levers of Yield Management, *Journal of Service Research*, November 1998, pp. 156–166.

Krajewski, L.J., Ritzman, L.P., Shift Scheduling in Banking Operations: A Case Application, *Interfaces*, April 1980, pp. 1–7.

LaForge, R., Lawrence, Craighead, C.W., Computer-based Scheduling in Manufacturing Firms: Some Indicators of Successful Practice, *Production and Inventory Management Journal*, First Quarter, 2000, pp. 29–34.

Lesaint, D., Voudouris, C., and Azarmi N., Dynamic Worforce Scheduling for British Telecomunications plc, *Interfaces*, January–February 2000, pp. 45–56.

Levitt, T., Production Line Approach to Service, *Harvard Business Review*, 50, September–October 1972, pp. 41–52.

Love, Robert, R. Jr., Hoey, J.M., Management Science Improves Fast-Food Operations, *Interfaces*, 20(2), March–April 1990, pp. 21–29.

Malhotra, M.K., Ritzman, L.P., Scheduling Flexibility in the Service Sector: A Postal Case Study, *Production and Operations Management*, 3, Spring 1994, pp. 100–117.

Melachrinoudis, E., Olafsson, M.A., Microcomputer Cashier Scheduling System for Supermarket Stores, *International Journal of Physical Distribution & Logistics Management*, 25(1), 1995, pp. 34–50.

Mentzer, J.T., Moon, M.A., Understanding Demand, *Supply Chain Management*, May–June 2004, pp. 38–45.

Mondschein, S.V., Weintraub, G.Y., Appointment Policies in Service Operations, *Production and Operations Management*, 12(2), Summer 2003, pp. 266–286.

Moss, S., Dale, C., Brame, G., Sequence-dependent Scheduling at Baxter International, *Interfaces*, 30(2), 2000, p. 70.

Radas, S., Shugan, S.M., Managing Service Demand: Shifting and Bundling, *Journal of Service Research*, 1, August 1998, pp. 47–64.

Ramani, K.V., Scheduling Doctors' Activities at a Large Teaching Hospital, *Production and Inventory Management Journal*, First/Second Quarter, 2002, pp. 56–62.

Salegna, G., Integrating the Planning and Scheduling Systerns in a Job Shop, *Production & Inventory Management Journal*, 37(4), 1996, pp. 1–7.

Sheu, C., Wacker, J.G., A Planning and Control Framework for Nonprofit Humanitarian Organizations, *International Journal of Operations and Production Management*, 14(4), 1994, pp. 64–77.

Skill Set Scheduling, www.pipkins.com/articles, 2005.

Smith, R., Find the Best Checkout Line, *Wall Street Journal*, December 8, 2011.

Trottman, M., Choices in Stormy Weather: How Airline Employees Make Hundreds of Decisions to Cancel or Reroute Flights, *Wall Street Journal*, February 14, 2006, pp. Bl–B3.

Chapter 7

Project Management

Reader's Choice—Do it right the first time

Davenport, T.H., De Long, D.W. and Beers, M.C., Successful Knowledge Management Projects, *Sloan Management Review,* Winter 1998, pp. 43–57. The authors, based on a study of 24 companies, identify eight characteristics of successful knowledge management projects.

Fleming, Q.W., and Koppelman, J.M., What's Your Project's Real Price Tag? *Harvard Business Review,* September 2003. The authors propose to use earned-value management (EVM) principles to track the actual performance of long-term capital projects.

Greiner, L.E., and Schein, V.E., The Paradox of Managing a Project-oriented Matrix: Establishing Coherence Within Chaos, *Sloan Management Reviews,* 22(2), Winter 1981, pp. 17–22. The authors examine several coordination issues for managing a project-oriented matrix.

Keil, M., and Montealegre, R., Cutting Your Loses: Extricating Your Organization When a Big Project Goes Awry, *Sloan Management Review,* April 2000. De-escalation of projects that are likely to fail is important before more organizational resources are committed. The authors focus on IT projects and suggest a four-stage de-escalation process.

Keil, M., and Hring, M.M., Is Your Project Turning into a Black Hole? *California Management Review,* November 2010. This article explains that a black-hole project results because of a drift from the original goal, treatment of symptoms, and managers' rationalization of the continuation of the project.

Lenfle, S., and Loch, C.H., Lost Roots: How Project Management Came to Emphasize Control Over Flexibility and Novelty, *California Management Review*, November 2010. The authors emphasize that the role of project management be expanded to include novel and uncertain projects rather than focusing only on projects that are definitive in their outcomes.

Macomber, J.D., You Can Manage Construction Risks, *Harvard Business Review*, 67(2), March–April 1989, pp. 155–161. The author describes various steps to avoid schedule slippages and cost overruns in constructions projects.

Randolph, W.A., and Posner, B.Z., What Every Manager Needs to Know about Project Management, *Sloan Management Review*, Summer 1988, pp. 65–73. The authors suggest 10 principles to manage projects effectively for the entire project life cycle.

Royer, I., Why Bad Projects Are So Hard to Kill, *Harvard Business Review*, 81(2), February, pp. 48–56. The author has analyzed failed innovations in two large French companies and has suggested ways to avoid such failures.

Sharpe, P., and Keelin, T., How SmithKline Beecham Makes Better Resource-Allocation Decisions, Harvard Business Review, March–April 1998, pp. 45–57. The authors describe the process at SmithKline Beecham to make investment decisions in various research and development projects and highlight the importance of information quality, credibility, and trust.

Slevin, D.P., and Pinto, J.K., Balancing Strategy and Tactics in Project Implementation, Sloan Management Review, Fall 1987, pp. 33–41. The authors use a project life-cycle framework to identify 10 strategic and tactical factors for project success. The project is likely to face problems if strategic and tactical factors are not well integrated.

Projects are special work configurations designed to accomplish singular or nearly singular goals such as putting on one play, writing new software, creating a mail-order catalog, and constructing a building. Bringing out a new product, building a factory, redesigning an established traditional hotel, and developing a new service belong to the same category of unique activities and qualify as projects.

After reading this chapter, you should be able to:

- Identify project characteristics and explain the unique work configurations known as projects.

- Describe the life-cycle stages of projects.
- Classify projects by their various types.
- Identify qualities of good project leaders, discuss project-management leadership and teamwork, and explain how project managers differ from process managers.
- Explain the basic rules for project management.
- Describe project management origins.
- Draw project network diagram.
- Find critical path and project duration.
- Calculate early start, early finish, late start, and late finish times of activities.
- Explain how to use forward-pass calculations to determine the shortest feasible time for project completion.
- Explain how to use backward-pass calculations to determine which project activities are on the critical path.
- Describe what slack means; explain how to derive it.
- Crash activities (including multiple paths) to reduce project duration; perform time-cost tradeoff analysis.
- Analyze probabilistic projects; explain when deterministic and probabilistic estimates for activity times apply.
- Show how to use optimistic and pessimistic activity time estimates to obtain a variance measure for activity times.
- Identify implications of limited resources.

7.1 Introduction

Projects consist of a set of goal-oriented activities that end when the goal is achieved. Such undertakings have a finite planning horizon. This is in contrast to the character of batch and flow-shop production. Projects have many attributes that are similar to custom work. However, the scale of projects is much greater, involving many participants and resources. The projects are time-based endeavors that bring together skills and technology to accomplish goals.

Building the Golden Gate Bridge in San Francisco epitomizes a major project. Imagine what it was like to be the project manager in charge of creating this magnificent bridge. The construction time (from January 5, 1933, to May 28, 1937) and bridge statistics can be found at http://www.goldengatebridge.org. Click on "How Long Did It Take to Build the Bridge." Use the search box to find "seismic retrofit," which is a continuing aspect of the project. The photographs are worth the visit to this web page. Many projects require unending updates, maintenance, and renewal. Consider the continuous updates of computer programs.

Projects often include some repetitive activities. Building several houses on one land subdivision is a project. Software programming is a project even though use is made of modular components (object-oriented programming). Projects may entail batch work and even some intermittent flow-shop work. However, the project itself integrates activities as it moves toward completion much as each additional chapter is written for a book or floors are added to buildings.

Projects can be classified by *degree of simplicity* to change things. Design changes, which result in engineering change orders (ECOs), may appear to be minor alterations in the product design. However, even simple changes require alterations of the process that can lead to systems complexities. A small design change can destroy the ability of fixtures to hold the parts for all downstream activities. Also ECOs can multiply in number and lead to severe quality problems. These problems are especially noticeable if there is insufficient time to test the interactions of the proposed changes with each other and with usage patterns. Having too many ECOs can disrupt the normal business of an organization.

Projects can be classified by *degree of complexity* reflecting the number of people, teams, components, and activities. Often, the number of issues to consider is very large. Building a new factory is complex. It requires doing a great variety of things that have not been done before. Building another McDonald's may seem, at first glance, to be highly repetitious. However, locations are different. Community officials and their rules are different. Time is different and things change over time (see "Select Country/Market" at http://www.mcdonalds.com).

Projects can be classified by *frequency of repetition*. Although NASA has launched many shuttle flights, they are not all the same. Challenger blew up on launching (January 28, 1986) because of special conditions. Seventeen years later (February 1, 2003), Columbia burned up during reentry. Again, unique conditions applied. Between these dates, hundreds of successful missions were flown. In space programs, it is essential to define what scenarios can be deemed repetitious; what parts are unique and unknown?

There are benefits from having repetitive activities within a project. Housing developments consist of the same house design being built many times.This allows parts to be purchased with quantity discounts. Training for repetitive activities is readily justified. The same activity plans (charts) can be used. As the project frequency increases, the project mindset must remain in place. When you plan for regularities, you must also plan for contingencies. We are reminded that object-oriented programming calls upon similar repetitive modules for computer software development. Project managers must deal with new ways to connect and combine the basic modules.

That mindset is goal-oriented with completion planned for a specific time. If the activities begin to be treated as a repetitive system, then the project orientation has been replaced by one of repetitive scheduling as used by job shops and intermittent flow shops. Note that even though many houses of the same design are being built, there are unique site considerations that must be taken into account. Nevertheless, some builders have produced houses in volume to reduce the costly project factors and replace them with lower manufacturing costs. Modularity of components is a supply chain factor that brings significant economies of scale to projects that are properly planned in this way. A balance is required between the repetitive similarities and the creative differences.

New product developments are not all the same. Bringing out a new automobile model may seem to be highly repetitious, but there are new elements to deal with every time. For example, the hybrid concept was essentially theoretical a decade ago. The same applies to bringing a new movie to its marketplace. There are lessons learned from prior experiences and then there are the new and different factors to consider. The distinction that is being drawn is between making the last car of a run of 2000 made that day and the third animated movie brought to a world market by Pixar. It may be the same old thing to a few people who are skilled at planning a dinner party, but for most people, that remains a daunting project. Although everyone would agree that a dinner party is simpler than launching a shuttle, for many, if not most, it still qualifies as a challenging project.

Projects can be classified by how many really *new activities* are involved. Some projects have activities never done before. Examples of such projects might include NASA building an international space station together with Russia. The construction of a monorail train from Tampa to Orlando might not seem to be that different from the Shanghai monorail since the same technology will be used. Yet, totally different factors apply to the geological and architectural issues, relevant supply chains, and the politics involved.

7.2 Managing Projects

Competent project management methods keep track of what is required at start up, what has been done, and what still needs to be done. Also, good project methods point to activities that are critical for completion. Project managers expedite those activities that seem to be slipping. These points are part of the five project life-cycle stages described here:

1. Describing goals requires developing and specifying the desired project outcomes. (Architects lay out plans for building, cruise lines announce their schedules and destinations; everyone must set the goals of their projects.)

2. Planning the project requires specifying the activities that are essential to accomplish the goals. It involves planning the management of the project including the timing of the activities. (The project manager lays out the charts of sequenced activities and estimates how long it will take to do them. The time frame sets in motion the execution of the plan. The builder is usually the project planner.)
3. Carrying out the project requires doing the activities as scheduled. (Getting building permits, ordering materials, assembling different kinds of work crews needed at the right times, and constructing the building. The builder is usually the project manager.)
4. Completing the project can mean disbanding work groups and closing down the project-management team. However, firms that are in the business of project management, such as companies that build refineries, move their crews from project to project. Each project is goal-specific and finite. That is the mission of project-management companies when compared to organizations that need to use project management from time to time. The latter cannot avoid the fact that an ECO is a project and needs to be managed as such.
5. The use of continuous project teams is an increasingly attractive option. There is significant evidence that continuous project development is crucial to the success of twenty-first-century global organizations. Companies that do not have a project orientation might bring out a new product and then disband the project teams when the job is done. As will be discussed later, organizations increasingly opt to maintain continuous project capability.

7.3 Good Project Managers Are Leaders

Organizations encounter the need for project management whenever they consider introducing a new product or service. Often, they turn to their process managers and appoint them to deal with the project over its lifetime. The kinds of problems encountered in projects are different from those encountered in the job shop and flow shop. Time is money in several ways.

First, until the project is completed, there is seldom any return on investment. Second, when projects are new products, the first into the marketplace with a quality product gets a substantial market advantage. In the same way, when the project focuses on a major process improvement, there is likely to be a cost and/or quality advantage that also translates into a market differential.

The project manager is guided by strategic planning, which is tuned to windows of opportunity in the marketplace. This often means putting more resources to work to speed up project completion. Problems arise that slow the project. The costs of such delays can be many millions of dollars, whereas the production shop manager can accept delays that cost much less and are correctable the next time around. Usually, there is no next time around for the project manager.

The ability to manage under pressure and crises requires strong leadership, a fact that should be recognized when selecting project managers. Effective project managers are accustomed to living with great risk and the threat of large penalties. Their goals are strategic and usually vitally important to top management and the success of the company. Often, their goals are the change-management plans for the company. Thus, the profile of a successful project manager is different from that of job shop and flow-shop process managers. Further, project managers often require rapid systems-wide cooperation to resolve their problems quickly. This is a different kind of leadership than that required by process managers who are control-oriented.

7.4 Basic Rules for Managing Projects

The following basic rules apply to project management:

1. State project objectives clearly. They should be reduced to the simplest terms and communicated to all team members. There often are many participants in a project, and knowledge about objectives must be shared.
2. Expertise is required to outline the activities of the project and sequence them correctly. These activities are what must be done to achieve the goals. As a simple example of what happens if the right steps and sequences are not known, when the walls are plastered and painted before the electrical wiring and plumbing are done, the house will have to be unbuilt (going backward) and then rebuilt to achieve goal completion.
3. Accurate and achievable time and cost estimates for all project activities are essential. Slippage from schedule often means real trouble, whereas at other times it can be tolerated because there is sufficient slack. Slack is a time buffer that will be precisely defined in a later section. Project management requires knowing which activities to monitor and expedite.
4. Duplication of activities, in general, should be eliminated. Under some circumstances, however, parallel-path project activities are warranted, namely:
   a. If a major conflict of ideas exists and there is urgency to achieve the objectives, then it is sometimes reasonable to allow two or more groups to work independently on the different approaches. Preplanned evaluation procedures should exist so that as soon as it is possible the program can be trimmed back to a single path.

b. At the inception of a program (during what might be called the exploratory stage), parallel-path research is frequently warranted and can be encouraged. All possible approaches should be considered and evaluated before large commitments of funds have been made.
c. Parallel-path research is warranted when the risk of failure is high, for example, survival is at stake. When the payoff incentive is sufficiently great with respect to the costs of achieving it, then parallel-path activities can be justified for as long a period of time as is deemed necessary to achieve the objectives.

5. One system-oriented person should be responsible for all major decisions. The project manager must be able to lead a team that understands technological, marketing, and production constraints. Multiple project leaders are not advisable. The single project leader will have to be able to deal with many individuals reporting to him or her.
6. Project management methods are based on information systems utilizing databases that are updated on a regular basis.
   a. Project methods categorize and summarize a body of information that relates to precedence of activities, and their time and cost.
   b. Project methods can assess the effects of possible errors in estimates.

In this chapter, we will study techniques for planning and scheduling two types of projects. The first category of projects consist of activities for which the times and costs can be determined accurately and are assumed to be constant, whereas the second category of projects consist of activities whose times can be estimated but cannot be specified exactly. These two types of projects will be designated as deterministic and probabilistic projects respectively.

7.5 Project Management Origins

Starting about 1957, two similar approaches to large-scale project network planning and tracking were begun at separate locations and for different reasons. These were:

- PERT—program evaluation review technique
- CPM—critical path method

PERT was developed by the US Navy Special Projects Office in conjunction with Booz Allen Hamilton for the Polaris submarine launched missile project. This cold war project was considered urgent by the government and time was a critical variable. There were about 100,000 activities divided amongst thousands of suppliers. PERT set up activity networks, ideal for large projects, which could be systematically analyzed by computers.

CPM was a similar method developed by DuPont and Remington Rand, which later became Unisys. It was used to design and coordinate chemical plant operations. Even at the time of development, computers were crucial.

The essential difference between PERT and CPM is in specifying the times for performing various activities. This is primarily due to their origins and early projects for which they were developed and used. PERT was used for projects where the activity times were not certain because project managers were unfamiliar with activities. On the other hand, the projects and activities were familiar to the project managers in the case of CPM. These days the distinction between PERT and CPM seems to be disappearing and together these are called PERT/CPM or simply network techniques. These two methods share the notion of a critical path as discussed later in the chapter.

Both applications were very successful in reducing project time. Before network methods existed, project slippage was a fact of life. Projects often took 20% more time than expected and cost 20% more than budget estimates. With PERT and CPM, 20% reductions in expected values were experienced. The adoption of the new project methods was immediate within the United States. Many different kinds of software were developed that could be used for very large projects including year-end budget preparation. The Polaris submarine PERT was organized by the Navy Special Projects Office and also NASA used elaborate PERT charts to monitor and coordinate with vendors. Take a look at www.thebhc.org/publications/BEHprint/v022n1/p0210-p0222.pdf.

7.6 Project Network

Three steps are required to utilize these network models.

1. Detail all of the activities that are required to complete the project.
2. Establish the precedence relationships among activities and document the rationale for these relationships so that all teammates can share this information. The historical record is permanent and explicit. Draw a precedence diagram (a network) for the precise sequencing to be used based on technological feasibility, managerial objectives, administrative capabilities, equipment, and workforce constraints.
3. Estimate the time to perform each task or activity. The method of estimation for activity times needs to be detailed and related to project quality. For example, more time is required to use double-error checking—to be sure that no project defects occur. Double-error checking means that two different people (and/or methods) are used to verify that no errors occur. Two options are available for establishing times.

   Option 1: deterministic estimates for activity times.

   Option 2: probabilistic estimates for activity times.

7.6.1 Project Network Example

Consider the data for a project given in Table 7.1. This project consists of activities A through J as shown in the first column. Columns 2 and 3 show the relationships among the activities. The *immediate predecessor* of an activity is the activity (or activities) that must be done immediately before that activity, whereas the immediate follower is an activity (or activities) that must be done immediately after the given activity. For example, there are no immediate predecessors of activities A, B, and C. This means that these activities can be done as soon as the project starts. Activity A is the immediate predecessor of activity D, so activity D can be done only after activity A has been completed. Activities E and F can be done only after activity B (immediate predecessor of E and F) has been completed. Activity G must be done after activity C has been done. Activity H requires that activities D and E are completed before one can start on activity H. Similarly F, G, and H need to completed before activity I can start and finally activity I needs to be completed before activity J can start.

The relationships among activities can also be given by specifying the immediate follower(s). For example, D is an immediate follower of A. It is the same thing as saying that A is the immediate predecessor of D. Activities E and F are the immediate followers of B. Activity G is the immediate follower of C; activity H is the immediate follower of D and E; activity I is the immediate follower of F, G, and H; and J is the immediate follower of I. Activity J has no immediate follower since J is the last activity of the project. In any project, it is sufficient to specify either the immediate followers or the immediate predecessors.

Table 7.1 Project Data

Activity	*Immediate Predecessors*	*Immediate Followers*	*Time (Weeks)*
A	None	D	9
B	None	E, F	5
C	None	G	7
D	A	H	12
E	B	H	8
F	B	I	6
G	C	I	11
H	D,E	I	5
I	F,G,H	J	4
J	I	None	10

Note: Either immediate predecessors or immediate followers need to be specified.

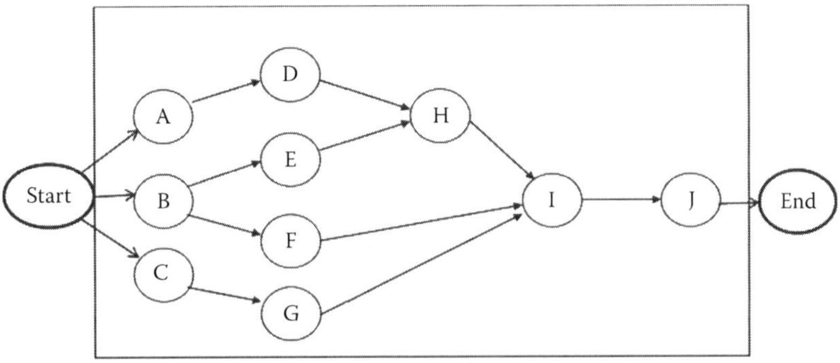

Figure 7.1 AON diagram.

The relationships given in Table 7.1 and discussed above are shown in Figure 7.1. Each activity is represented in a circle. The arrows show the relationships. By convention, the "Start" and "End" activities are added to make a single beginning and a single ending of the project. The times for the Start and End activities will be zero. This diagram is called an Activity on Node (AON) diagram because circles are the nodes and each activity is represented as a node. This diagram is also known as a network or network diagram.

7.7 Critical Path and Project Duration

Critical path is defined in terms of time. PERT/CPM is a time-based method. Time and cost estimates will be related at a later point. For now, time is the crucial parameter. The goal is to determine the project duration with the aim in mind to achieve the shortest project completion time and then, control of project-cycle time to meet the goals of the plan.

The project duration can be found by listing all the paths in the network. This project network shown in Figure 7.1 has four paths. These are listed in Figure 7.2. The Start and End activities are not included while listing the paths. The length of

Paths					*Length*	
A	D	H	I	J	40	Critical Path
B	E	H	I	J	32	
B	F	I	J		25	
C	G	I	J		32	

Figure 7.2 Path-length (weeks).

a path is equal to the sum of all activities that belong to that path. For example, the time of path, A–D–H–I–J = 9 + 12 + 5 + 4 + 10 = 40 weeks. The time of path B–F–I–J = 5 + 6 + 4 + 10 = 25 weeks. Refer to Figure 7.2 for the time of the other paths.

The project duration is equal to the length of the longest path, that is, 40 weeks in this case. The longest path is also called the critical path. There can be more than one critical path in a network. It may be noted that the total time of all activities is not the project duration because some activities can be done simultaneously. For example, activities A, B, and C can be done simultaneously (also called as—done in parallel) when the project starts. A path that is shorter than the critical path is said to have slack time which is the difference between the length of the critical path and the length of the path having the slack time. For example, the path B–E–H–I–J has a slack time of 8 (= 40 – 32) weeks. The concept of slack time is further discussed later in this chapter.

7.8 Early Start and Early Finish Times

The critical path gives the project duration. However, it does not tell when to start an activity and when to finish an activity. We find out the Early Start (ES) time and the Early Finish (EF) time of each activity. These times are given in Table 7.2 and are calculated as described below.

Activities A, B, and C do not have any predecessors, so they can all start at time zero as shown in Table 7.2. ES times of activities A, B, and C are zeros. We can now calculate the EF times of these activities by using

$$\text{EF time} = \text{ES time} + \text{Activity time}.$$

Table 7.2 ES and EF Times for the Activities

Activity	*Immediate Followers*	*Time (Weeks)*	*ES Time*	*EF Time*
A	D	9	0	9
B	E, F	5	0	5
C	G	7	0	7
D	H	12	9	21
E	H	8	5	13
F	I	6	5	11
G	I	11	7	18
H	I	5	21	26
I	J	4	26	30
J	None	10	30	40

Activity A starts at time 0 and its processing time is 9, so it will be finished at time 9 (= 0 + 9). Similarly, B will finish at time 5 and C will finish at time 7. Activity D can start when its immediate predecessor (A) has been completed. A is completed at time 9. Therefore, the ES time of D is 9. Its EF time is 21 (= 9 + 12), where 12 is the processing time of activity D. Activities E and F can start after activity B is finished at time 5. So the ES time of both E and F is 5. The EF time of E is 13 (= 5 + 8) and the EF time of F is 11 (= 5 + 6). Similarly, the ES and EF times of activity G are 7 and 18, respectively. Activity H can start only after activities D and E are completed since both D and E are the immediate predecessors of H. Activity D is completed earliest at time 21 and activity E is completed earliest at time 13. Therefore, the ES time of activity H is 21. *The rule is that if an activity has several immediate predecessors, then the ES time of that activity is equal to the largest of the EF times of all preceding activities.* The EF time of H is, therefore, 26 (= 21 + 5). The ES time of activity I is 26 which is the largest of the EF times of F, G, and H (immediate predecessors of I). The EF time of activity I is 30 (= 26 + 4). Finally, the ES of activity J is 30 (ES of I) and the EF of J is 40 (= 30 + 10).

7.9 Late Start and Late Finish Times

In this section, we discuss how to find the Late Start (LS) and the Late Finish (LF) times of all activities in the project. The calculations are shown in Table 7.3.

Table 7.3 LS and LF Times for the Activities

Activity	*Immediate Followers*	*Time (Weeks)*	*LS Time*	*LF Time*
A	D	9	0	9
B	E, F	5	8	13
C	G	7	8	15
D	H	12	9	21
E	H	8	13	21
F	I	6	20	26
G	I	11	15	26
H	I	5	21	26
I	J	4	26	30
J	None	10	30	40

The calculations start by fixing the deadline for project completion which is the LF time by which the last activity of the project has to be finished. In this project, activity J is the last activity. Activity J can be completed earliest at time 40 as discussed above and shown in Figure 7.2. Let us set the due date for completion of the project as 40. It could have been a different due date. A different due date will generally be greater than the calculated project due date (40 in this case). A smaller due date can also be specified in which case durations of some of the activity have to be reduced (see Section 7.11 on crashing activities). Therefore, the LF time of activity J is 40. The LS time of an activity is calculated by using

$$\text{LS time} = \text{LF time} - \text{Activity time.}$$

For activity J, the LS time is, therefore, 40 – 10 = 30. Next, we need to fix the LF times of all activities that precede activity J. Only activity I precedes activity J. Therefore, the LF time of activity I is equal to 30. Using the equation given above, the LS time of activity I is 26 (= 30 – 4). Activities F, G, and H precede activity I. Therefore, the LF times of F, G, and H are set equal to 26. An explanation of fixing these LF times is in order. Activity I must start latest by time 26 for the project to be completed in time. For I to start at 26, all its immediate predecessor must have been completed by 26. Using the equation, we calculate the LS times of F, G, and H as 20 (= 26 – 6), 15 (= 26 – 11), and 21 (= 26 – 5), respectively. Similarly, the LF times of D and E are 21. Note that D and E are the immediate predecessors of H. The LS time of D is 9 (= 21 – 12) and that of E is 13 (= 21 – 8). The LF time of C is 15 which is the LS time of G, and the LF time of A is 9 which is equal to the LS time of D. The LS time of A is 0 (= 9 – 9) and the LS time of C is 8 (= 15 – 7). The LF time of B is 13 and it needs some explanation. It may be noted that activities E and F are the immediate followers of activity B. E and F cannot start unless B has been finished. E must start latest by time 13 and F must start latest by time 20 as discussed earlier. Therefore, B must finish by time 13, otherwise E cannot be started at time 13. The LS time of B is 8 (= 13 – 5). *The rule to find the LF time of an activity when there are multiple followers is: the late finish time (LF) of that activity is equal to the smallest value of the latest start time (LS) of all the followers.*

7.10 Slack Time

The slack time of an activity is the time by which the activity can be delayed without delaying the project completion time. Slack is the allowable slippage in time. Slack time, by definition, can be wasted without changing project-completion time. It is calculated by using one of the following equations. Both of them give the same answer.

$$\text{Slack time} = \text{LS time} - \text{ES time.}$$

or

$$\text{Slack time} = \text{LF time} - \text{EF time}.$$

The slack times of all activities are shown in Table 7.4. Consider, for example, activity B. The slack time is 8 (= 8 – 0 or 13 – 5). This means that activity B can start at any time between 0 (ES) and 8 (LS) without delaying the project. Activities A, D, H, I, and J have zero slack time. It means that if these activities do not start at their ES times, the project will be delayed. In the network, the activities with zero slack time belong to the critical path. A–D–H–I–J is a critical path (see Figure 7.2). It is the set of activities that define the shortest time in which the project can be completed without slippage.

The slack time is very useful when scheduling activities under resource constraints. From the above discussion, we know that activities A, B, and C can start at time zero. However, if the number of workers or any other resource (like machines, money or material) is unexpectedly limited, one or more of these activities may have to be delayed. The project manager needs to make a decision to delay one or more activities. We know, in this case, that activity A should not be delayed. B and/or C can be delayed up to 8 time units (weeks, days, etc.). The slack of an activity describes the amount of time that can be viewed as a safety buffer for slippage. The project manager views small amounts of slack and zero slack as activities that demand constant attention because slippage there results in project delay. Also, reallocation of

Table 7.4 Slack Times (Weeks)

Activity	*Immediate Followers*	*Time (Weeks)*	*ES Time (Weeks)*	*EF Time (Weeks)*	*LS Time (Weeks)*	*LF Time (Weeks)*	*Slack Time (Weeks)*
A	D	9	0	9	0	9	0
B	E, F	5	0	5	8	13	8
C	G	7	0	7	8	15	8
D	H	12	9	21	9	21	0
E	H	8	5	13	13	21	8
F	I	6	5	11	20	26	15
G	I	11	7	18	15	26	8
H	I	5	21	26	21	26	0
I	J	4	26	30	26	30	0
J	None	10	30	40	30	40	0

resources can affect slack, which puts discretionary powers for what is critical, and what is not, in the hands of project planners.

7.11 Reducing Project Duration—Crashing Activities

We continue with the example discussed above. The project completion time is 40 weeks because the length of the critical path is 40 weeks (see Figure 7.2). Suppose we want to reduce the project duration, say, to 39 weeks. The length of the critical path must be reduced to 39 weeks. This requires that we reduce the time of one of the critical path activities by 1 week. For simplicity, we assume that the time can be reduced by the whole integer and not reduced in fractions such as reduce one activty by 1/2 and another activity by 1/2. Suppose we can reduce the time of activity A by 1 week by bringing in additional resources. As a result, activity A will be completed in 8 weeks instead of 9 weeks. The length of the critical path will become 39 weeks, whereas the lengths of other paths remain the same. The project duration will be 39 weeks. See the results in Figure 7.3.

Suppose we now want to further shorten the project duration to 38 weeks. Again, we must reduce the time of one of the activities on the critical path. Activity A whose time is now 8 weeks may be picked again (if it is still alterable with additional resources or by redesigning what must be done or one of the other activities can be selected for crashing). In the project-management literature, reducing the time of an activity is known as crashing the activity. Suppose we choose activity J (with activity time of 10 weeks) for crashing. The activity of time J becomes 9 weeks by calling upon additional resources. The length of the critical path becomes 38 weeks. The lengths of all other paths also shorten by 1 week because activity J belongs to all paths. The lengths of all paths after crashing J are shown in Figure 7.4.

How do we decide which activity (activities) should be crashed? To make that decision, additional time and cost analysis needs to be done.

Paths					*Length (Weeks)*
A	D	H	I	J	39
B	E	H	I	J	32
B	F	I	J		25
C	G	I	J		32

Figure 7.3 Reduction of the length of critical path.

Paths					Length (Weeks)
A	D	H	I	J	38
B	E	H	I	J	31
B	F	I	J		24
C	G	I	J		31

Figure 7.4 Length of paths after crashing activities A and J.

7.12 Cost Analysis

To find out which activities should be crashed, we have to find out the number of weeks by which an activity can be crashed (shortened) and the cost of crashing per week.

7.12.1 Example

We will use the example given in Table 7.5. The AON network for this example will be the same as given in Figure 7.1. The project consists of 10 activities. The list of activities (A through J) and their immediate predecessors, the normal time, the

Table 7.5 Example for Crashing Activities of a Project

Activity	*Immediate Predecessor (S)*	*Normal Time (Weeks)*	*Crash Time (Weeks)*	*Normal Cost ($)*	*Crash Cost ($)*	*Cost of Crashing per Week ($)*	*Maximum Crashing Possible (Weeks)*
A	None	9	6	13,000	15,550	850	3
B	None	5	4	7000	7900	900	1
C	None	7	5	15,000	15,800	400	2
D	A	12	8	12,000	14,800	700	4
E	B	8	5	9000	10,500	500	3
F	B	6	4	5000	6200	600	2
G	C	11	9	13,000	14,000	500	2
H	D, E	5	4	8000	9000	1000	1
I	F, G, H	4	3	3000	3500	500	1
J	I	10	8	12,000	15,000	1500	2

crash time, the normal cost, and the crash cost for each activity are also given in this table.

An activity is generally completed in its normal time at the normal cost; but it can be expedited and completed in less than the normal time at an additional cost. By how much time an activity can be crashed is limited by the nature of that activity. The crash time is the absolute minimum time to which an activity's time can be reduced. The time of an activity cannot be reduced below its crash time, no matter what resources are available. For example, let us consider activity A. The normal time of A is 9 weeks and it will cost \$13,000 to complete this activity in 9 weeks. However, the activity can be crashed to 6 weeks and the cost will be \$15,550. Activity A can be crashed by a maximum of 3 weeks at an additional cost of \$2550.

7.12.2 Cost of Crashing an Activity

> The maximum possible time by which an activity can be crashed is the difference between its normal time and crash time.

Based on the time and cost information we can calculate the cost of crashing per week (or any other time unit being used for the project) by using the following formula.

$$\text{Cost of crashing per week} = \frac{(\text{Crash cost} - \text{normal cost})}{(\text{normal time} - \text{crash time})}.$$

For activity A, the cost of crashing per week = \$850 = (15,550 – 13,000)/(9 – 6). This means that if the time of activity A is reduced by 1 week, its cost will increase by \$850. In other words, activity A can be crashed up to a maximum of 3 weeks and the cost of crashing per week is \$850. What are the time and cost implications for activity A? This means that activity A can be done in 9 weeks (normal time) at \$13,000 (normal cost). The time of activity A will be 8 weeks if it is crashed by 1 week at a cost of \$13,850; the cost will increase by \$850. If activity A is crashed by 2 weeks, its cost will increase by \$1700 (= 2 × 850), and the cost will increase by \$2550 (= 3 × 850) if A is crashed by 3 weeks. Activity A will be completed in 6 weeks if it is crashed by 3 weeks at a cost of \$15,550.

7.12.3 Reducing Project Duration

The project duration can be reduced by crashing activities as discussed above. However, only the activities that belong to the critical path must be crashed.

Consider a project with two paths. Path 1 consists of activities P–Q–R and Path 2 consists of activities X–Y–Z. Suppose the length of Path 1 is 20 days and the length of Path 2 is 18 days. Path 1 is the critical path and the project duration is 20 days. Suppose one of the activities on Path 2 is crashed by 1 day incurring some additional cost. The length of Path 2 will become 17 days but the project duration (20 days) remains unchanged since Path 1 is still the critical path, and the project cost has increased due to the crashing of an activity on Path 2. If an activity on Path 1 was crashed by 1 day, then the project duration will become 19 days; Path 1 is still critical but its length is 19 days. If Path 1 is crashed by an additional 2 days, then the length of both paths will be 17 days. At this stage, activities on both paths must be crashed to reduce project duration. Crashing activities on multiple paths is discussed later.

Crash activities only on the critical path(s) to reduce project duration.

Let us return to our prior example (see data in Table 7.5 and Figure 7.1). We want to reduce the duration of the critical path (A–D–H–I–J). We begin by crashing the activity with the minimum cost of crashing per week which is activity I ($500 per week). The order in which activities on the critical path will be crashed is: I (1 week at $500 per week), activity D (4 weeks at $700 per week), activity A (3 weeks at $850 per week), activity H (1 week at $1000 per week), and finally activity J (2 weeks at $1500 per week). Once all activities on the critical path have been crashed, the project duration cannot be reduced further.

In the case of multiple critical paths, activities on all critical paths must be crashed simultaneously by the same amount.

Table 7.6 gives the sequence of crashing various activities and the resulting length of each path and the cost after crashing the activities.

Schedule 1 is the normal schedule in which all activities are done in normal time. The activity cost is $97,000 which is the sum of normal costs of all activities. In Schedule 2, activity I is crashed by 1 week at a cost of $500. The activity cost increases to $97,500. The critical path is still A–D–H–I–J and its length becomes 39 weeks. It may be noted that the length of each path has reduced by 1 week since activity I belongs to all the paths.

Next, we crash activity D which has the minimum cost of crashing per week of all activities that belong to critical path. Activity D can be crashed by up to 4 weeks.

Table 7.6 Schedule of Crashing Activities

	Schedule 1	*Schedule 2*	*Schedule 3*	*Schedule 4*	*Schedule 5*	*Schedule 6*
Path	Normal schedule	Crash activity I by 1 week	Crash activity D by 4 weeks	Crash activity A by 3 weeks	Crash activity H by 1 week	Crash activity J by 2 weeks
A–D–H–I–J	40	39	35	32	31	29
B–E–H–I–J	32	31	31	31	30	28
B–F–I–J	25	24	24	24	24	22
C–G–I–J	32	31	31	31	31	29
Cost of crashing ($)	0	500	2800 = 700*4	2550 = 850*3	1000	3000 = 1500*2
Activity cost ($)	97,000	97,500	100,300	102,850	103,850	106,850

We can crash D by 1 week at a time or by the entire 4 weeks. In Schedule 3, activity D has been crashed by 4 weeks. The length of the path A–D–H–I–J becomes 35. The lengths of all other paths remain unchanged since activity D does not belong to the other paths. The cost of crashing D by 4 weeks is $2800 (= $700*4). The cost of this schedule increases to $100,300 ($97,500 + $2800).

Activity A has been crashed by 3 weeks in Schedule 4. The length of the critical path A–D–H–I–J becomes 32, whereas the lengths of all other paths remain unchanged. The cost of crashing is $2550 (= $850*3) and the total cost of this schedule is $102,850 ($100,300 + $2550).

In Schedule 5, activity H is crashed by 1 week at a cost of $1000. The lengths of paths A–D–H–I–J and B–E–H–I–J are reduced by 1 week since H belongs to both of these paths. The lengths of the other two paths remain unchanged. The total cost of Schedule 5 increases to $103,850 (= $102,850 + $1000). There are now two critical paths in the network A–D–H–I–J and C–G–I–J.

To reduce the project duration further, we must crash one activity on each path or a common activity. The only activity left to be crashed on path A–D–H–I–J is activity J which can be crashed by 2 weeks. Activity J belongs to path C–G–I–J also. Therefore, activity J is crashed by 2 weeks at a cost of $3000 (= $1500*2). The lengths of all the four paths are reduced by 2 weeks since activity J belongs to all the paths. Paths A–D–H–I–J and C–G–I–J continue to be critical paths with a length of 29 weeks each. However, we cannot reduce the project duration further because all activities on path A–D–H–I–J have been crashed. Path C–G–I–J can still be crashed but the project duration will not decrease and the cost will increase—not very wise. The total activity cost for Schedule 6 is $106,850 (= $103,850 + $3000) and the project duration is 29 weeks. The project manager has to choose a particular schedule. How is this decision made? Why should a project manager spend more money to expedite the project? These questions can be answered by taking into account the fixed costs of the project.

7.12.4 Fixed Costs

The project costs that have been discussed above are the costs of completing an activity. However, there are other project costs that are not activity-dependent but are time-dependent. These are called the fixed costs. For example, the salary of the project supervisor will be a fixed cost if the supervisor is paid per week. Similarly, if a builder, who is building an apartment complex, completes the project 2 months late, he loses the rent that could have been earned if the project was completed on time. The fixed cost is the rent lost for 2 months. Similarly, in the case of a new product development project, if a company can bring the product earlier to the market, it will start getting the returns on investment sooner.

For the example that we have been discussing (Tables 7.5 and 7.6), suppose the fixed cost of the project is $800 per week. In this case, the fixed cost will be $32,000 ($800*40) if the project is completed in 40 weeks, and $23,200 ($800*29) if the project is completed in 29 weeks. Table 7.7 gives the activity costs, fixed costs, and the total costs for all project durations from 29 to 40 weeks.

Some of the important observations from this table include the following. Activity cost decreases as the project duration increases; it goes down from $106,850 (for 29 weeks) to $97,000 (for 40 weeks). The fixed cost increases as the project duration increases; it increases from $23,000 (for 29 weeks) to $32,000 (for 40 weeks). The total cost first decreases as the project duration increases, reaches a minimum, and then increases again. The minimum cost for this project is $128,300 for the project duration of 35 weeks (Schedule 3 in Table 7.6).

It may be observed that as an activity is crashed by 1 week, the total cost of the project goes up by an amount equal to the crashing cost of that activity and decreases by an amount equal to the fixed cost per week. To observe this phenomenon, let us review various schedules in Table 7.6. Schedule 1 requires crashing of I at a cost of $500. The fixed cost is $800. So crashing I by one week will reduce the total project cost by $300. Observe in Table 7.7 that the total cost has gone down from $129,000 (for a 40-week schedule) to $128,700 (for a 39-week schedule). Schedule 3 (Table 7.6) requires crashing D at a cost of $700 per week. The cost advantage will be $100 since the fixed cost is $800. Total cost is reduced by $100 per week as we move from the 39-week schedule to 35-week schedule. To make the project duration 34, activity A has to be crashed by 1 week at a cost of $850. This will increase the total project cost by $50 as can be seen in Table 7.7. Total cost

Table 7.7 Cost Analysis of All Project Durations

Project Time	*Activity Cost ($)*	*Fixed Cost ($)*	*Total Cost ($)*
29	106,850	23,200	130,050
30	105,350	24,000	129,350
31	103,850	24,800	128,650
32	102,850	25,600	128,450
33	102,000	26,400	128,400
34	101,150	27,200	128,350
35	100,300	28,000	128,300
36	99,600	28,800	128,400
37	98,900	29,600	128,500
38	98,200	30,400	128,600
39	97,500	31,200	128,700
40	97,000	32,000	129,000

for a 34-week schedule is $128,350 when compared to $128,300 for the 35-week schedule. This trend continues until the minimum project duration is reached. The guiding principle for crashing process time is: crash an activity (or several activities in the case of multiple critical paths) if the total cost of crashing the activities per week is less than the fixed cost per week.

> Crash an activity (or group of activities in the case of multiple critical paths) if the total cost of crashing the activities per unit time is less than the fixed cost per unit time.

7.13 Crashing Multiple Paths

As discussed above, in the case of multiple critical paths, activities on all critical paths have to be crashed. We may crash an activity common to all critical paths or different activities on each path. The activity (or activities) to be chosen depends on the crashing costs. Consider the data given in Table 7.8. This project consists of two paths. Path 1 consists of activities a–b–c and Path 2 consists of activities a–p–q. The length of each path is 20 days. Both paths have to be crashed by 1 day each to reduce the project duration by 1 day. The costs of crashing all activities

Table 7.8 Crashing Multiple Paths

Path 1	a–b–c	20 days	Crash b	$400
Path 2	a–p–q	20 days	Crash p	$700
Cost of crashing per day				
	Activity	Cost ($)		
	a	1200		
	b	400		
	c	600		
	p	700		
	q	800		

are given in Table 7.8. The minimum cost activity to be crashed on Path 1 is activity "b" at a cost of $400, whereas the minimum cost activity to be crashed on Path 2 is activity "p" at a cost of $700. The total cost of crashing will be $1100 if we crash activities "b" and "p." We could have crashed the common activity "a" whose cost is $1200. For this project, crashing "a" is not desirable in comparison to crashing "b" and "p" together. However, suppose the cost of crashing activity "a" is less than $1100, say $1000. In this case, it becomes desirable to crash activity "a" rather than crashing activities "b" and "p." This kind of analysis has to be done where multiple critical paths have to be crashed to reduce project completion time.

7.14 Probabilistic Projects

We have done the analysis for the deterministic projects so far where the activity times were known and fixed. However, there are projects where the activity times are not fixed but they can be described using a probability distribution. In this section, we are going to study such projects using methods developed for the PERT system.

The activity times for these projects have to be estimated. For each activity, the project manager or the project team makes three time estimates that are intended to capture: optimistic time (*o*), most likely time (*m*), and pessimistic time (*p*). The activity will generally be completed in the most likely time. However, if circumstances are very favorable, the activity can be completed in less time. This is called the optimistic time. If things go wrong with an activity than the activity may be delayed, pessimistic time is an estimate of the activity time in such a situation. The expected completion time for an activity is the weighted sum of optimistic, most likely, and pessimistic time estimates. The most likely time is given four times more weight than the optimistic and pessimistic times. The expected time for an activity is then estimated by using

$$\text{Expected time} = \frac{o + 4m + p}{6}.$$

7.14.1 Probabilistic Projects Example

Consider a project with 10 activities. The list of activities (A through J) and their immediate predecessors are given in Table 7.9. In addition, the optimistic times, the most likely times, and the pessimistic times for all activities are given in this table.

The expected time for each activity can be calculated using the formula given above. For example, the expected time for activity A is 8 (= (5 + 4*8 + 11)/6) days. The project duration can be found by listing all paths in the network. It may be noted that we have used the same list of activities and precedence relationships as given in Table 7.1. The network diagram will be the same as shown in Figure 7.1. There are four paths in the network (same as shown in Figure 7.2). However, the length of each path will be calculated using the expected times. The list of paths and the length of each path are given in Table 7.10.

Table 7.9 Data for Probabilistic PERT Analysis

Activity	*Immediate Predecessor (s)*	*Optimistic Time (o) (Weeks)*	*Most Likely Time (m) (Weeks)*	*Pessimistic Time (p) (Weeks)*	*Expected Activity Time (Weeks)*	*Variance of Activity Times*
A	None	5.00	8.00	11.00	8.00	1.000
B	None	2.00	6.00	10.00	6.00	1.778
C	None	3.00	3.00	3.00	3.00	0.000
D	A	8.00	9.00	10.00	9.00	0.111
E	B	3.00	5.00	10.00	5.50	1.361
F	B	2.00	4.00	7.00	4.17	0.694
G	C	3.00	5.00	10.00	5.50	1.361
H	D, E	3.00	4.00	5.00	4.00	0.111
I	F, G, H	6.00	9.00	11.00	8.83	0.694
J	I	2.00	5.00	8.00	5.00	1.000

Table 7.10 Analysis of Probabilistic Paths

Project Due Date					*36*
Paths	*Expected Time of the Path*	*Variance of the Path*	*Standard Deviation of the Path*	*z-Value*	*Probability of Completing the Path by Due Date*
A–D–H–I–J	34.83	2.92	1.71	0.684	0.753
B–E–H–I–J	29.33	4.94	2.22	2.998	0.999
B–F–I–J	24.00	4.17	2.04	5.879	1.000
C–G–I–J	22.33	3.06	1.75	7.818	1.000
Probability of completing the project (critical path) by due date					0.753

The longest path is A–D–H–I–J (critical path). The expected length of this path is 34.83. Therefore, the expected project completion time is 34.83 weeks. Once we say that the "expected" time is 34.83 weeks, the project manager will be interested in knowing the probability to complete the project in 34.83 weeks or any other due date. To calculate the probability of completing the project by a given due date, we need to calculate the variance of each activity which is given by

$$\text{Activity variance} = \text{square of } [(\text{pessimistic time} - \text{optimistic time})/6] = [(p\text{-}o)/6]^2$$

For example, for activity A, the variance = 1 = square of ((11 – 5)/6). The variance may be zero for some activities when all three time estimates are the same (activity C, for example). Variances of all activities are given in Table 7.9.

Based on the activity variances, the variance of each path can be calculated by adding the variances of all activities that belong to a particular path. From variances, the standard deviation of each path can be calculated using the known result in statistics that standard deviation (σ) = square root of variance. The standard deviations of all paths are also given in Table 7.10. Now we can calculate the probability of completing the project by a given due date. Suppose the project due date is D. We need to calculate the probability of completing each path by the due date. For doing this, we calculate the z-value for each path using the formula z-value = $(D - T_e)/\sigma$, where T_e is the expected completion of the path and σ is its standard deviation.

Suppose, the project due date is 36 weeks. The z-value for path A–D–H–I–J, whose standard deviation is 1.71, will be 0.684 = (36 – 34.83)/1.71. Similarly, the z-values of all paths are calculated and are given in Table 7.10. From the z-value, the probability of completing each path within 36 weeks can be calculated by using the z-tables (see Appendix B) or the Excel function "normsdist." The probability of completing path A–D–H–I–J in 36 weeks is 0.753. The probabilities of completing all paths by the due date are also given in Table 7.10. The probability of completing the project by the due date is *assumed* to be the probability of completing the critical path by the due date. For this example, the probability of completing the project in 36 weeks is 0.753. In more advanced calculations, the probability of completing the project is obtained by multiplying the completion probabilities of all paths. In that case, the probability of completing the project is $0.7521 = (0.753 \times 0.999 \times 1.00 \times 1.00)$.

Note: The probabilities of completing the noncritical paths by the due date are larger than the probability of completing the critical path by the due date. Therefore, these probabilities can be ignored. However, the probability of a noncritical path whose length is very close to the critical path and which has a large variance may impact the probability of project completion.

7.15 Resource Management

Resource management is an important issue for project design. The fundamental idea of resource management is to switch extra resources from places where they are not essential to places where they could be used immediately. Alternatively, resource management aims to balance resource assignments across activities over time. It also provides some control for time-based management of project life cycles and facilitates shortening the critical path. It is a method for improving the speed of project achievements. It can also help improve quality of achievements and then rapid attainment.

Resource management seeks to move people from overstaffed activities to those that are understaffed. It attempts to reallocate money from where there is overspending to where there is under-spending. These efforts must make sense in technological and process terms. Similarly, the project manager would prefer smooth demand for cash instead of sporadic cash outflows. If, among a set of simultaneous activities, a few are receiving the greatest percentage of project expenditures, it often is desirable to level these allocations.

The beginning game is thoughtful and deliberate. The design of many projects is such that they start slowly while many ideas are being considered. There is much slack in activities that are off the critical path as reports are written and approvals are sought. There seems no need for urgency. The critical path is itself stretched out by bureaucratic organization of the project. This point of view has been undergoing change because of competitive pressures.

The end game has everything coming to a head very quickly. The end game has a time deadline that begins to appear to be looming. Then, it is decided that time must be made up. There is a surge in spending to speed up the project. Often, in the rush, mistakes are made, and damage control uses up time that was meant for thoughtful and deliberate means to bring the project to a successful conclusion. It is well known to project managers that this type of pattern is not desirable. Unless caught by circumstances, present-day project management avoids this damaging scenario.

In search of shorter project cycles, the newer approach to projects uses a multi-functional team with rapid communication to secure approvals. The cash-flow pattern is far more balanced from start to finish. Resources are assigned along all paths so that the critical path can be shortened and slack imbalances can be corrected before the beginning, or early on in the project's time line.

Resource management has two functions—resource leveling and resource scheduling. In resource leveling, the goal is to minimize the fluctuations in resources required from one period to another over the life of the project. On the other hand, in resource scheduling, it is assumed that there is an upper limit on the resources available and all activities are to be scheduled within the resource constraints. Study of the techniques for resource leveling and scheduling is beyond the scope of this chapter. However, we will illustrate the impact of shifting resources over the project completion time.

The existence of slack is an important basis for resource leveling and scheduling. To illustrate, consider Figures 7.1 (network) and 7.2 (list of all paths) which were obtained from the data given in Table 7.1. No resources were considered while doing this analysis. Suppose a part of one of the resources, say a team of workers, is shifted from activity B to activity A. This may result in increasing the time of activity B, say by 1 week (to 6 weeks), and decreasing the time of activity A by 1 week (to 8 weeks). The lengths of the four paths will now be: A–D–H–I–J (39), B–E–H–I–J (33), B–F–I–J (26), and C–G–I–J (32). Path A–D–H–I–J is still the critical path but its length is 39 weeks and hence the project duration has been reduced by 1 week. This was accomplished by shifting resources from one activity to another rather than spending more resources. Project manager and clients will be pleased. Leveling should not be confused with crashing of activities that increases the cost. In addition, due to a decrease in the length of critical path, slacks of individual activities will also change. For example, the slack of activity B will become 7 weeks instead of 8 weeks.

Resource management provides the means to withdraw resources, shift resources, or to add new resources. This approach changes budgetary constraints. Changing the budget is a strategic decision. The tactical aspect of resource leveling is that it can affect the entire system without changing the budget. All project team members are impacted when resources are shifted and they should be kept informed of project alterations. Good resource-allocation decisions are a project-management responsibility. This is another example of where tactical changes have far-reaching impact on strategic issues.

Summary

Chapter 7 starts by identifying project characteristics and explains the unique work configurations known as projects. The life-cycle stages of a project and classification of projects are studied next. The chapter then delves into managing projects. The qualities of a good project leader and team work are emphasized. Basic rules of project management are discussed. The origins of project management discipline are traced.

The chapter then moves on to scheduling of projects and describes how project network diagrams are drawn. The chapter discusses the identification of the critical path and its importance. The chapter explains how project duration is calculated.

Calculation of ES, EF, LS, and LF times of activities are fully explained. These require forward-pass and backward-pass calculations. The chapter continues by describing what slack means and how to derive it. Crashing of activities (including multiple paths) to reduce project duration is explained next. The time-cost tradeoff analysis is described and discussed.

The chapter moves on to an analysis of probabilistic projects. It details when deterministic and probabilistic estimates for activity times apply and shows how to use optimistic and pessimistic activity time estimates to obtain expected activity times and a variance measure for activity times. The chapter ends with a discussion of the implications of limited project resources.

Review Questions

1. What are the unique attributes of project management? Frame the answer in terms of other P/OM work configurations.
2. Classify projects by type and describe project life cycles.
3. Why is a project manager considered a leader?
4. What is team work and why is team work repeatedly mentioned when discussing good project management?
5. What is PERT's relationship to critical-path methods?
6. What does the forward-pass procedure accomplish in critical-path methods?
7. What does the backward-pass procedure accomplish in critical-path methods?
8. What is a critical path and how does knowing it in detail help project managers?
9. What is slack and how does knowing exactly what it is and where it resides help project managers?
10. What are the differences between deterministic and probabilistic activity time estimates?
11. What are the strengths and weaknesses of PERT?
12. Describe how cost/time trade-off methods can be used given the decision to spend an additional 10% on the project.
13. Describe project crashing and contrast it to normal time.
14. What advantages can be gained by using crashing?
15. What are the dangers of, using crashing?

Problems

1. Consider the following AON network and the data given in the following table to answer the next four questions.
   a. Identify the critical path.
   b. Find the earliest completion time of the project.
   c. Find ES, EF, LS, and LF of each activity.
   d. Find the slack for each activity.

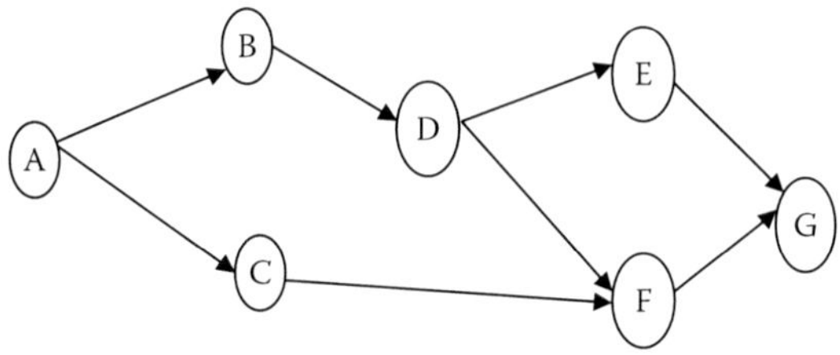

Activity	*Time (Days)*
A	5
B	6
C	7
D	4
E	9
F	3
G	4

2. The Delta Company manufactures a full line of cosmetics. A competitor recently developed a new skin rejuvenating cream that appears to be successful and potentially damaging to Delta's skin care position in the marketplace. The sales manager has asked the operations manager what the shortest possible time would be for Delta to reach the marketplace with a new product packaged in a competitively redesigned container. The operations manager has drawn up the following table.

Activity Description	*Activity Symbol*	*Immediate Follower*	*Duration (Days)*
Design product	A	H and G	30
Design package	B	E and C	15
Test market package	C	D	20
Distribute to dealers	D	None	20
Order package materials	E	I and F	15

Fabricate package	F	D	30
Order materials for product	G	I and F	3
Test-market product	H	J	25
Fabricate product	I	J	20
Package product	J	D	4

a. Construct the AON diagram.
b. Find the critical path.
c. Find the ES, LS, EF, and LF.
d. What is the project duration?
e. Determine the slack of each activity.
f. Neither the sales manager nor the P/OM is satisfied with the way the project is designed. However, the P/OM insists that because of the pressure of time, the company will be forced to follow this plan. In what ways does this plan violate good practice?

3. Consider the data given in the following table. The fixed cost is $900 per week. Perform a time-cost tradeoff analysis. What is the minimum cost to complete the project and what is the corresponding time?

Activity	*Immediate Predecessor (s)*	*Normal Time (Weeks)*	*Crash Time (Weeks)*	*Normal Cost ($)*	*Crash Cost ($)*
A	None	9	6	13,000	15,550
B	None	5	4	7000	7900
C	None	7	5	15,000	15,800
D	A	12	8	12,000	14,800
E	B	8	5	9000	10,500
F	B	6	4	5000	6200
G	C	11	9	13,000	14,000
H	D, E	5	4	8000	9000
I	F, G, H	4	3	2500	3500
J	I	10	8	12,000	15,000

4. Consider the data given in the following table and answer the questions that follow.

Activity	*Immediate Predecessor (s)*	*Normal Time (Days)*	*Crash Time (Days)*	*Normal Cost ($)*	*Crash Cost ($)*
A	None	4	2	200	400
B	A	4	3	300	600
C	A	1	1	200	200
D	B and C	3	2	600	650
E	D	2	1	500	900

   a. What is the normal project time?
   b. Identify the critical path?
   c. What is the normal cost of the project?
   d. If the project time is to be reduced by one day, which activity should be crashed first? What is the cost of crashing per day of activity E?
5. There are two paths in a network (see below). The length of Path 1 is 30 days and that of Path 2 is 28 days. You are now told that activity F which currently takes 4 days will actually take 3 days more, that is, 7 days. What will be the project duration with this revised time?
   Path 1: A–B–D–E–G
   Path 2: A–C–D–F–G
6. There are two paths in a network (see below). The length of Path 1 is 30 days and that of Path 2 is 28 days. You are now told that activity D which currently takes 2 days will actually take 3 days more, that is, 5 days. What will be the project duration with this revised time?
   Path 1: A–B–D–E–G
   Path 2: A–C–D–F–G
7. Suppose the length of critical path in a project is 40 days. There are four activities A, B, C, and D on the critical path. The variances of these activities are: A (1.4), B (1.2), C (0.9), and D (0.4). What is the probability of completing the project within 42 days?
8. Suppose the length of critical path in a project is 50 days and the standard deviation is 2 days. What due date should be set for the project so that the probability of completing the project is 90%? Round the answer to the nearest higher integer.

9. Suppose the length of critical path in a project is 35 days. The probability of completing the project within 34 days is:
   a. Less than 50%
   b. Greater than 50%
   c. Equal to 50%
   d. Cannot be determined from the above data

Readings and References

Baker, B.M., Cost/Time Trade-Off Analysis for the Critical Path Method: A Derivation of the Network Flow Approach, *Journal of the Operational Research Society*, 48(12), December 1997, pp. 1241–1244.

Chen, M.T., Simplified Project Economic Evaluation, *Cost Engineering*, 40(1), January 1998, pp. 31–35.

Clayton, E.R., and Moore, L.J., PERT vs. GERT, *Journal of Systems Management*, 23, February 1972, pp. 11–19.

Conkright, T.D., So You're-Going to Manage a Project, *Training*, 35(1), January 1998, pp. 62–64.

Davenport, T.H., De Long, D.W. and Beers, M.C., Successful Knowledge Management Projects, *Sloan Management Review*, Winter 1998, pp. 43–57.

Eden, C., Williams, T., and Ackermann, F., Dismantling the Leaming Curve: The Role of Disruptions on the Planning of Development Projects, *International Journal of Project Management*, 16(3), June 1998, pp. 131–138.

Fleming Q.W., and Koppelman, J.M., What's Your Project's Real Price Tag? *Harvard Business Review*, September 2003.

George, G., Construction Projects Span the Globe, *Transmission & Distribution*, 50(I), January 1998, pp. 38–48.

Greiner, L.E., and Schein, V.E., The Paradox of Managing a Project-oriented Matrix: Establishing Coherence Within Chaos, *Sloan Management Review*, 22(2), Winter 1981, pp. 17–22.

Helgadottir, H., The Ethical Dimension of Project Management, *International Journal of Project Management*, 26(7), 2008, p. 743.

Icmeli-Tukel, O., and Rom, W.O., Ensuring Quality in Resource Constrained Project Scheduling, *European Journal of Operational Research*, 103(3), December 16, 1997, pp. 483–496.

Karlos, A., et al., Foundations of Project Management, *International Journal of Project Management*, 27(1), 2009, p. 1.

Keil, M., and Hring, M.M., Is Your Project Turning into a Black Hole? *California Management Review*, November 2010.

Keil, M., and Montealegre, R., Cutting Your Loses: Extricating Your Organization When a Big Project Goes Awry, *Sloan Management Review*, April 2000.

Kolisch, R., Resource Allocation Capabilities of Commercial Project Management Software Packages, *Interfaces*, 29, July–August 1999, pp. 19–31.

Kumar, P.P., Effective Use of Gantt Chart for Managing Large-Scale Projects, *Cost Engineering*, 47(7), July 2005, pp. 14–21.

Larson, M., Manage Your Projects Before They Manage You, *Quality*, September 1997, pp. 64–67.

Lenfle, S., and Loch, C.H., Lost Roots: How Project Management Came to Emphasize Control Over Flexibility and Novelty, *California Management Review*, November 2010.

Levy, F., Thompson, G., and Wiest, J., The ABC's of the Critical Path Method, *Harvard Business Review*, 41(5), October 1963.

Ling, F.Y., et al., Key Project Management Practices Affecting Singaporean Firms' Project Performance in China, *International Journal of Project Management*, 27(1), January 2009, p. 59.

Lomas, D.W., Leadership Styles of Project Managers in Hong Kong: An Empirical Study, *International Journal of Management*, 14(4), December 1997, pp. 667–672.

MacCrimmon, K.R., and Ryavec, C.A., An Analytical Study of the PERT Assumptions, *Operations Research*, 12, January–February 1964, pp. 16–37.

Macomber, J.D., You Can Manage Construction Risks, *Harvard Business Review*, 67(2), March–April 1989, pp. 155–161.

Matta, N.F., and Ashkenas, R.N., Why Good Projects Fail Anyway? *Harvard Business Review*, September 2003, pp. 109–114.

Miller, R.W., How to Plan and Control within PERT, *Harvard Business Review*, 40, March–April 1962, pp. 93–104.

Newmark, H.R., Auditing Construction Projects, *The Internal Auditor*, 54(6), December 1997, pp. 36–41.

Oates, D., Understanding and Solving the Causes of Project Failure, *Knowledge Management Review*, 9(5), May–June 2006, p. 5.

Randolph, W.A., and Posner, B.Z., What Every Manager Needs to Know about Project Management, *Sloan Management Review*, Summer 1988, pp. 65–73.

Rodrigues, A.G., and Williams, T.M., System Dynamics in Project Management: Assessing the Impacts of Client Behaviour on Project Performance, *Journal of the Operational Research Society*, 49(1), January 1998, pp. 2–15.

Royer, I., Why Bad Projects Are So Hard to Kill, *Harvard Business Review*, 81(2), February 2003, pp. 48–56.

Sharpe, P., and Keelin, T., How SmithKline Beecham Makes Better Resource-Allocation Decisions, *Harvard Business Review*, March–April 1998, pp. 45–57.

Shenlar, A.J., From Theory to Practice: Toward a Typology of Project-Management Styles, *IEEE Transactions on Engineering Management*, 45(1), February 1998, pp. 33–48.

Slevin, D.P., and Pinto, J.K., Balancing Strategy and Tactics in Project Implementation, *Sloan Management Review*, Fall 1987, pp. 33–41.

Smith-Daniels, D., Teaching Project Management to MBAs: The Means to How Many Ends? *Decision Line*, May 1997, pp. 11–13.

Thamhaim, H.J., and Wilemon, D.L., Conflict Management in Project Life Cycle, *Sloan Management Review*, Spring 1975, pp. 31–50.

Tighe, G., From Experience: Securing Sponsors and Funding for New Product Development Projects—The Human Side of Enterprise, *Journal of Product Innovation Management*, 15(1), January 1998, pp. 75–81.

Chapter 8

Quality Management

Readers' Choice—"Quality means doing it right when no one is looking."—Henry Ford

Apte, U.M., and Reynolds, C.C., Quality Management at Kentucky Fried Chicken, *Interfaces*, 25(3), 1995, p. 6. The program developed by Kentucky Fried Chicken (KFC) Corp. to improve service quality is used as a benchmark for continuous process improvement by all KFC stores. The reduced service time as a result of this program is one of the measurements of quality.

Crosby, P.B., *Quality is Free (The Art of Making Quality Certain)*. McGraw-Hill, 1979. Crosby (1979) demanded a zero-defects goal which treats any failures as intolerable.

Harris, C.R., and Yit, W., Successfully Implementing Statistical Process Control in Integrated Steel Companies, *Interfaces*, 24(5), 1994, p. 49. Implementation processes of statistical process control (SPC) projects were analyzed at 12 integrated steel companies to identify key success (and failure) factors.

Hosseini, J., and Fard, N.S., A System for Analyzing Information to Manage the Quality-Control Process, *Interfaces*, 21(2), 1991, p. 48. The system developed by Hay and Forage Industries to manage quality process cuts across all levels and departments in the organization. The system can track quality characteristics of hundreds of parts.

Kumar, S., and Gupta, Y.P., Statistical Process Control at Motorola's Austin Assembly Plant, *Interfaces*, 23(2), 1993, p. 84. The total quality management (TQM) program implemented at Motorola's Austin, Texas assembly plant includes: training, appropriate use of SPC for process control, design of experiments, coordinating problem-solving teams, and certifying operators and machines.

Roethlein, C.J., and Mangiameli, P.M., The Realities of Becoming a Long-term Supplier to a Large TQM Customer, *Interfaces*, 29(4), 1999, p. 71. This paper describes the efforts of one of the long-term suppliers of Whirlpool Corporation, Stanley Engineered Components, to meet Whirlpool's TQM requirements.

Rust, R.T., Keiningham, T., Clemens, S., and Zahorik, A.Z., Return on Quality at Chase Manhattan Bank, *Interfaces*, 29(2), 1999, p. 62. This paper evaluates the impact of improving service quality by calculating the net present value and return on investment of the quality improvement project at Chase Manhattan bank.

Seawright, K.W., and Young, S.T., A Quality Definition Continuum, *Interfaces*, 26(3), 1996, p. 107. This paper describes various definitions of quality and their impacts on achieving competitive advantage through total quality management (TQM).

This chapter provides an understanding of the quality of goods and services; and builds the conceptual foundation of quality management. We discuss seven well-known and widely practiced quality achievement methods for providing quality assurance which is comprised of a set of systems-wide activities aimed at establishing confidence that quality goals will be achieved.

After reading this chapter, you should be able to:

- Explain why quality is a fundamental factor in strategic planning.
- Define and analyze quality in terms of its many dimensions.
- Explain how to set quality standards.
- Use various quality control methods.
- Construct control charts.
- Distinguish between producers' and consumers' quality concepts.
- Develop acceptance sampling plans and operating characteristic curves.
- Explain how both tangible and intangible quality dimensions are measured.

- Detail the things to consider when developing a rational warranty policy.
- Discuss ISO 9000 standards in an international context.
- Apply the costs of quality to determine rational product strategies.
- Describe the control monitor feedback model.
- Explain why quality competitions and prizes are given worldwide.

8.1 Introduction

There are two viewpoints of quality that coexist and cooperate. Producers (manufacturers or service providers) view quality as a set of standards and specifications that must be met (called conformance). On the other hand, customers view quality as attributes that please them. Finally, there are organizational measures of quality that combine the two views in various ways. These include global ISO quality standards, the Malcolm Baldrige National Quality Award system (see Garvin, 1991), the Deming Prize, and other competitions for prizes (see Francis and Thor, 1994).

Quality assurance starts by assessing the requirements of the customer. The quality goals are determined by the preferences of the customer. The product has to satisfy a variety of customer types (called market segments). For example, qualities that are associated with a great vacation or a restaurant of choice differ between market segments. Since September 11, 2001, safety and security have joined the list of qualities that cannot be taken for granted. Quality of life is a growing concern as world and urban populations increase, global warming heats up, and global economic interdependencies (such as outsourcing) become a bone of contention.

While the consumer's perception of "good" quality is important, the perception of poor or terrible quality is a calamity. Perceived quality is a critical variable in any business assessment. If it is not included, there can be serious repercussions. When it comes to gaining competitive advantage, better quality has leverage with new and old customers. Better quality increases customer loyalty. An investment in quality can be called an expense for improved customer holding. Trade-offs between the cost of getting new customers and the additional price paid for better quality to hold onto existing customers need to be evaluated. Current research indicates that the cost of holding onto an existing customer is significantly lower than the cost of obtaining a new customer (which requires obtaining a switch from a competitive brand).

Alienating existing customers with product weaknesses and failures has a host of other side effects, which are undesirable. Word spreads quickly about product failures. There are a growing number of consumer protection publications. Government agencies pursue a number of quality factors to assure consumer protection. Recalls are used to remove defective products from stores. Large-scale auto, food, and battery callbacks are given prime-time coverage on TV. A strong quality program decreases the probability of callback situations arising.

Better product quality obtained at a reasonable price generally goes hand-in-glove with growth in market shares and increases in revenues. Sometimes having a good quality program and achieving quality improvements result in decreases in court costs for claims of harm caused by malfunctions and other types of liability. Another advantage of better quality is that people who work in companies that really have better quality enjoy an environment of higher morale.

Quality Assurance requires a team effort—the Olympic perspective. The Olympic perspective calls for team play with everyone striving to achieve the company's goals in the best possible manner. To the managers of the firm, "going for the gold" is no less a meaningful objective than for Olympic teams. In achieving effective team effort for gold medal quality, not all competitors strive to be the best in the world. Some are happy to be silver or bronze. Others consider being at the Olympics a sufficient reward. Companies have similar goal differences. It is important to realize that not all companies strive to be the best. Nevertheless, because quality failures are widely recognized to have negative impact on long-term performance, all companies with long-term objectives consider quality improvement to be a common goal. Companies with short-term objectives do not care at all. Striving for quality unites all of the people, components and constituencies of the system. That does not mean that everyone strives with equal vigor and equivalent knowledge. The Olympic credo of striving to be better is readily translated into dynamic goals of continuous improvement for management.

Quality starts with a zero-error mind-set. The appropriate zero-error mind-set is to abhor defectives and do as much as feasible to prevent them from occurring. When they occur, learn what caused them and correct the situation. Meanwhile, raise the goals and improve the standards. Tougher hurdles are part of the quality framework, which seeks continuous improvement. Do it right the first time is the motto of the approach that champions zero errors. Those who say that no defects are permissible are at odds with those who say that a few defects are to be expected. The latter group goes on to say that one should learn from mistakes and take the necessary measures to prevent them from recurring. The safest position is to be flexible. Good advice is to go for a "no defects" policy. If and when it fails to work, switch to a "learn from mistakes" policy. After corrective action has been taken to correct the mistakes, switch back to the zero-errors goal.

8.2 How Much Quality

The two definitions of quality discussed above are held and exercised by different interests. One person can have both points of view, depending on what hat he/she is wearing at the time. Everyone has an intrinsic sense of what quality means to them as a consumer. They want the best that can be had, so they are always judging the "degree of excellence."

As a producer, it is often necessary to compromise and set quality standards that are not the best in their class. If the level of quality that is set by management is viewed as unacceptable by the market, then the strategic planners have failed. Inexpensive automobiles do not have the same set of qualities as expensive ones. There are different markets for SUVs and compacts. It is the producer's problem to choose appropriate quality standards so that the market judges them to be fitting for that price class.

Producers strive to balance the market forces for high quality with consumers' cost preferences and the company's production capabilities. Figure 8.1a illustrates a common economic concept of the optimal quality level.

There are valid exceptions to the relationships that Figure 8.1a depicts. The cost of quality does not necessarily rise as the quality level increases. The fact that many times quality can be improved without spending money will be discussed at a later point. There also is the issue of how improved quality is obtained, allowing

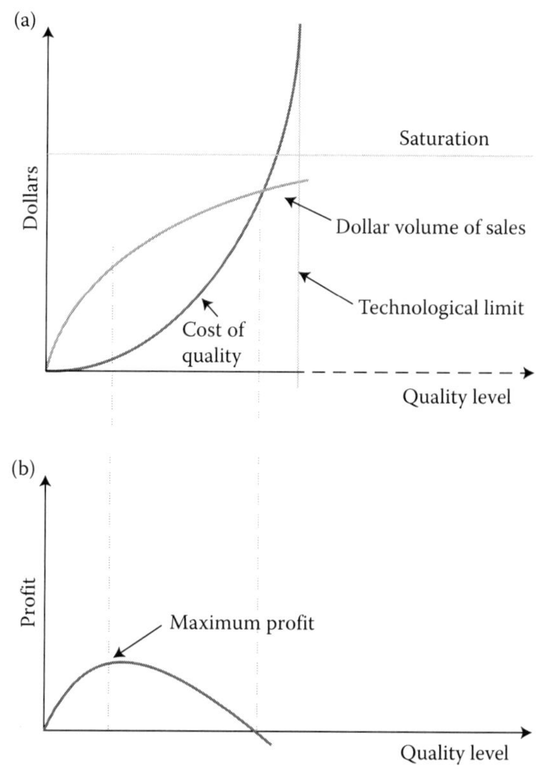

Figure 8.1 (a) Quality level versus cost of quality and dollar volume of sales. (b) Quality level versus profit.

that money to be spent on training and technology. Figure 8.1a also depicts a limit to how much quality can be improved. This concept can be faulted as not taking scientific possibilities and the ingenuity of creative people into account. There also is a question about the dollar volume of sales approaching saturation no matter how much better the quality level is made. Given such issues, it is difficult to know where the quality level of maximum profit occurs. Nevertheless, quality level standards should be set in accord with the notion of maximum profit. Figure 8.1b shows how the total profit may vary as the quality improves. One should strive to achieve the quality level at which the profit is maximized.

Some producers may be endowed with an innate sense of how to set quality standards so as to best balance costs and benefits. Most producers have to learn how to deal with such issues. This is an essential part of what P/OM does since process design determines what can be gained from the application of total quality management (TQM).

8.3 Dimensions of Quality

The dimensions of quality are the descriptors that must be examined to determine the quality of a product. The wine business is an enormous industry on a global scale. How do the presidents of wine companies characterize the quality of their products? They measure their production output by bouquet, color, and taste plus chemical analyses. Cars are evaluated within categories of cost, by their power, safety features, capacities, fuel efficiencies, and style, among other things. Appearance and style dimensions, so often important, are difficult to rate. Experts at ladies' fashion shows are at no loss for words. Yet this billion-dollar industry is only able to explain successes and failures after the fact. When rating the quality of cities to live in, the authors of such research base their studies on dimensions that define quality of life. The complex set of demographically sensitive criteria include amount of crime, cost of living, job availability, transportation, winter mildness, and the quality of schools for families with children.

The starting point for managing quality is to define the relevant set of quality dimensions, recognizing the special needs of market niches and segments. Not everyone will agree about what should be on the list or about the importance of the dimensions that are on the list. Individuality accounts for different perceptions about "what counts."

A sample of customers was asked what qualities could be improved in the service they received from their bank. Their answers included the following: length of time waiting for tellers should be decreased; availability of officers for special services should be increased; banks should be open longer hours; rates on interest-bearing accounts should be raised; and service charges should be dropped. The list was long and not everyone agreed on the relative importance of these points. At the same bank, a group of officers was asked to define the quality of the services their bank offered or should offer. The officers' answers show what a different perspective prevails between customers and producers. They wanted to increase the number of different products that the bank offers (CDs, checking accounts, pass-books, mortgages, loans, investment services); the average time spent by tellers per transaction should be small; variability of tellers' times should be decreased; the percentage of times that customers wait longer than five minutes should be small; and the average length of waiting lines for tellers should be about three people. Waiting lines of zero or one signify that something is wrong with staffing assignments. What could bring these two groups closer together? Customers describe quality in terms of personal ideals. The bankers describe quality in terms of how well they meet the economically sound standards that have been established by their bank. They also talk about changing the standards. The balance between these two positions is related to the costs and benefits of providing more of what the customer wants.

8.3.1 Models of Quality

Start to model quality in the form of lists of generic categories of quality. The lists will enable producers and consumers to check off and define all of the dimensions of quality that each deems applicable to the products. Eight categories, derived by David Garvin (see Garvin, 1987), a Harvard professor who researched quality issues, are listed below.

1. Performance
2. Features
3. Reliability
4. Conformance
5. Durability
6. Serviceability
7. Aesthetics
8. Perceived quality

These quality dimensions are discussed below. The automobile is used as a manufacturing example and a resort hotel is used as a service example. Describing qualities provides excellent practice in taking the first steps necessary to install a quality program.

- The *performance dimension* relates to the quality of the fundamental purpose for which the product is purchased. How well does the car do what it is supposed to do? Are the rooms quiet and the beds comfortable?
- The *features dimension* refers to product capabilities not considered to be part of normal performance expectations. These might be GPS and satellite radio for the car, or a spa and access to the internet (either wired or wireless) for the resort guests.

- The *reliability* dimension relates to performance that can be depended upon with a high level of assurance. The car starts, drives, and does not break down. If the windshield wipers do not work, there is a reliability problem. For the resort, if the room key does not work all of the time, there is a reliability problem.
- The *conformance* dimension alludes to the degree to which the measured production qualities correspond to the design quality standards that have been specified. Windshield wipers are not supposed to fail (say) for the first five years and the room key is expected to work for a normal stay. Conformance is definitely the producer's responsibility (auto and resort management). P/OM is charged with meeting the conformance dimensions of quality. These quality standards are specified by the management.
- The *durability* dimension deals with how well the product endures in the face of use and stress. Some cars are roadworthy after being driven more than 100,000 miles. Some room phones fail within a few weeks. Rooms get quite shabby with constant use in a few years and customers switch to other hotels when things get shabby unless prices drop.
- The *serviceability* dimension is related to how often service is required; and how difficult, and how costly it is to service and repair the product. Serviceability for both cars and resort rooms involves the combination of preventive and remedial maintenance.
- The *aesthetics* dimension refers to the appearance of the product. For both autos and resorts, design styling counts initially, and maintenance is crucial.
- The *perceived quality* dimension relates to the customers' perceptions of the product's quality and value received for monies paid. This dimension integrates the prior seven dimensions with the customers' sense of value for them. Market research is one of the most important means for determining the customers' perceived quality.

For different individuals, certain dimensions are more important than others. Frequently, there are regional differences in quality perceptions. Hard water areas have their own special quality dimensions for soap. Snow tires have no impact in the Southern states of the U.S.A. Such effects are amplified in the international marketplace. P/OM may have to set different standards for region A as compared to region B. Overall, it should be evident that quality definition is enhanced by requiring and enabling a systems perspective. Industrial designers and operations managers form a powerful team to apply the systems approach to all of the linked factors that relate to the eight quality dimensions.

Assumptions must be made about the way in which the measurable, physical factors relate to the consumers' evaluations of the "ilities" of the product. The "ilities" include utility, dependability, reliability, durability, serviceability, maintainability, repairability, and warrantability. Manufacturability, which includes assembly, is a producer's concern although it has its effects on the consumers' "ilities."

Quality definition is not simple. The measurement of how well a product or service performs its conforming functions is essential if standards are to be determined. There is no point in measuring the wrong things. When the standards are established to everyone's satisfaction, then P/OM can make the right products, test them, and improve the process. The same statement applies to both goods and services (e.g., cars and hotels).

Some other items that are important for quality evaluation include (a) product and service failure, (b) warranty policies (including how long a period is covered), (c) repairability, (d) human factors also called ergonomic factors to describe the efficiency of human beings in their work environments, (e) aesthetic aspects, and (f) product variety. These are discussed below.

8.3.1.1 Failure—Critical to Quality Evaluation

Failure occurs when a product ceases to perform in an acceptable fashion. Sometimes it is in the mind of the customer. Almost everyone has had a favorite restaurant that has fallen from grace. With respect to physical failure, there are engineering as well as logical conventions that define it. The car does not start. The fish is raw. The light bulb is burned out. The room is not made up. If the light source is considered "failed" after its output falls below a certain threshold, then P/OM needs to consider that aspect of quality standards.

With globally outsourced producers, problems are compounded. A rash of quality problems with products made in China and exported to the United States has made apparent the difficulty of controlling the quality of non-domestic production. Tainted pet food, toys made with leaded paints, contaminated food products, and unsafe tires are examples of what has been called "The Quality Fade." (see knowledge@wharton.upenn.edu, 2007). The Quality Fade issue has occurred repeatedly with high costs to consumers as well as producers. Quality fade is defined as "The initial qualities of production output cannot be maintained." Among the questions to be raised is "Why was there no overseeing function for P/OM at the outsourcing site?" P/OM should be fully aware of how it can influence the failure rates and characteristics of the products that it is producing. P/OM should also be working closely with R&D and engineering design to develop new production capabilities that can increase the product's expected lifetime or mean time between failures (MTBF). The measure MTBF is often used to depict reliability. A competitive benchmark of expected lifetime is a useful guide. However, customers' willingness to put up with failures is the ultimate standard for acceptability.

There are many reasons why failure and reliability, as definitions of quality, play an extremely important role. Some types of failure are life threatening. This consideration plays an important part in the costs of protection through insurance and litigation. Every effort must be made to ensure safety and to document that this effort has been honest. The courts of law expect such ethical behavior. Some types of failure do not permit repair. The definition and specification of quality should also be concerned with the ease of maintenance and the cost of replacement parts. These factors affect the consumer's judgment of quality.

What level of failure is acceptable in each circumstance? Crosby (1979) demanded a zero-defects goal which treats any failures as intolerable. Another point of view is to learn from failures so that no failure is ever repeated. In this regard, it may be important to note that certain systems must fail from old age and repeated use. Good quality managers protect their systems from adverse effects of such failures. Therefore, it is the system that avoids failures and with proper maintenance components that fail are removed and replaced before they degrade systems performance.

8.3.1.2 Warranty Policies

A product warranty is a guarantee by the producer to protect the customer from various forms of product failure. The specifics are spelled out in contractual fashion. Thus, it is typical to state for how long a time, and for how much use, the product is covered. The conditions of use are generally stated. One topic of common concern is whether companies are capable of offering warranty periods that are calculated to provide minimum costs because product failures start right after the covered period expires. For the most part, this assumption is bogus. Companies may wish that they could be so smart. The extent of product recalls demonstrates that coping with the delivery of superb quality is a failed ambition that costs companies far more than they would like to spend.

8.3.1.3 The Service Function—Repairability

Repairability and maintainability are quality dimensions that require a fully functional service capability. Speed of service is an important auxiliary attribute. A service policy is an agreement between the company and its customers that spells out how much service will be rendered, how fast that service will be provided, what service steps will be taken, and what charges will be borne by the customer (warranty contract).

Nikon has a service policy of repairing the entire camera and not just the parts that are responsible for the immediate cause of failure. Their service policy includes furnishing an estimate by mail or phone before beginning work. Service policies are taken very seriously by customers who require service. Organizations distinguish themselves from one another by the care that they show and the fairness of their service policy.

8.3.1.4 Functional Human Factors

Quality management must focus on the importance of human factors such as safety, security, comfort, and convenience. The human factors area (also as we have previously pointed out... called ergonomics) relates equally well to office or factory conditions. It concerns dangers from products in use and services provided such as plane trips and taxi rides. Many human factor qualities are overlooked until they exceed reasonable limits and are then rejected. For example, the comfort of a chair in a restaurant might be considered acceptable until it passes a threshold, and then it becomes a factor for evaluation of the restaurant. Using market segmentation according to body types, only some customers may find the restaurant chair uncomfortable.

Many safety factors cannot be seen. Starting in 1996, Firestone ATX tires got a reputation for tread separation causing SUV rollovers. In 2006, Bridgestone Firestone Corp. was still trying to reach the 5% of original tire owners for replacement. Word of mouth in such situations has a large negative effect. Proper P/OM heads off the existence of such issues. Food safety is another invisible factor. Customers rely on food companies to take proper measures to ensure that salmonella is not present. Contaminant control is a P/OM responsibility. Why are there so many *warnings* and *recalls*? The answer is the P/OM is not given the authority to monitor and control at all stages of product design, development, and production. Management saves money and time on a short-term basis only when there is no one to guard the system (i.e., negligence overrides the production control system).

The health quality aspects of foods related to processing and ingredients such as trans-fatty acids and partially hydrogenated oils is a quality issue. Nutritionists decry the use of too much fat, too many calories, too much sodium, etc. P/OM makes the product that uses these ingredients. However, the recipes are part of the product design, which relates to management strategy, marketing planning, and market research information. Because the underlying properties of food are invisible to the consumer, labeling has taken on importance that is proportional to the consumer's ability to use that kind of knowledge. Problems of many kinds can be traced to inadequate and incompetent strategic planning.

8.3.1.5 Nonfunctional Quality Factors—Aesthetics and Timing

Nonfunctional qualities play a major role in the consumer's judgment of quality. Appearance and style are intangibles, and customer satisfaction with aesthetics is costly to measure. There can be high variability in what constitutes preferred designs, and there are ambiguous design criteria for what works. Because appearance, style, and other non-functional qualities are intangibles, expert opinion and market research are the only ways to measure satisfaction. Nevertheless, these dimensions are as important to the definition of product quality as any that are found in the functional categories of quality.

Market research often attempts to determine the image that consumers conjure up of themselves to justify the purchase of items. Self-image is significant for high-ticket items such as Rolls Royces or Mercedes. Self-image applies to a broad range of acquisitions such as health club memberships, art purchases, top-of-the-line cameras, maid service, and trips into space on Russian satellites.

How consumers interpret intangible qualities involves sociological and psychological dimensions of quality. Thinking along these lines is not easily associated with P/OM, which strives to meet the standards that are specified. That is why industrial designers play a major role in product design decisions. There is broad systems responsibility to achieve effective nonfunctional attributes for a product. The Apple Industrial Design Group (IDG) was established in April 1977 by Steve Jobs as an integral part of the company. There is an interesting history about how Apple's "uniform design language" came into existence. However, designs are created, P/OM is ultimately charged with making exactly what designers fashion.

Timeliness is another elusive nonfunctional quality dimension. It is most evident in the fashions and styles of clothing, athletic footwear, iTune downloads, and the latest fad of youthful school-age market segments. The extent to which styles go in and out of fashion is never overlooked by those who are in the seller's chair. If vibrant laptop colors have real lasting power, the inventory of drab computers will have to be sold at a loss. Otherwise, there will be a glut of colored laptop cases. Managing inventories of home furnishings products requires an understanding of the dynamics of style shifts. For example, the furniture suit business (e.g., Thomasville) requires being on top of inventory shifts of types and colors of wood.

8.3.1.6 The Variety Dimension

Marketing can assist strategic planners in their evaluation of the importance of variety of choice. Variety is defined as the number of product alternatives (in a given product class) that are made available to the customer. Variety can include different quality dimensions at different prices. The key to defining variety is that customers see the various choices as substitutable alternatives.

The number of flavors that Jello offers is an illustration of variety as a quality. Does the customer want to buy green or red Jello? If green, will it be lime- or kiwi-flavored? In Brazil, it might be avocado-flavored. Having set down what the customer seems to want, what does P/OM have to do to switch the production line from green to red? How much does that cost? Will it be less expensive to switch from red to green? Answers to such questions are based on operational considerations. For example, it is easier to switch from white to black paint than the reverse.

Variety often increases a brand's market share because people like to switch flavors from time to time. They might switch away from their favored brand to get a different flavor occasionally. Similarly, some people like white cars and other like red ones. A dealer with only one color car is likely to sell fewer cars. Variety of the product line is dependent upon process capabilities to shift inexpensively from one color, flavor, or size to another. Can different prices be charged for choices among varieties? If it costs a lot to add another variant, customers must be willing to pay for the additional choice, and that is a matter of price elasticity. For example, to get decaffeinated coffee requires an extra step since the coffee beans contain caffeine naturally.

Customized product is the apex of variety. Suits made to fit a special customer are one of a kind. Suits off-the-rack and then altered to fit are still customized but at a lower level. Industrial consumers can also choose between altered to fit or tailor-made equipment. The highest variety levels fall within that range of customization. P/OM produces the level of variety determined and set by marketing strategy. Because variety entails additional production costs, marketing has to factor these expenses into the equation. P/OM is in charge of the knowledge base in the business unit that can inform marketing about the costs of variety. Together, marketing and P/OM examine the trade-offs between higher costs and greater sales. At the same time, variety develops loyal customers in special market niches. Some customers prefer regular cola. Others prefer diet drinks. Some like vanilla or cherry added to the drink. Others like more caffeine. Some choose bottles and others choose cans. Choice is a quality that is a competitive factor. It must be understood by marketing and P/OM alike.

The above factors that define the quality of a product or service lead to the determination of the cost of the product or service as discussed below.

8.4 The Costs of Quality

A good approach to understanding quality is to analyze the costs associated with achieving, or failing to achieve it. There are three basic costs of quality. These include the costs of prevention, appraisal, and failure. The nature of these costs is discussed below.

8.4.1 The Cost of Prevention

Prevention involves the use of conscious strategies to reduce the production of defective product, which by definition does not conform to agreed-upon quality standards. The entire system must be designed, coordinated, and controlled to prevent defectives. This includes the materials and equipment used, appropriate skills, and the correct process to deliver product conforming to standards. Generally by spending more, the percent of defectives can be reduced. Figure 8.2 shows this kind of relationship, although the real shape of the curve would have to be determined for a specific situation.

If some mistake causes defectives to occur, and the problem is noticed and corrected, then it is fair to say that quality improvement was free. Defectives have been prevented from occurring by remedying a wrong. In service organizations, as in manufacturing, there are many examples of cost-free quality improvements. Philip Crosby wrote a book about quality being free (see Crosby, 1979) and established a quality school for executives in Winter Park, FL that emphasized this focus as part of its curriculum. The cost of prevention goes up when quality standards are raised meaning a lower level of defectives. It is a reasonable policy to raise the

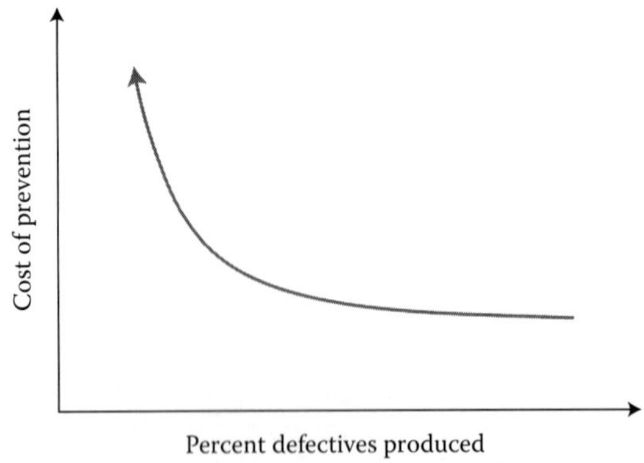

Figure 8.2 Percent defectives versus cost of prevention.

quality standards whenever the prior standards are being consistently met. The costs involved in achieving these ever more stringent specifications will increase, gradually at first, and then markedly as the technological limitations of the materials and the process are approached. This is shown in Figure 8.3.

In manufacturing, the tolerance limits define the range of acceptable product. As an example, the lead that fits into a mechanical pencil might be specified as 0.5 mm. This describes the diameter of the lead. No one supposes that each piece of lead that comes in a container marked 0.5 mm will be exactly 0.5 mm. The tolerance limit sets the standard. For example, it could be specified that each

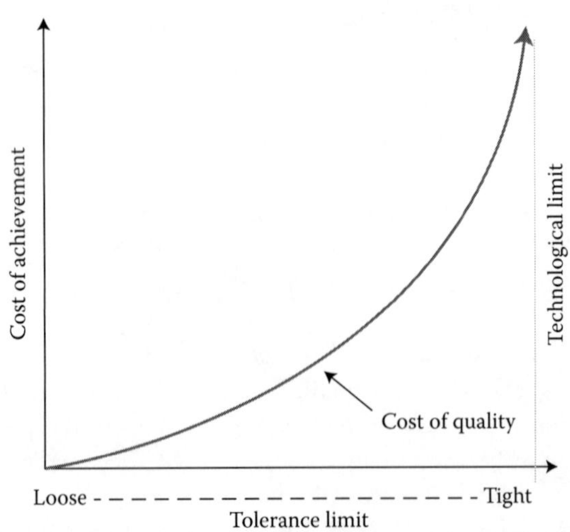

Figure 8.3 Tolerance limits versus the cost of achievement.

piece of lead must fall within the range of 0.5 mm plus or minus one twentieth of a millimeter (0.5 ± 0.05). This means that the lead diameter could be as wide as 0.55 mm and as thin as 0.45 mm. In comparison, if the stated tolerance limit allowed a range of 0.51–0.49, that range would be much tighter and more costly to achieve.

Figure 8.3 illustrates the exponential increase in costs as the accepted tolerance range is tightened and approaches the technological limits of the process. When the tolerance range becomes tighter (i.e., 0.005 is tighter than 0.05), the cost of getting all of the units produced to fall within that range becomes greater. When designers narrow the defined acceptable range, then existing equipment may not be able to do the job. A greater percentage of output will not conform to the new standard. Everything that falls outside the tolerance range limit is called defective. These defectives must either be scrapped or reworked. Both are considered part of the cost of failure.

8.4.2 The Cost of Appraisal (Inspection)

When product is examined to see whether it conforms to the agreed-upon standards, it is undergoing inspection and appraisal. Both terms are used interchangeably and have to do with the evaluation of whether or not the product conforms to the standards. It should, however, be understood that inspection and quality control (QC) are not synonymous. Inspection is one of the steps in quality assurance. Product that is not judged to conform because it fails to fall within the tolerance limits is sorted out. There must be a policy regarding what to do with product that does not conform, for each way that it may not meet specifications. Some types of defectives have to be scrapped; others can be reworked and sold for a discount.

The usual way to sort out the items that do not fall within the tolerance limits is to inspect all of the items. However, it is possible to use *acceptance sampling* (AS) methods, which, as the name indicates, consist of inspecting a sample of the production lot (see Section 8.11 on AS for details). If the sample fails to pass, then the entire lot is inspected and detailed. Detailing means removing the defectives so that each item in the lot conforms to the specifications. Therefore, when detailing, the inspector separates the bad from the good.

Figure 8.4 shows the three basic costs of quality. One is the cost of prevention of defects (previously seen in Figure 8.2). Second is the cost of inspection, which increases as the percent defective in the production lots increases. The reasons for the cost increase are that more inspectors are needed and an increased amount of detailing is necessary. If sampling is used, more samples of greater size will be taken in the same time period. All other things being equal, the increase in inspection costs would be almost linear, and not too steep, as a function of increasing percent defectives. The inspection cost line is shown in Figure 8.4.

Figure 8.4 puts together Figures 8.2 and 8.3. It also shows the cost of failure, as discussed below, which tends to increase exponentially as the percent defectives

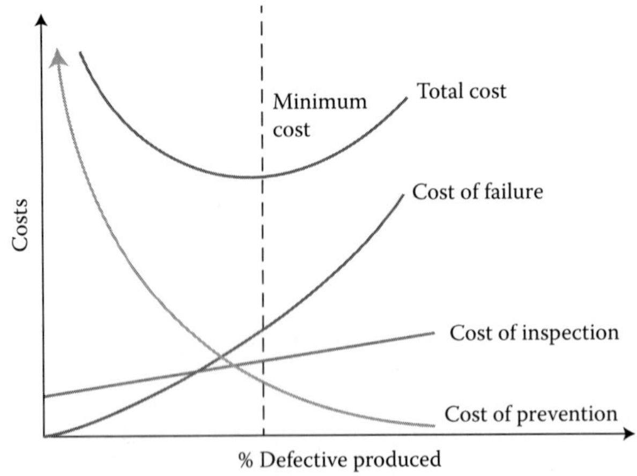

Figure 8.4 Percentage defectives produced versus costs (total, failure, inspection, and prevention).

produced (and not caught by inspection) increases. The fourth curve in Figure 8.4 is the total cost of quality, which is the sum of the other three curves.

8.4.3 The Cost of Failure

In Figure 8.4, the cost of failure rises linearly, at first, and then accelerates as the percentage of defectives increases. This might reflect a replacement cost for failed items that are under warranty. Then, as the percent of defectives continues to rise, product failure costs can be far more severe. Word of mouth among customers causes customer defections and so the cost of lost business escalates. The curve starts to move up geometrically. Severe costs of failures occur when customers begin to defect to competitors in large numbers as a result of serious product failures.

This lost revenue stream, often called the lost *lifetime value* (LTV) of customers, has to be taken into account as a significant cost of failure. An estimate of the average LTV of a customer is an important guide for deciding how much to spend to prevent failures that damage customer loyalty. LTV can be a large amount of revenue. For example, it can be thousands of dollars lost if a regular pizza customer of a particular fast-food take-out service defects to another one because of dissatisfaction. It can be hundreds of thousands of dollars if a loyal luxury car customer defects to another brand. Lost revenue streams are markedly significant when a business traveler permanently shifts hotel chains.

Further, serious failures could involve liability of very large sums of money. The expense of litigation in court trials, as well as the accompanying bad publicity, has costs that are difficult to estimate. Additional costs of failures are related to product callbacks, which require rework involving labor and material costs. Product callbacks often carry other penalties beyond the cost of repairing or replacing failed product. These include possible legal damage claims, bad publicity, and the loss of customers. Curves that chart the cost of failure are likely to have steeply ascending, exponential, or geometric shapes.

8.4.4 The Total Cost of Quality

The three kinds of costs in Figure 8.4 permit derivation of a total cost curve. This curve is U-shaped. Total costs are minimized at some level of percent defectives. However, there can be disagreement with this proposition. Situations exist in which, for small increases in prevention costs, percent defectives will be dramatically reduced. Also, there are conditions when the cost of failure does not rise exponentially because the impact of failure is trivial (the pencil point broke) or because inspection catches 100% of the defectives. With effective inspection, the percent defectives produced are substantially higher than the percent defectives shipped and there is a cost of scrapping the product.

On the other hand, if the cost of failure rises very fast, then the indicated minimum cost point is pushed towards the zero level of percent defectives. The total cost of quality increases geometrically as the percentage defectives rises. Six-Sigma methodology assumes that this situation of serious consequences for failure often exists, for example, when the airplane control system does not recognize that the course setting is heading into a mountain (google American Airlines flight 965). The curves for the costs of quality are logical in certain circumstances; they are conjectural and plain wrong in others. However, modeling quality costs remains an important effort and worthwhile endeavor for P/OM. The extra effort to ascertain the correctness of the cost model is justified when prevention is viable and/or failure involves an extreme penalty.

8.5 QC Methodology

The body of knowledge that relates to quality attainment starts with the detection of problems. This is followed by diagnosis based on the analysis of the causes of the problems. Next comes the prescription of corrective actions to treat the diagnosed causes. This is followed up through observations and evaluations to see how well the treatment works. The entire set of steps is called QC.

Quality originates with a thorough understanding of the process. P/OM is involved in getting everyone in the company motivated to know all about the processes being used and changes under consideration. Process QC methods are used to analyze and improve process quality. There are seven time-honored and well-established methods. These are (1) data check sheets, (2) bar charts, (3) histograms, (4) Pareto analysis, (5) cause-and-effect charts, (6) statistical QC (SQC) charts, and (7) run charts. The time period from which these methods are derived goes back to one of the greatest quality pioneers, Shewhart (1931). The first five methods are discussed in this section.

SQC charts are very powerful quantitative tools and are widely used. The following four major charts are discussed in this chapter: x-bar Charts, R-Charts, p-Charts and c-Charts. The SQC methodology also includes Acceptance Sampling (AS). We have devoted one section to each of these charts and to AS because of the importance of SQC to organizational success. A discussion of run charts is included in the section where we explain the construction of control charts.

8.5.1 Data Check Sheets (DCSs)

Data check sheets (DCSs) primarily organize the data. They are ledgers to count defectives by types. They can be spreadsheets, which arrange data in a matrix form when pairs of variables are being tracked. These are used for recording and keeping track of data regarding the frequency of events that are considered to be essential for some critical aspect of quality. For a particular product, there may be several data points required to keep track of different qualities that are being measured. It is useful to keep these in the same time frame and general format so that correlations might be developed between simultaneous events for the different qualities.

A data check sheet, such as shown in Figure 8.5, could record the frequency of power failures in a town on Long Island in New York State. The data check sheet shows when and how often the power failures occur. The DCS is a collection of information that has been organized in various ways, one of which is usually chronologically. The information about when power failures have occurred can be used in various ways. First, it provides a record of how frequently power failures arise. Second, it shows where the failures occur. Third, it can reveal how quickly power is restored. Fourth, it can show how long each type of failure (A–F) has to wait before receiving attention, usually because work crews are busy on other power failures. After the data check sheet is completed, problem identification follows. Then causal analysis by type of failure and location can begin. Organized collection of data is essential for good process management.

8.5.2 Bar Charts

Bar charts represent data graphically. Often data are converted from check sheets to bar charts and histograms. Figure 8.6 shows a bar chart for the recorded number of power failures that were listed on the data check sheet (Figure 8.5). The six types of failures listed as A through F on the data check sheet are represented with separate bars in Figure 8.6. The number of incidents is shown by the y-axis. Bar graphs compare the number of single-event types (e.g., types of complaints). A separate bar is used for each data set. The types should be chosen to be operationally useful, that is, something can be done to reduce the number of incidents.

Date	*A*	*B*	*C*	*D*	*E*	*F*	*Comments*	*Location*
12/6/12	X						Crew delay	P21
1/02/13			X				6 hr. service	P3
2/15/13				X			3 hr. service	P22
3/07/13				X			Snowstorm	P21
4/29/13			X				3 hr. service	P3
5/15/13						X	Flooding	R40
6/07/13				X			Crew delay	P5
7/29/13		X					R40/no service	P21-R40
8/21/13			X				Road delays	B2
9/30/13			X				Crew delay	R40
11/02/13					X		Partial crew	P3
12/14/13		X					Flat tire	P5
Total	1	2	4	3	1	1		

Figure 8.5 Data check sheet with types of failures A through F.

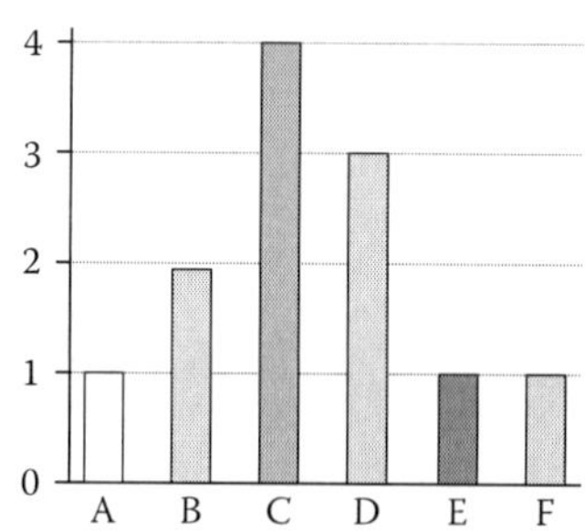

Figure 8.6 Bar chart for the type of failure.

8.5.3 Histograms

Histograms are frequency distributions. When data can be put into useful categories, the relative frequencies of occurrence of these categories can be informative. For example, in Figure 8.7, the *X*-axis describes the number of paint defects ($i = 0, 1, 2, \ldots, 8$) that have been found per hour during an inspection of autos coming off the production line. The *Y*-axis records the observed frequency for each number of paint

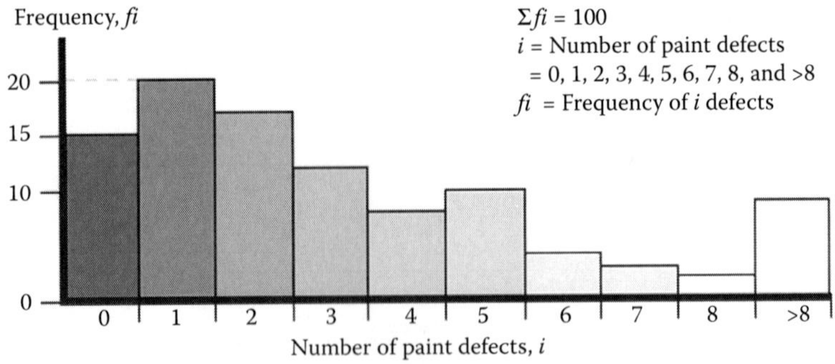

Figure 8.7 Histogram of the number of paint defects per hour.

defects. The total number of observations is 100. A review of this histogram provides important information. The relative frequency for the number of times that no paint defects are found is 15 out of 100 (15%). This is quite low and needs an explanation. In many auto companies, the standards for paint and finish are so high that few cars are found faultless. The definition of a defect is often set by such stringent standards that a customer would not be able to detect the imperfection. Figure 8.7 shows that one defect is found 20% of the time. More than one defect occurs 65% of the time. There is a notable decrease in the number of paint defects, yet more than eight paint defects occur nine times. Even with rigorous standards, this chart should not please the production and operations Manager in charge of quality.

8.5.4 Pareto Analysis

Pareto analysis seeks to identify the most frequently occurring categories; for example, what defect is the most frequent cause of product rejections, next to most, etc. This facilitates identification of the major problems associated with process qualities. It rank orders the problems so that the most important causes can be addressed. The procedure to get Pareto-type information can take many forms. First, it helps us to develop a comprehensive list of all the problems that can be expected to occur. To do this, many people should be consulted both inside and outside the company. Surveys, telephone interviews, and focus groups can be used to find out what problems people say they have experienced.

Complaints are a valuable source of information. Complaints are so vital to P/OM managers that they should encourage marketing to keep Pareto data on complaints. They should monitor changes in complaints both by type and frequency. Clusters of complaints at certain times provide important diagnostic information for process managers. Problems mentioned in the text range from power failures, to scratches or bubbles on an auto's finish, to complaints about restaurant service.

Some problems arise during inspection of the product while it is on or leaving the production line. These are internally measurable problems. Other problems

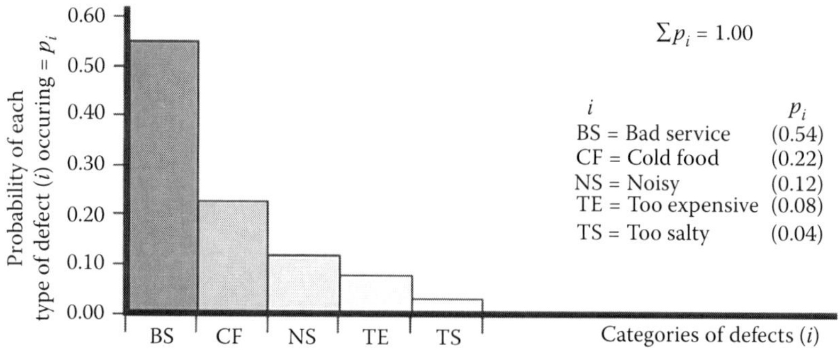

Figure 8.8 Pareto chart of complaints in a restaurant.

must wait for customer experiences. They are learned about through complaints from customers who write or call toll-free numbers set up by the company to provide information and receive complaints. Surveys also pick up complaints or they become apparent when customers seek redress within the warranty period. Retailers, wholesalers, and distributors know a lot and can communicate what they hear from their customers.

Figure 8.8 shows how a Pareto chart can be used to analyze a restaurant's most frequent quality problems. "Bad service" is the number one complaint. Among all five complaints, it occurs 54% of the time. "Cold food" is the second most frequent criticism with 22% of the objections. The third grievance is about ambiance—"Noisy" (12%). The fourth criticism is that the restaurant is "Too expensive" (8%). Excessively salty food is the fifth complaint (4%). The rank order of the hierarchy of defect causes is evident in Figure 8.8. The probabilities reflect a high degree of skew (exaggeration) for the top two sources of complaints. This kind of skew is typical of Pareto frequencies that are found in actual situations. Here, the first two causes of defects account for 76% of the complaints which is fairly typical.

Often, 20% of the total number of problems and complaints occur 80% of the time. This is called the 20:80 rule. Some things occur far more frequently than others. In the restaurant case, 2 out of 5 quality problems account for 76% of the complaints.

What is important about these numbers is the need to be sensitive to the real issues of "what counts." If the most frequent complaint is trivial (such as "too popular"), then, P/OM moves on to complaints that matter and where something can be done to improve the situation. Pareto charts do not separate the critical few from the unimportant many.

8.5.5 Cause and Effect Charts (Ishikawa aka the Fishbone)

Cause and effect charts organize and depict the results of analyses concerning the determination of the causes of quality problems. They show at a glance what factors affect quality and thereby what may be the causes of quality problems. One of these cause and effect charts is the fishbone (or Ishikawa) chart which is based on listing variables that are known to affect quality. These charts are organized with the quality goal shown at the right. Scatter diagrams, studied in the chapter on forecasting, also help in determining causality. To determine the cause of a quality problem, start by listing everything in the process that might be responsible for the problem. A reasonable approach is to determine the factors that cause good (or bad) quality. These potential causes are grouped into categories and subcategories, as is done for the following example, which lists many of the factors responsible for the quality of a cup of coffee. The processes used to grow coffee beans, harvest, roast, store, and ship the coffee would be included in this cause-and-effect diagram if there was control over these factors. Here, the coffee-making process begins with the goal of making a good cup of coffee. All participating processes are part of the causal system.

8.5.5.1 Example: A Good Cup of Coffee

The quality of a cup of coffee is a big business with growth opportunities, challenges, and concerns. Chains of coffee shops have opened thousands of stores all over the United States, Europe, and Asia. Starbuck's (http://www.starbucks.com) and Dunkin' Donuts (http://www.dunkindonuts.com) have enormous numbers of stores all over the world. With billions of dollars at stake, companies find it worthwhile to scrutinize various business models each with its own concept of how consumers perceive quality. Table 8.1 captures many of the multidimensional factors that determine the quality of a cup of coffee. Specifically, Table 8.1 presents a list of causal factors that should be considered by all participants in the system that affects the quality of a cup of coffee. After making the list, these variables can be charted using the fishbone construction method developed by Ishikawa (1987).

There are nine main variables and different numbers of subcategory variables. The latter are shown in Figure 8.9 as fins radiating from the nine lines of the main variables, connected to what looks like the spine of a fish. The subcategories are listed in Table 8.1 under the nine main variables.

The fishbone diagram is conducive to systems-wide discussion, involving everyone, concerning the completeness of the diagram. Have all the important factors been listed? When the fishbone diagram is considered complete, there follows the determination of what variations should be tested. Market research is used to test consumer reactions to the new product and process alternatives.

Returning to the power failure situation, the household consumer of electricity is unlikely to know which cause is responsible for any failure. The electric utility, on the other hand, can do the necessary detective work to determine what has caused each failure. These include but are not limited to: lines are knocked down by cars and trucks as well as by wind storms; transformers fail more often under

Table 8.1 Variables for the Quality of a Cup of Coffee

Coffee Beans–Type Purchased
Source of beans (locations in Sumatra, Colombia, Jamaica, Costa Rica, Hawaii, etc.) • Grade of beans • Size and age of beans • How stored and packed (how long stored before grinding)
Roasting Process • Kind of roaster • Temperature used • Length of roasting
Grinding Process • Type of grinder • Condition of grinder • Speed and feed used • Fineness of grind setting • Length of grinding • Quantity of beans ground • Stored for hours long before use
Coffee Maker • Drip type or other; exact specs for coffee makers including k-cup Keurig and Verisimo • Size and condition of machine (how many prior uses) • Method of cleaning (vinegar, hot water, detergents) • How often cleaned • Number of pounds of coffee used
Water • Amount of water used • Type of water used (chemical composition) • Storage history before use • Temperature water is heated to

continued

Table 8.1 (continued) Variables for the Quality of a Cup of Coffee

Filter
• Type of filter used (material: paper (and type), gold mesh, other metals)
• Size of filter
• How often used
• How often replaced
Coffee Server
• Type of container
• Size of container
• How often cleaned
• How long is coffee stored?
Coffee Cup
• Type of cup
• Size of cup
• Cleaned how (temperature of water, detergent)
• Cleaned how often
Spoon in Cup–Sweetener–Milk or Cream
• Type of spoon (material/how often cleaned and how)
• Type of sweetener used
• Type of milk: 1%, 2%, half and half, skim, light cream, heavy cream, etc.

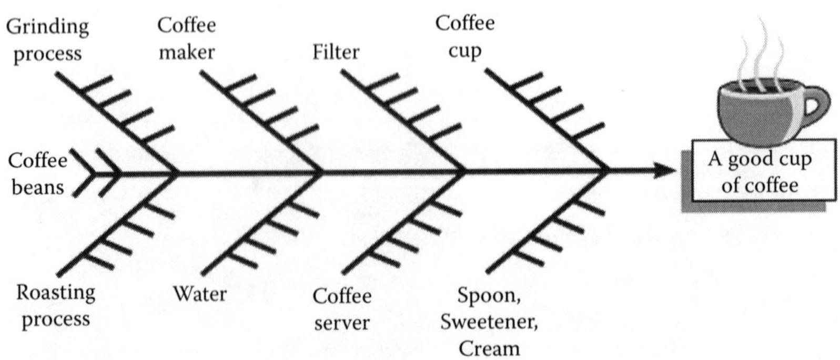

Figure 8.9 Ishikawa (fishbone) diagram—making a good cup of coffee.

peak demands; and fires result from overloaded power generation equipment correlated with summer heat spells. Time of day, day of the week, date of the year may all be useful data for patterns to be found. The Pareto charts can provide a quick approach to identify the most frequent types of power failures, probably by region and extent. Ishikawa diagrams now help with the search for causes of these problems. To sum up, Ishikawa diagrams enumerate and organize the potential causes of various kinds of quality failures.

Knowing the most frequently occurring types of failures might lead to solutions that could reduce their frequency. Overhead wires in certain locations could be buried at a cost that might be much lower than the expensive remedial actions taken after storm-caused failures. Power-grid sharing could be initiated before overloads occur. Strong correlations provide information that connects types of failures with the times and places that they occur, which may facilitate problem solving (and the reduction of failures).

8.6 Control Charts for Statistical Process Control

In this section, we are going to discuss various methods for process control to make sure that the product of the right quality is being produced; and service of the right quality is being provided. This is accomplished through constructing statistical process control (SPC) charts. These charts are used for detecting quality problems, much as early warning detection systems. SPC charts also can help discern the causes of process quality problems. The SPC charts are extremely powerful for monitoring conditions of the process and for stabilizing processes. SPC is part of the broader QC methodology based on statistical concepts. This is called SQC. SQC consists of SPC and AS. We discuss AS later in this chapter. The SPC charts are generally called QC charts also. We will use the terms SPC charts and QC charts interchangeably.

Control charts are constructed by inspecting the process output. The mean of the output and standard deviation of the output are calculated. Inspection can be done by variables or attributes as discussed below.

8.6.1 Inspection by Variables versus Attributes

Output from a process (manufacturing or service) is measured with respect to qualities that are considered to be important. The output can be measured in two different ways: by variables and by attributes. In the case of measurement by variables the numbered measurements are continuous, as with a ruler for measuring inches or with a scale for measuring pounds. The variables can have dimensions of weight, temperature, area, strength, electrical resistance, loyalty, satisfaction, defects, typing errors, complaints, accidents, readership, and TV viewer ratings. The fineness of the units of measure is related to the quality specifications for the outputs. The process control charts used for variables include the charts for the output mean (*x*-bar charts) and the output range (*R*-charts).

In the case of the measurement by attributes, the output units are classified as accepted or rejected, usually, by some form of inspection device that only distinguishes between a "go" or a "no go" situation. Rejects result from measurements falling out of the range of acceptable values that are pre-set by design. Rejects are defective product units. Under the attribute system of classification, process quality is judged by the number of defective units discovered by the pass–fail test. It is charted as the percentage of defects discovered in each sample and is called a *p*-chart. Also under the attribute system of measurement is the *c*-chart, which counts the number of nonconformities of a part. For example, in the eye of the specialist, a particular diamond can be categorized as having various imperfections, which are counted. Distribution of the number of imperfections (*c*) is the basis of the *c*-chart.

Monitoring by variables is more expensive than by attributes. The inspection by variables requires reading a scale and recording many numbers correctly (for example, successive measurements of the weight of cans of peanuts that must be at least 16 ounces). Measurement error must be avoided. A reading of 15.99 is unacceptable whereas 16.01 is acceptable. Numbers have to be recorded arithmetically Their significance must be understood to allow proper actions to be taken.

In comparison, the attributes used by the *p*-chart (from a go/no-go gauge) are simpler. Go/no-go is an unequivocal pass–fail exam. Many times it can be performed by a visual check—good or bad? Attribute measurement methods build in the acceptable and unacceptable ranges. Monitoring by attributes can often be automated. The engineering of the inspection process is an investment in simplifying the task. When error prevention is critical, assessment by attributes often is safer. We should note that there is more information inherent in analysis of variables than in analysis of attributes which means that *x*-bar charts have greater sensitivity than *p*-charts to critical changes that signal fundamental system instabilities.

Inspection, whether by variables or attributes, leads to measurement of the quality of the process output. However, variability exists in all systems. The causes of process variation are discussed below.

8.6.2 Causes of Process Variation

Two seemingly identical manufactured parts made by the same process have their own signatures, like fingerprints, when measured to whatever degree of fineness is required. Systems with less variability may differ with respect to measurements in thousandths of an inch whereas systems with more variability may be noticeably different in terms of hundredths of an inch. The causes of process variation can be classified in two categories: (a) chance causes and (b) assignable causes. Both of these causes are discussed below.

8.6.2.1 Chance Causes

Chance causes result from inherent (intrinsic and innate) properties of the system. It is the variability of a stable system that is functioning properly. That variability is said to be caused by chance causes (also called simple causes). These causes cannot be removed from the system, which is why they are called inherent systems causes. They are also called chance causes because they are predictable and stable in statistical terms, as are the number of heads from an honest coin throw. A process that experiences only chance causes, no matter what its level of variability, is called a stable process. The variability that it experiences is called random variation.

8.6.2.2 Assignable Causes

The second kind of variability has traceable and removable causes, which are called assignable or special causes. Unlike chance causes, these causes are not background noise. They produce a distinct trademark, which can be identified and traced to its origins. They are called assignable because these causes can be designated as to type and source, and removed.

The process manager identifies the sources of quality disturbances—a tool that has shifted, a gear tooth that is chipped, an operator who is missing the mark, an ingredient that is too acidic—and does what is required to correct the situation. Thus, tool, die, and gear wear is correctable.

Broken conveyor drives can be replaced. Vibrating loose parts can be tightened. Human errors in machine setups can be remedied. These are examples of assignable causes that can be detected by the methods of SPC. An alternative name that might be preferred to assignable causes is identifiable systems causes.

8.6.3 What Are QC Charts?

QC charts are the means for plotting the quality of the process output measured through inspection whether by variables or attributes. QC charts can spot identifiable causes of system variation. Once identified, knowledge of the process leads to the cause, which is then removed so that the process can return to its basic, inherent causes of variability. To construct a control chart we must find the mean and the standard deviation of the process output. We will use the symbol μ for the mean and σ for the standard deviation in all our discussions in this chapter. Suppose we are producing steel rods. The diameter of the rods is the variable of interest and defines the quality of the output. To construct the control chart, we calculate the mean (average) diameter and the standard deviation of the process output. The methods to find mean and standard deviation depend on the type of chart and will be discussed later.

In a control chart, the *Y*-axis represents the observed process output measured at successive intervals (or for successive samples) and the *X*-axis represents the time interval (or successive sample numbers). A control chart consists of a center line

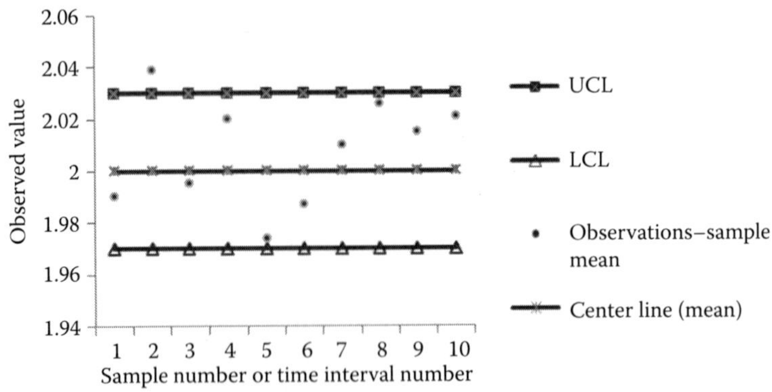

Figure 8.10 Example of statistical process control chart.

representing the mean of the process output and a line above the center line called the upper control limit (UCL) and a line below the center line called the lower control limit (LCL). The measured values of the quality variable are plotted as individual points. An example of a control chart is shown in Figure 8.10. In this chart 10 observations are plotted. Point 2 lies outside the control limits—above the UCL. This is an indication that the process was out of control when that measurement was taken. It points to the need for causal analysis. All other points are within the control limits and their variation is primarily due to the inherent process variability. It may also be noted that if the spread between UCL and LCL is made smaller some other points may go outside the control limits. Point 5 is a prime candidate to go outside the control limits because it is very close to the LCL. Point 8 is another prime candidate on the upper limit.

The other phenomenon to be noted is that the last four points are all above the center line which warrants some investigation. The process may be heading out of control. This phenomenon is discussed below under the title analysis of statistical runs.

8.6.3.1 Analysis of Statistical Runs

Run charts can help spot the occurrence of an impending problem. Thus, run analysis is another early warning detection system. It can be of great help in diagnosing the causes of problems.

A run of numbers on a control chart is successive values that all fall either above or below the mean line. Consider the control chart given in Figure 8.10. The first six points are well distributed around the center line. However, the last four points are all above the mean line. What is the probability of four heads (or tails) in a row? It is $(1/2)^4 = 1/16 = 0.0625$ which is relatively small. Certainly, a heads-up is warranted.

The process is currently within control limits but does not look normal and may go out of control if this trend continues. Look for the problem. Is there something changing in the process or is it a statistical fluke? Sample the process again, as soon as possible. Make the interval between samples smaller than usual until the quandary is resolved. Sometimes a separate run chart is kept to call attention to points that are falling on only one side of the mean value of the chart. Separate charts are not required if the regular control charts are watched for runs. The charting decision is dependent upon the likelihood that runs will play a major role in controlling quality. A monotonic run (continuously increasing or decreasing) signals even greater urgency to consider the likelihood that an assignable cause has changed the process's behavior.

The important decision to be made in constructing the control charts is to establish the UCL and LCL. We will illustrate this with the help of an example.

8.6.3.2 Example: Control Charts for a Manufacturing Process

Suppose a manufacturing process can produce steel rods with an average diameter of 2.00″ ($\mu = 2.00$). The standard deviation (σ) representing process variation is 0.01. This information can be used to construct the control chart. It is assumed that the process output follows a normal distribution. Based on the well-known properties of the normal distribution we can say that 99.72% of the output will lie between ± 3 standard deviations; and 95.44% of the output will lie between ± 2 standard deviations. The percentages 99.72% and 95.44% are known as confidence intervals. The values 3 and 2 are standard normal variables and are represented by the symbol z. The table in Appendix B gives the values of z for various confidence intervals. Appendix B is included at the end of this textbook. The UCL and LCL are established as follows.

$$\text{UCL} = \mu + z\sigma \quad \text{and} \quad \text{LCL} = \mu - z\sigma.$$

For 99.72% confidence interval ($z = 3$), the limits are:

$$\text{UCL} = 2.00 + 3 * 0.01 = 2.03,$$

$$\text{LCL} = 2.00 - 3 * 0.01 = 1.97.$$

For 95.44% confidence interval ($z = 2$), the limits will be:

$$\text{UCL} = 2.00 + 2 * 0.01 = 2.02,$$

$$\text{LCL} = 2.00 - 2 * 0.01 = 1.98.$$

Figure 8.10 shows the control chart for this example with a confidence interval of 99.72%. If two sigma limits are used, the space between the control limits will be narrower and point 5 will fall outside the LCL (1.98).

The number of standard deviations used will control the sensitivity of the warning system. Three-sigma charts remain silent when two-sigma charts sound alarms. Alarms can signal out-of-control as well as false alarms. Understanding how to set the control limits is related to balancing the costs of false alarms against the dangers of missing real threats. One-sigma limits will sound more alarms than two-sigma ones. The judicious choice of the number of standard deviations to be used is an operations management responsibility (in conjunction with other participants) based on the assessment of the two costs stated earlier. There must be an evaluation of what happens if the detection of a real problem is delayed. The systems-wide solution is the best one to use.

The probability of a subgroup mean falling above or below three-sigma limits by chance rather than by cause is 0.0028 (= 1.00 – 0.9972), which is pretty small, but still real. For two-sigma, it would be 0.0456 (= 1.00 – 0.9544), which is more than 16 times larger than the three-sigma result. In some systems, two successive points must go out of control for action to be taken. There are fire departments that require two fire alarms before responding because of the high frequency of false alarms. When a response is made to out-of-control signals, knowledge of the process is the crucial ingredient for remedying the problem.

We will discuss four control chart examples—two charts for variables and two charts for attributes in the following sections. Control charts are used to stabilize a process and are then used to monitor the process once it has been stabilized.

8.6.4 Stable Process

For quality to be consistent, it must come from a process that is stable, that is, one with fixed parameters. This means that the process average and its standard deviation are not shifting. In the case of measurement by attributes a stable system will give a consistent proportion of defective items. The process capability studies are used to establish the limits within which a process can operate. It is important to note that in real production situations, enough data are obtained to create reliable control charts. Then, more data are collected to examine whether the process is stable. Initially a sample of observations is taken and control charts are constructed. Evidence of temporary instability often is thrown out as being associated with start-up conditions. The process is reexamined to see if it has stabilized after the start-up. The initial sample size is often 25–30 sets of data, followed by another 25–30 sets of data. Typically, a data set would consist of 2–10 successive observations from which a mean and standard deviation would be determined. A process that is operating without any assignable causes of variation is stable by definition, although its variability can be large. Stability does not refer to the degree of variability of a process. Instability reflects the invasion of assignable causes of variation, which create quality problems. Once the process is considered stable, then regular observations are made at scheduled intervals.

8.6.5 Selecting a Manufacturing Process

After the design of a product is complete, an appropriate manufacturing process is to be selected. Returning to the example of manufacturing steel rods whose desired diameter is 2″. The product design specifies the tolerances on the diameter as ± 0.05″. This means that the required output should be between 2″ ± 0.05″. In other words any rod whose diameter is between 1.95″ and 2.05″ meets the required quality level. 1.95″ is called the lower specification limit (LSL) and 2.05″ is called the upper specification limit (USL).

Now we have to select an appropriate process which can produce rods within the USL and LSL. Suppose we have two processes A and B which can produce the required steel rods. The average diameters and the standard deviations of the rods produced by the two processes could be different. However, we assume that the average diameters (μ) of the rods produced by both processes are 2″ but variations are different. Suppose the process variation, measured as standard deviation (σ) are 0.01 and 0.025, respectively, for processes A and B. Based on the discussion of the control charts that we had earlier, we can say that for a 99.72% confidence level, the UCL and LCL for process A will be:

$$\text{UCL (Process A)} = 2.00 + 3 * 0.01 = 2.03,$$

$$\text{LCL(ProcessA)} = 2.00 - 3 * 0.01 = 1.97.$$

The USL, LSL and UCL and LCL for process A are shown in Figure 8.11a.

The UCL and LCL for 99.72% confidence interval for process B will be

$$\text{UCL(Process B)} = 2.00 + 3 * 0.025 = 2.075,$$

$$\text{LCL(Process B)} = 2.00 - 3 * 0.025 = 1.925.$$

The USL, LSL and UCL and LCL for process B are shown in Figure 8.11b.

The UCL (2.03) and LCL (1.97) of process A lie within the USL (2.05) and the LSL (1.95). This means that at least 99.72% of the output will be of acceptable quality. On the other hand, the UCL (2.075) and LCL (1.925) of process B lie outside the USL (2.05) and LSL (1.95).

Therefore, some items produced by process B will fall outside the USL and LSL. By definition, these items will be considered defective. Process A is more accurate and precise as compared to process B and hence process A is likely to be more expensive. Process B is less expensive but produces more defective units. An appropriate process is selected after doing a careful analysis of the costs of the processes and the cost of rejected units which might have to be scrapped if they cannot be reworked. Note that the items falling above the USL might be re-machined whereas the item falling under the LSL will have to be scrapped. More detailed process capability studies can be conducted to find an appropriate process.

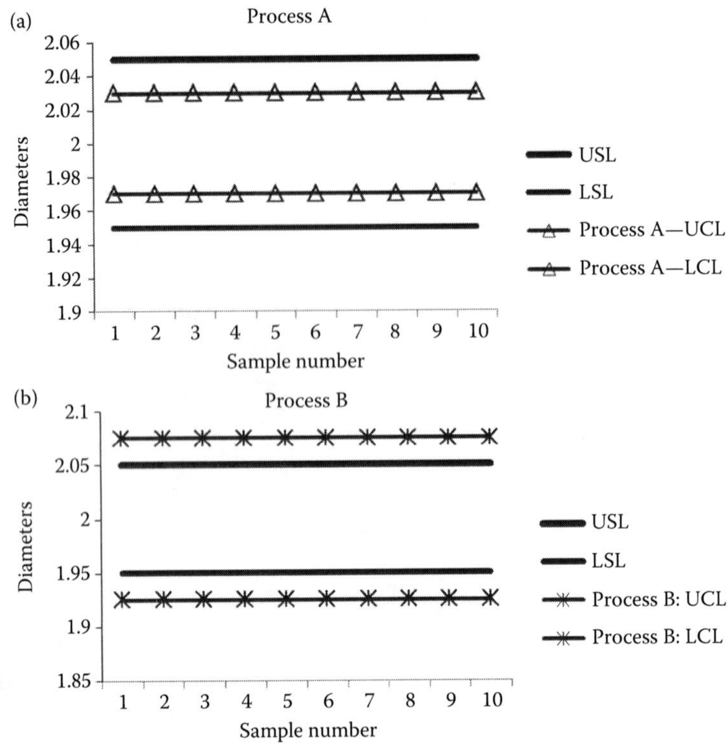

Figure 8.11 USL, LSL and UCL, LCL (a) for process A; (b) for process B.

8.7 Control Charts for Variables: *x*-Bar Charts

The *x*-bar chart uses measurement by variables. A sample of size n of the product is taken at pre-defined intervals. Each unit in the sample is measured along the appropriate scale for the quality being analyzed. The sample mean (*x*-bar; $\bar{x}$) is calculated. After that the grand mean (*x*-double bar; $\bar{\bar{x}}$) which is the mean of all the sample means is calculated. The grand mean represents the center line of the control chart. The following formulas are used to find the UCL and LCL:

$$\text{UCL} = \bar{\bar{x}} + z * \sigma,$$

$$\text{LCL} = \bar{\bar{x}} - z * \sigma,$$

where σ is the standard deviation of the distribution of the sample means and z is the standard normal variable as discussed earlier. The calculations for σ are laborious. Therefore, an alternative method that uses the range of sample observations, R, will be used to represent the variability of the process. The control limits UCL and LCL in this method are computed by using the following formulae:

$$\text{UCL}(\bar{x}) = \text{Grand mean} + A_2 * \bar{R},$$

$$\text{LCL}(\bar{x}) = \text{Grand mean} - A_2 * \bar{R},$$

where $\bar{R}$ (R-bar) is the mean of the ranges of the samples and A_2 is a factor which is dependent on the sample size n and z = three standard deviations. Values of A_2 for various values of n are given in Table 8.2. We illustrate the construction of x-bar charts with an example entitled The Belgian Chocolate Truffle Factory (BCTF).

Table 8.2 Factors for Determining Three-Sigma Control Limits Using *R*-Bar

Number of Observations in Subgroup (n)	A_2	D_3	D_4
1	n.a.	n.a.	n.a.
2	1.88	0	3.27
3	1.02	0	2.57
4	0.73	0	2.28
5	0.58	0	2.11
6	0.48	0	2
7	0.42	0.08	1.92
8	0.37	0.14	1.86
9	0.34	0.18	1.82
10	0.31	0.22	1.78
11	0.29	0.26	1.74
12	0.27	0.28	1.72
13	0.25	0.31	1.69
14	0.24	0.33	1.67
15	0.22	0.35	1.65
16	0.21	0.36	1.64
17	0.2	0.38	1.62
18	0.19	0.39	1.61
19	0.19	0.4	1.6
20	0.18	0.41	1.59

8.7.1 Example: Control Charts for a Manufacturing Process

The famous truffle bonbons made by the Belgian Chocolate Truffle Factory (BCTF) are marketed all over the world as being hand-dipped using the best ingredients. The standard weight per piece set by BCTF is 30 grams of the finest chocolate. A perfect pound box would contain 16 chocolates, each weighing 28.35 grams. Company policy, based on principles of ethical practice, is to err on the side of giving more, rather than less, by targeting 1.65 extra grams per piece. However, extra chocolate is expensive. Worse yet, if weight and size are inconsistent, the larger pieces create dissatisfaction with the truffles that meet the standard. Also, consistency of size is used by customers to judge the quality of the product. Even though the company claims customization as an advantage, complaints are received that the candies are uneven in size and weight. This is an ideal opportunity to use an *x*-bar chart to find out if the chocolate-making process is stable or erratic.

Weight is easier to measure than size and is considered to be a good surrogate for it. The company wants to test for consistency of product over a production day. Samples were taken over the course of one day of production at 10:00 A.M., 11:00 A.M., 1:00 P.M., 3:00 P.M., and 4:00 P.M. The hours chosen are just before the line is stopped for tea break and/or for cleanup. The sample subgroup size is of four consecutive days with the observations made at the specific times. This yielded 20 measures and five sample subgroup means.

For instructional purposes, this sample is about the right size. For real results, all five days of production for a typical week should be examined. Except as an example, a 1-day sample is too small. There is a rule of thumb widely used: Practitioners consider 20–25 subgroup means as appropriate for a start-up or diagnostic study.

Table 8.3 shows five sample subgroup sets of data as the columns with one for each time period. Measurements are in grams. There is also a range measure

Table 8.3 Data for Belgian Chocolate Truffle Factory (BCTF)

Subgroup	*I*	*II*	*III*	*IV*	*V*	*Sum*	*Grand Mean*
Time	10 A.M.	11 A.M.	1 P.M.	3 P.M.	4 P.M.		
Day 1	30.50	30.30	30.15	30.60	30.15		
Day 2	29.75	31.00	29.50	32.00	30.25		
Day 3	29.90	30.20	29.75	31.00	30.50		
Day 4	30.25	30.50	30.00	30.00	29.70		
Sum	120.40	122.00	119.40	123.60	120.60		
Average	30.10	30.50	29.85	30.90	30.15	151.50	30.30
Range	0.75	0.80	0.65	2.00	0.80	5.00	1.00

in this chart that will be used to find UCL and LCL. It is the difference between the largest and the smallest number in the column. The grand mean value is (151.50)/5 = 30.30, which indicates that, on average, these chocolate truffles are 1% heavier than the standard. This average value seems well suited to the ethical considerations of the firm because it is above, but not too far above, the target value of 30.00 grams. BCTF's policy is ethical but expensive. Customers get more than a pound of chocolates on a regular basis.

In Table 8.3 individual sampled values fall below the 30-grams target five times, even though in all cases they are well above 28. The lowest value sampled was 29.50. The highest value is 32. The three highest values all appear at 3:00 P.M., which also has the largest sample mean value. This information may be useful to the process manager who can search for the explanations.

We can now determine the UCL and LCL for this process. The control chart drawn in Figure 8.12 is based on plotting the sample means for each of the five time periods. The mean is the average of observations for all four days for a given time period. For example, for time period I (10.00 A.M.) the average is 30.10 = (30.50 + 29.75 + 29.90 + 30.25)/4 from Table 8.3 It is important to note that the underlying distribution is that of the sample means for the five time periods. It is necessary to estimate the variability of this distribution of the sample means in order to construct control limits around the grand mean of this process.

The UCL and the LCL, shown in Figure 8.12 could have been determined by calculating the standard deviation for the sample means given in Table 8.3. However, an easier calculation of variability uses the range, *R*, values. It is well suited for repetitive computations that control charts entail. Column values of *R* for BCTF for each

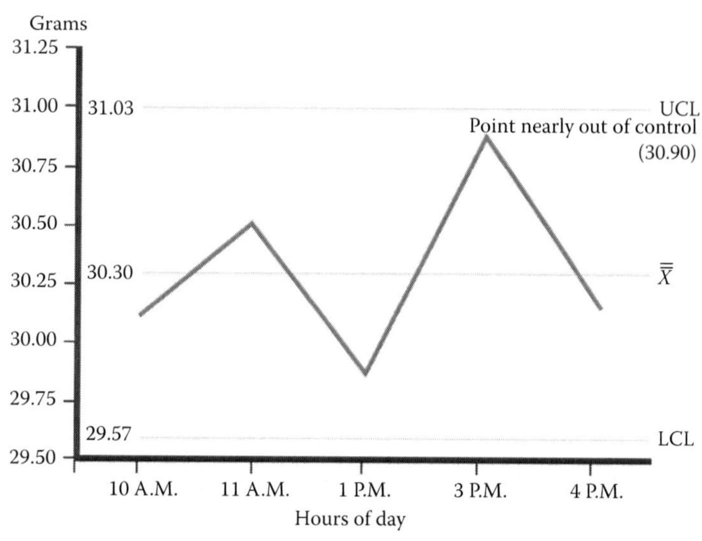

Figure 8.12 Control chart for Belgian Chocolate Truffle Factory (BCTF).

time period are given in Table 8.3. The range within each sub-group is defined as: R = Maximum value – Minimum value. For example, the largest value in the 10:00 A.M. sample subgroup is 30.50 and the smallest is 29.75, so R = 0.75. UCL and LCL will be computed using the average value of R for the subgroup samples.

The calculations of UCL and LCL for the Belgian Chocolate Truffle Factory are given by the following equations.

$$\text{UCL}(x\text{-bar}) = 30.30 + 0.73 * 1.00 = 31.03,$$

$$\text{LCL}(x\text{-bar}) = 30.30 - 0.73 * 1.00 = 29.57,$$

where 0.73 is the value of A_2 for a sample size of 4 (see Table 8.2).

8.7.2 Discussion of UCL and LCL for BCTF

Control limits are thresholds designed statistically to signal that a process is not stable. The *x*-bar chart for BCTF shows no points outside the control limits. However, note that the 3:00 P.M. sample can be identified on the chart as being nearly out of control. The 3:00 P.M. sample, when it is backed up by a lot more data, is the kind of visual warning that usually warrants attention. Perhaps, the 3:00 P.M. sample is taken during the afternoon tea break, during which the chocolate mixture becomes thicker and the molding machine tends to make heavier pieces. The cleanup comes after 3:00 P.M. Tracking and discovering such causes will lead to elimination of the problems.

Although the process is stable, it appears to have too much variability (32.00 is max–29.50 is min). It may need to be redesigned. The chocolate mold-filling machines might need replacement or rebuilding. Workers using the filling machines might need additional training. From the previous comment on variability, it becomes apparent that there is a need to study the range measure. That will be done shortly.

If two standard deviations were used the UCL and the LCL will be 30.79 and 29.81, respectively (calculations not shown here). In that case the 3:00 P.M. point would be outside the control limit. The 3:00 P.M. point was 30.90. This outlier could trigger a costly process evaluation, which might not be justified. The *x*-bar value of 29.85 for the third subgroup is uncomfortably close to the LCL value of 29.81 leading to the possibility of another costly trigger. For this process, where a safety margin has been built into the system, three-sigma seems more satisfactory than two-sigma.

8.8 Control Charts for Variables: *R*-Charts

The *x*-bar chart is accompanied by another kind of control chart for variables. Called an *R*-chart, it monitors the stability of the range. In comparison, the *x*-bar chart checks for the stationarity of the process mean, that is, it checks that the mean of the distribution does not shift about—while the *R*-chart checks that the spread of the distribution around the mean of the process stays constant. In other words, *R*-charts control for shifts in the process standard deviation.

The *x*-bar and *R*-charts are best used together because sometimes an assignable cause creates no shift in the process average, or in the average value of the standard deviation but a change occurs in the distribution of *R* values. For example, if the variability measured for each subgroup is either very large or very small, the average variability would be the same as if all subgroup *R* values were constant. Remembering that the sampling procedure is aimed at discovering the true condition of the parent population, the use of *x*-bar and *R*-charts provides a powerful combination of analytic tools.

The formulas for the control limits for three standard deviations for the *R*-chart are:

$$\text{UCL}(R\text{-bar}) = D_4 * R\text{-bar},$$

$$\text{LCL}(R\text{-bar}) = D_3 * R\text{-bar},$$

where D_4 and D_3 are given in Table 8.2.

The Belgian Chocolate Truffle Factory data from Table 8.3 can be used to illustrate the control chart procedures. The mean value of the range is 1.0. Continuing the use of three-sigma limits, the values of D_4 and D_3 are 2.28 and 0.00, respectively (see Table 8.2). The control limits for the *R*-chart are given below.

$$\text{UCL}(R\text{-bar}) = 2.28 * 1.00 = 2.28,$$

$$\text{LCL}(R\text{-bar}) = 0.0 * 1.00 = 0.00.$$

Figure 8.13 shows that the *R*-chart has no points outside the control limits. The same pattern prevails as with the *x*-bar chart. The 3:00 P.M. point is close to the upper limit. This further confirms the fact that the process is stable and is unlikely to do much better. However, the sample is small, and good chart use would throw out the 3:00 P.M. sample and take about 25 more samples before making any decision.

8.9 Control Charts for Attributes: *p*-Charts

The control chart for attributes, the *p*-chart, shows the proportion (or percent) defective in successive samples. It has UCL and LCL similar to those found in the *x*-bar and *R*-charts. The attribute used is good (accepted) or bad (rejected). Only one chart is needed for the attribute, *p*, which is the percent defective. Also, only the UCL really matters. The LCL is always better as it gets closer to zero defectives. The value of *p* is defined as follows: *p* = number rejected/number inspected.

Assume that the Belgian Chocolate Truffle Factory can buy a new molding machine that has less variability than the current one. BCTF's operations managers

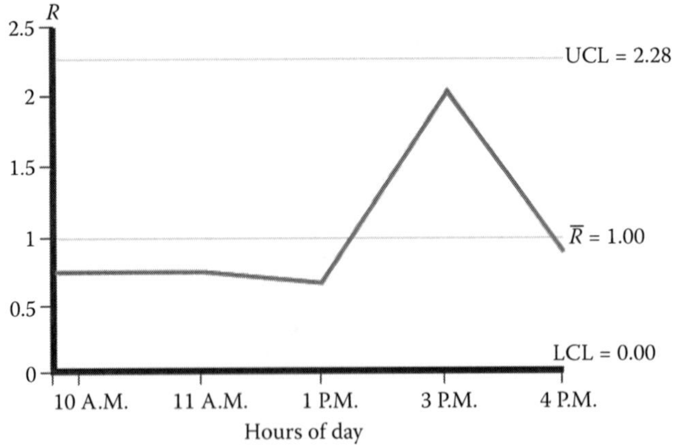

Figure 8.13 The three-sigma limits *R*-chart for Belgian Chocolate Truffle Factory (BCTF).

in cooperation with the marketing managers and others have agreed to redefine acceptable standards so that profits can be raised while increasing customer satisfaction. Also, it was decided to use a p-chart as the simplest way to demonstrate the importance of the acquisition of the new molding machine.

Let us assume that previously, a defective product was defined as a box of truffles that weighed less than a pound. What was in the box was allowed to vary, as long as the pound constraint was met. Now the idea is to control defectives on each piece's weight. The new definition of a reject is as follows: A defective occurs when weight of a chocolate <29.60 g or weight of a chocolate >30.40 g. This standard set on pieces is more stringent than the earlier one set on box weight. The new standards will always satisfy BCTF's criteria to deliver nothing less than a pound box of chocolate.

It has been suggested that the data of Table 8.3 be tested against the p-chart criteria. Rejects are marked with an R in Table 8.4.

The number of rejects (NR) is counted, and p is calculated for each subgroup, p = number of rejects/sample size. See Table 8.4. For example, for the 10:00 A.M. sample, the number of rejects is 1 and the sample size is 4. Therefore, $p = 0.25$. The values of p are 0.50, 0.25, 0.75, and 0.25 for samples II, III, IV, and V, respectively. The average number of rejects (p-bar) is $0.40 = (0.25 + 0.50 + 0.25 + 0.75 + 0.25)/5$. The value of p-bar could have been calculated by dividing the total number of rejects (8) by the total number inspected (20). The total number inspected is number of days (4) * number of samples (5).

The p-chart requires the calculation of the standard deviation σ which is given by the following formula:

$$\sigma = \sqrt{\frac{(\bar{p})(1-\bar{p})}{n}}.$$

Table 8.4 Rejects (R) Due to Excessive Weight for BCTF

Subgroup	*I*	*II*	*III*	*IV*	*V*	*Sum*	*Grand Mean*
Time	10 A.M.	11 A.M.	1 P.M.	3 P.M.	4 P.M.		
Day 1	30.5R	30.30	30.15	30.6R	30.15		
Day 2	29.75	31.00R	29.5R	32.00R	30.25		
Day 3	29.90	30.20	29.75	31.00R	30.5R		
Day 4	30.25	30.50R	30.00	30.00	29.70		
NR	1	2	1	3	1	8.00	
p	0.25	0.50	0.25	0.75	0.25		0.40

The UCL and LCL are given by the following formulas:

$$\text{UCL}(p\text{-bar}) = p\text{-bar} + z * \sigma,$$

$$\text{LCL}(p\text{-bar}) = \text{p-bar} - z * \sigma,$$

where z is the confidence interval as discussed for x-bar and R-charts. The values of z for various confidence intervals are: $z = 1.96$ for 95%, $z = 2.00$ for 95.44% and $z = 3.00$ for 99.72%. Other values can be read from the table in Appendix B.

For the BCTF example, p-bar = 0.40 (see Table 8.4). Using the formula given above, the value of $\sigma = 0.0245$.

$$\sigma = \sqrt{((0.04 * 0.06)/4)} = 0.0245.$$

The control limits for a confidence interval of 95% with $z = 1.96$ are determined as follows:

$$\text{UCL} = 0.4 + 1.96 * 0.0245 = 0.8801,$$

$$\text{LCL} = 0.4 - 1.96 * 0.245 = -0.08 = 0.$$

The LCL cannot be less than zero, so it is set to zero when the calculated value of LCL is negative.

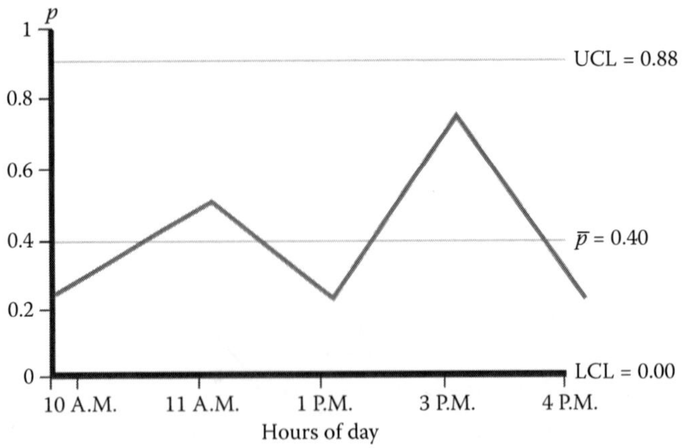

Figure 8.14 The *p*-chart for Belgian Chocolate Truffle Factory.

It may be noted that UCL and LCL are functions p, n, and σ. The control limits can step up and down according to the sample size, which is another degree of flexibility associated with the p-chart. The appropriate control chart for percent defectives for the data given in Table 8.4 for the Belgian Chocolate Truffle Factory is shown in Figure 8.14. There are no points outside the control limits, although the 3:00 P.M. value of 0.75 is close to the upper limit's value of 0.88. The 3:00 P.M. sample can be dropped if it is considered an outlier. It can be eliminated from the calculations and 25 more samples of four should be taken to recalculate the control limits.

Calculations in Table 8.4 show that the p-values are rather high. Companies that subscribe to the concepts of TQM are committed to obtaining defective rates that are less than 1%; and for BCTF the average defective rate is 40%. It is evident that the new criteria and the present process are incompatible. Constructing the p-chart confirms this evaluation. BCTF either has to go back to the old standard or get the new machine. This analysis might provide the motivation for changing machines.

8.10 Control Charts for Attributes: *c*-Charts

In many situations, the number of defects in an item, rather than percentage of defectives, is of interest. For example, the finish of an automobile or the glass for the windshield is typical of products that can have multiple defects. It is common to count the number of imperfections for paint finish and for glass bubbles and marks. The c-chart is constructed in such cases. It shows the number of defects in successive samples. The count c of the number of defects per part lends itself to control by attributes. The expected number of defects per part is c-bar. The standard deviation of the distribution of c is the square root c-bar. The UCL and LCL are calculated by the following formulas (similar to p-charts).

$$\text{UCL}(c\text{-bar}) = c\text{-bar} + z * \sigma,$$

$$\text{LCL}(c\text{-bar}) = c\text{-bar} - z * \sigma,$$

where σ = square root of c-bar and z-values are calculated just like the calculation for p-charts.

The best explanation is achieved by means of an example with data from Table 8.5 charted in Figure 8.15.

The total number of defects in 10 samples is 38. Therefore, c-bar = 3.8 (38/10) and $\sigma = 1.95$. For a two-sigma limit ($z = 2$),

$$\text{UCL}(c\text{-bar}) = 3.8 + 2 * 1.95 = 7.70,$$

$$\text{LCL}(c\text{-bar}) = 3.8 - 2 * 1.95 = -0.098 = 0.$$

A negative LCL is interpreted as zero. The UCL is breached by the fourth sample as shown in Figure 8.15. This violation seems to indicate that all is not well with this process. It is possible to throw out the fourth sample and recalculate the charts. In reality, larger samples would be taken to draw up the initial charts, and then further samples would determine the stability of the process.

Table 8.5 Data for c-Charts

Sample Number	*Number of Defects*
1	3
2	2
3	5
4	8
5	2
6	5
7	3
8	4
9	3
10	3
Total	38

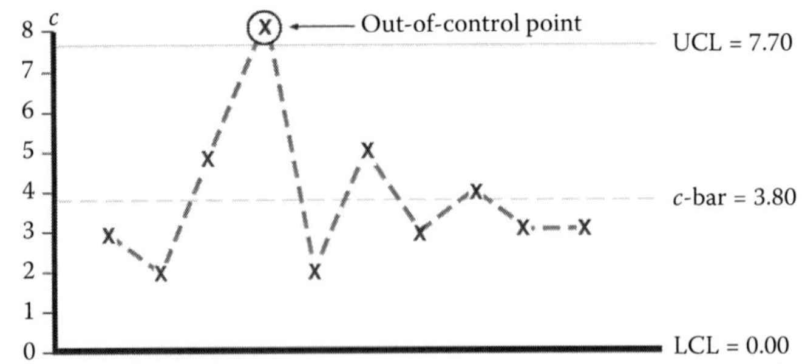

Figure 8.15 c-chart for number of defects per part.

8.11 Acceptance Sampling

In the previous sections, we discussed process control that ensures that we are producing the product or providing the service within the acceptable quality ranges. However, when a product has been produced and is shipped to the buyer, the buyer wants to ensure that the shipment contains the product of the right quality. The buyer initiates the inspection process to assess the quality of the shipment. The inspection may be done by variables or by attributes depending on how the quality has been defined for that particular lot. This process of inspection continues through the entire supply chain. In general there is a supplier (manufacturer and/or seller) and a buyer (customer). Suppliers' goods are checked for adherence to quality standards. After the inspection the buyer decides whether the submitted lot is of the right quality or not. The decision may lead to two types of errors: Type I and Type II errors.

Type I errors occur when an innocent person is found guilty by a jury of peers or when a medical test falsely proclaims illness in a healthy patient. Type II errors occur when a guilty person is found innocent by a jury of peers or when a medical test erroneously declares an ill person to be healthy. In the operations management context Type I errors occur when work is rejected on false premises. Type II errors occur when work is accepted that should be rejected (according to mutually agreed upon standards). These two types of errors are interdependent. Generally, as one gets larger the other gets smaller. It is useful to be conscious of these errors as they affect general management decision making—as well as when negotiating quality sampling plans between buyers and sellers. In a continuing effort to improve quality, the standards by which quality is measured are raised and the criteria are tightened. With very high standards and the desire to have zero defects, a six-sigma program may be warranted.

Type I errors are called errors of commission because an action follows the test result. When a supplier's shipment has been rejected by a test sample, that supplier may call for a 100% inspection. Defectives can be removed from the shipment by a procedure called detailing. Type II errors are called errors of omission. The shipment is accepted and the buyer may or may not discover that the percent defectives were above the contract specification. How does one discover that he or she did not marry the person best suited to be a lifelong companion? The effects of Type I and Type II errors can be tactical or have immensely significant strategic impact for competitive success. In this regard, balancing errors of commission and omission requires strategic thinking that unites all of the components and constituencies of the system. It is dangerous to set goals based on thinking about only one type of error.

The inspection process may involve a 100% inspection or the decision may be based on inspecting a sample of items taken from the lot. Acceptance Sampling (AS) is the process of using samples at both the input and the output stages in a manufacturing process. AS uses statistical sampling theory to determine if the supplier's outputs shipped to the producer meet standards. Then, in turn, AS is used to determine if the producer's outputs meet the producer's standards for the customers. Usually, the purchased materials are shipped in lot sizes, at intervals over time. The buyer has set standards for the materials and inspects a sample to make certain that the shipment conforms. When the sample has too many defectives, the lot will be rejected and returned to the manufacturer—or it can be detailed. Detailing means using 100% inspection of rejected lots to remove defectives. Sometimes the supplier delivers the sample for approval before sending the complete shipment. Sampling plans are often part of the buyer–supplier contract. AS is particularly well suited to exported items. Before the products are shipped to the buyer, they are inspected at the exporting plant. Only if they pass are they shipped. The same reasoning makes sense for shipments that move large distances, even within the same country. In the case of sampling, the sample is assumed to represent the quality of the shipment submitted for inspection. It is also assumed that the process that made the product is under control in the SQC sense. Each method has its pros and cons as discussed below.

There are times when 100% inspection is the best procedure. When human checking is involved, it may be neither cost-efficient nor operationally effective. Automated inspection capabilities are changing this scenario. The use of 100% inspection makes sense for parachutists and for products where failure is life-threatening. Inspections that are related to life-and-death dependencies deserve at least 100% and, perhaps, even 1000% inspection. For parachute inspections and surgical cases, a second opinion is a good idea. It is human nature that it is easier to spot defects in the works of others than in one's own work. NASA continuously monitors many quality factors at space launchings. Aircraft take-off rules are a series of checks that constitute 100% inspection of certain features of the plane. There have been occasions when pilots forgot to do a checking procedure with dire consequences.

An economic reason to shun 100% inspection is that it is often labor-intensive. This means it is likely to be more expensive than sampling the output. A common criticism of 100% inspection is that it is not fail-safe because it is human to err. Inspectors' boredom makes them very human in this regard. Sampling puts less of a burden on inspectors. Inspection procedures performed by people are less reliable and have lower accuracy when the volume of throughput is high. Imagine the burden of trying to inspect every pill that goes into a pharmaceutical bottle. If the throughput rate is one unit per second and it takes two seconds to inspect a unit, then two inspectors are going to be needed and neither of them will have any time to go to the bathroom. With present scanner technology, the dimensions of the problem are changed. In such a situation, sampling inspection is a logical alternative. If destructive testing is required (as with firecrackers), sampling is the only alternative. Otherwise, the fire-cracker manufacturer will have no product to sell. Destructive testing is used in a variety of applications such as for protective packaging and for flammability. The sampling alternative is a boon to those manufacturers.

8.11.1 Single Sampling Plans

A single sampling plan is defined by the following three numbers:

N = Lot Size (shipment size). This is the total number of items received in a lot. n = the sample size. The items to be inspected should be a representative sample drawn at random from the lot size N. Inspectors know better than to let suppliers choose the sample.

c = the acceptance number which is the maximum number of defective items allowed in the sample. When defectives exceed c, reject the lot, which is shown below in the formula.

The sampling process consists of inspecting n items picked up randomly from the lot of N items. Suppose d items are found to be defective. The criteria to accept/reject the lot are given below.

- Accept the lot if $d \leq c$.
- Reject the lot if $d > c$.

EXAMPLE

Five items ($n = 5$) are drawn from a lot size of 20 ($N = 200$). The acceptance number is set at $c = 2$. Then if three items are found to be defective ($d = 3$), reject the entire lot because $d > c$. If two or fewer items are found to be defective, d is either equal to or less than c; therefore, accept the lot. It makes sense to remove the defectives from the lot. This means that ($n - 2$) units are returned to the lot. The defectives may be able to be repaired or require scrapping.

8.11.1.1 Operating Characteristic Curves

As stated above, N, n, and c represent a sampling plan. A unique operating characteristic (OC) curve can be drawn for the sampling plan. The OC curve shows the probability of accepting a lot as the actual percent defective in the lot (p) goes from zero to one. Consider the three OC curves shown in Figure 8.16. It is assumed that N and n are constant for the three OC curves; however, the values of c are 0, 1, and 2. Let us consider the OC curve for $c = 1$ which is the middle curve. The X-axis gives the proportion of defectives (p) in the lot. Suppose $p = 0.15$. It means the production process (assuming a stable process) used to produce this lot produces on an average 0.15 proportion of defectives (15%); see the discussion of

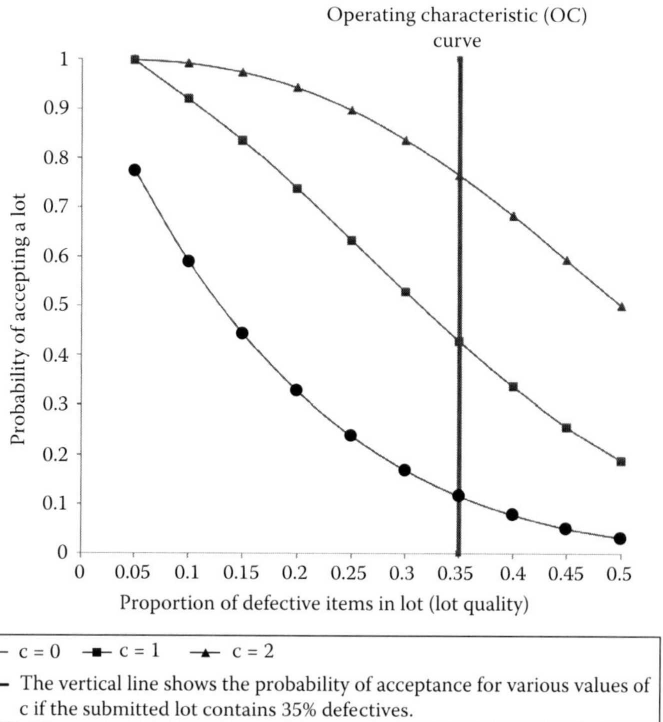

Figure 8.16 OC curves for various values of c

p-charts. The probability of accepting the lot (given on the Y-axis) for $p = 0.15$ is about 0.84 (84%). The probability of acceptance will be about 95% for $p = 0.1$; and it will be about 45% for $p = 0.35$. This means that as the quality of the lot decreases (an increase in the value of p) its acceptance probability decreases. This phenomenon may be observed for all the three OC curves.

Now let us compare the probability of acceptances of the three OC curves for a given value of p, say $p = 0.35$. See the vertical line at 0.35 and read the probabilities where it intersects the OC curves on the Y-axis. The probability of acceptances are approximately 12%, 43%, and 76% for $c = 0$, 1, and 2, respectively. This means allowing more defectives in sample (larger c) increases the probability of acceptance.

It may be noticed that when c is increased there is a marked decrease in the discriminating ability of the sampling plan. This is because the probability of accepting a lot with a given value of p rises significantly when the acceptance number (c) increases. For example, when $c = 2$, the probability of acceptance at $p = 0.05$ is almost one and it decreases to about 0.69 (at $p = 0.4$). This shows a decrease of 0.31 in the probability of acceptance. For $c = 0$ the change in probability is from 0.78 (for $p = 0.05$) to 0.09 (for $p = 0.4$). This shows a decrease of 0.69 in the probability of acceptance. Clearly, $c = 0$ (smaller value of c) has more discriminating ability.

The probability values stated here are approximations because these are read from the lines in Figure 8.16. The buyer will prefer to have a lower value of *c* because for a given *p* it reduces the probability of accepting the lot whereas a supplier will prefer a large value of *c*. A supplier does not want the lot to be rejected.

The acceptance number *c* has a big effect on the probability of acceptance as discussed above.

The other controlling parameter is *n*. The larger the value of *n*, the more discriminating is the sampling plan. The sample size *n* drives the value of the probability of acceptance. The effect of *N* is not great except when the value of *N* is quite small. The selection of sample size is based on statistical considerations (not discussed in this chapter). However, it may be mentioned that it has been established that OC curves with the same sampling proportion (*n*/*N* ratio) do not provide the same probability of acceptance. They are not even similar OC curves.

Designing a sampling plan requires determining what the values of the probability of acceptances should be for different levels of *p*. This is to provide the kind of protection that both the supplier and the buyer agree upon. Then, the OC curve can be constructed by choosing appropriate values for *n* and *c*. Note that an agreement is required. The design is not the unilateral decision of the buyer alone. The situation calls for compromise and negotiation between the supplier and the buyer. A sampling plan that minimizes the cost of inspection and gives the desired results needs to be designed.

8.11.2 Multiple-Sampling Plans

Multiple-sampling plans are utilized when the cost of inspection required by the single-sampling plan is too great. Single-sampling plans require that the decision to accept or reject the lot must be made on the basis of the first sample drawn. With double sampling, a second sample is drawn if needed. Double sampling is used when it can lower inspection costs. The double-sampling plan requires two acceptance numbers $c1$ and $c2$ such that $c2 > c1$. Then, if the observed number of defectives in the first sample of size $n1$ is $d1$:

Accept the lot if $d1 \leq c1$.
Reject the lot if $d1 > c2$.
If $c1 < d1 \leq c2$, then draw an additional sample of size $n2$.

The total sample is now of size $n1 + n2$; and the total number of defectives from the double sample is now $d1 + d2$. Then the decision rule is

Accept the lot if $(d1 + d2) \leq c2$.
Reject the lot if $(d1 + d2) > c2$.

Double sampling saves money by hedging. The first sample, which is small, is tested by a strict criterion of acceptability. Only if it fails are the additional samples taken. Double-sampling costs have to be balanced against the costs of the single large sample.

Multiple sampling plans follow the same procedures as double sampling where more than two samples can be taken. In the same sense, the costs of multiple-sampling plans can be compared to single or double sampling. OC curves can be drawn for double and multiple sampling plans also.

8.12 International Quality Standards

Each country has its own quality standards. For example, The American National Standards Institute (ANSI) and the American Society for QC (ASQC) have jointly published a series of quality specifications—called the Q90 series. However, it is important to establish global quality standards because of increasing world trade and global supply chains. The International Organization for Standardization (ISO) was set up to meet the need for international quality standards. ISO officially began operation on February 23, 1947. ISO is constantly engaged in setting standards for products, processes, information systems, and the environment with respect to every type of organization and activity. The organization, ISO has developed a comprehensive system for certifying that companies in any part of the world have voluntarily met the standards and qualify as ISO certified. The most well-known ISO standards are the ISO 9000 (quality series) and the ISO 14000 (environmental series).

ISO's work program ranges from standards for traditional activities, such as agriculture and construction, through mechanical engineering, to medical devices, to the newest information technology developments. Standardization of screw threads helps keep chairs, children's bicycles, and aircraft together with minimum wasted time. Try to solve the repair and maintenance problems that would be caused by a lack of standardization. Standards establish common parts and international consensus on terminology, which makes technology transfer easier. Without the standardized dimensions of freight containers, international trade would be slower and more expensive. Without the standardization of telephone and banking cards, life would be more complicated and more expensive. A lack of standardization may even affect the quality of life itself: for the disabled, for example, when they are barred access to consumer products—public transportation and buildings—because the dimensions of wheelchairs and entrances are not standardized. Standardized traffic symbols provide danger warnings no matter what the spoken language is. Consensus on grades of various materials gives a common reference for suppliers and clients in business dealings. Agreement on a sufficient number of variations of a product to meet most current applications allows economies of scale with cost benefits for both producers and consumers.

An example of failure is the lack of standardization of paper sizes. ISO 216 defines A and B paper series. ISO 269 defines a C paper series, and there is more. This is not a happy situation for printers and faxes being interchangeable around the globe. This failure is costing almost every user a lot of money that is wasted.

Standardization of performance or safety requirements of diverse equipment makes sure that users' needs are met while allowing individual manufacturers the freedom to design their own solution on how to meet those needs. Standardized protocols allow computers from different vendors to "talk" to each other. Standardized documents speed up the transit of goods, or identify sensitive or dangerous cargoes that may be handled by people speaking different languages. Standardization of connections and interfaces of all types ensures the compatibility of equipment of diverse origins and the interoperability of different technologies.

Agreement on test methods allows meaningful comparisons of products, or plays an important part in controlling pollution—whether by noise, vibration, or emissions. Safety standards for machinery protect people at work, at play, at sea ... and at the dentist. Without the international agreement contained in ISO standards on quantities and units, shopping and trade would be haphazard, science would be—unscientific—and technological development would be handicapped.

Standards contribute to making the development, manufacturing, and supply of products and services more efficient, safer, and cleaner. ISO generic management standards can deal with quality (ISO 9000) and the environment (ISO 14000) though the vast majority are highly specific to a particular product, material, or process.

Standards enhance global competitiveness of businesses and quality of life by promoting and facilitating voluntary consensus standards and conformity assessment systems. The existence of divergent national or regional standards can create barriers to trade. Examples include different electric voltages (110 and 220) and different plug sizes all over the globe. International standards are the technical means by which political trade agreements can be put into practice. Initially these standards are voluntary but may become market requirements if adopted by regulatory agencies of the governments.

The original ISO materials were written in great detail with guideline books that contain policies, rules, and principles. They applied to both manufactured and service products. Being familiar with the statements of these standards in their original 9000 to 9004 form is helpful in understanding the transformation and growth of global-based quality requirements.

ISO 9000: Provides overall guidance for prospective users of the ISO standards. It explains the ISO system, in general, and gives directions for using the other components of the ISO 9000 series as described in the following standards.

ISO 9001: Applies to firms that are engaged in the pre-market stages including design and development, and also in the in-market stages including production, installation, and servicing.

ISO 9002: Applies to firms that are only engaged in the in-market stages of production, installation, and servicing and not engaged in the pre-market stages of design and development.

ISO 9003: Applies to firms that deal with testing and inspection of products when they are acting as an inspection agent or as a distributor.

ISO 9004: Describes the accepted quality management system and serves as a guide to the application of that system to production systems for goods and services.

Many U.S. companies require their suppliers to adhere to ISO 9000 standards or the ANSI/ASQC equivalent standards Q90–Q94. In the same sense, U.S. companies wanting to do business in Europe and in Asia are committing resources to the considerable work that needs to be done for certification. Environmental standards ISO14000 are being globally adopted in growing numbers. There is much to be learned by going to http://www.iso.org and clicking on various items. For example, under Popular standards there are: ISO 31000 Risk management, ISO 9000 Quality management, and ISO 26000 Social responsibility.

8.13 Industrial Recognition of Quality

There are several national and international awards that recognize companies that excel in quality. Two of the most important quality awards include The Deming Prize www.juse.or.jp/e/deming/ and The Malcolm Baldrige National Quality Award www.nist.gov/baldrige/

The Deming Award: Since 1951, the Japanese Union of Scientists and Engineers (JUSE) has been awarding the Deming Prize to companies (from any country) that have achieved outstanding quality performance. Deming, as an American statistician, advised Japanese manufacturers about various principles of quality including his work with Dr Shewhart on SPC. The Japanese felt that Deming's influence was so great that they named the prize for him.

The companies that compete for this award include both manufacturing and service industries. For example, Takenaka Komuten, an architectural and construction firm, in Japan won the Deming Prize in1979 and Florida Power and Light (FPL) in the USA won the award in 1989.

8.13.1 The Malcolm Baldrige National Quality Award

During the 1980s, US companies felt the power that Japanese competitors exercised because of their devotion to quality. A variety of government initiatives were undertaken in the USA to spur improvement of the quality of American products. The Malcolm Baldrige National Quality Award was conceived as a program to promote competitiveness based on high standards of quality management among US companies. There is a special small business category; the other two categories are manufacturers and service companies. Only companies located in the USA can compete for this award.

Summary

The reason that better quality is strategically important is explained. TQM—the systems management approach to quality—is the way to obtain better quality.

This chapter states the case that before better quality can be achieved, quality has to be defined. There is a difference in the way that consumers and producers think about quality. Various dimensions of quality are explored. One of many is the warranty policies that companies offer. Another is the kind of services that are offered.

National and international standards are influential. How these standards are set is discussed, including US and international ISO 9000 as a set. There are important Japanese criteria, and other quality standards. Prizes and awards for best quality are detailed.

The costs of quality are explored, including the costs of prevention, appraisal, and failure. Control of output quality is not possible without a feedback system, which is explained. This chapter describes a broad range of SQC techniques including control charts and AS.

Overall, this chapter presents a systems approach to quality attainment which is critical because quality is the most critical variable for the success of any organization. The effects of quality failures range from the catastrophic (with loss of life, severe trauma, property damage and long recovery periods) to reduction of consumer loyalty and a decrease in the LTV of customers.

Review Questions

1. What is the role of quality in strategic planning within any company?
2. What is the competitive role of quality externally (in the marketplace)?
3. Since quality is important in different ways to different providers of services, why do we need to teach what is so obvious?
4. What is TQM and why is it important for attaining better quality?
5. Friends have said that TQM means *the quickest method*. Is that correct?
6. What are the differences between consumer and producer definitions for quality? Sometimes the consumer is called the buyer and the producer is called the supplier. What are the differences between them?
7. How does the systems approach apply to the following statement: Quality—like a chain—is betrayed by its weakest link.
8. Discuss the statement "Although the name TQM may be viewed by some as faddish, the fundamental concept is not."
9. What quality dimensions apply to the software industry?
10. Set down appropriate quality dimensions for buying or designing a hybrid automobile.
    a. Explain the differences in the systems approach.
11. What problem exists because quality standards can age?
12. Explain P/OM's relationship with market research.
13. Explain how consistency of conformity translates into low reject rates.

14. Discuss the strengths and weaknesses of quality circles. Are they similar to student cohort work groups?
15. What are the ways to reduce the cost of detailing?
16. Explain what the House of Quality does and how it works. (Look up House of Quality in Wikipedia. Follow up on Quality Function Deployment before answering this question.)
17. What prize competitions exist for quality and why are these prizes given?

Problems

1. After setup in the job shop, the first items made are likely to be defectives. If the order size calls for 125 units, and the expected percent defective for start-up is 7%, how many units should be made?
2. A subassembly of electronic components, called M1, consists of five parts that can fail. Three parts have failure probabilities of 0.03. The other two parts have failure probabilities of 0.02. Each M1 can only be tested after assembly into the parent VCR. It takes a week to get M1 units (the lead time is one week), and the company has orders for the next five days of 32, 44, 36, 54, and 41. How many M1 units should be on hand right now so that all orders can be filled?
3. Use SPC with the following table of data to advise this airline about its on-time arrival and departure performance. These are service qualities highly valued by their customers. The average total number of flights flown by this airline is 660 per day. This represents three weeks of data. It is suggested that a late flight be considered as a defective. Draw up a *p*-chart and analyze the results. Discuss the approach.

   The number of late flights (NLF) each day

Day	1	2	3	4	5	6	7	8	9	10	11	12	13	14	15	16	17	18	19	20	21
NLF	31	56	65	49	52	38	47	43	39	41	37	48	45	33	22	34	29	31	35	44	37

   Assuming that days 1, 8, and 15 are Mondays; 2, 9, and 16 are Tuesdays, etc., Use the concepts from chapter on forecasting to determine if there is any correlation between day of the week and the number of late flights.
4. New data have been collected for the BCTF. It is given in the following table. Start your analysis by creating an *x*-bar chart and then create the *R*-chart. Provide interpretation of the results.

Weight of Chocolate Truffles (in Grams)

Subgroup →	I	II	III	IV	V
Time- →	10 A.M.	11 A.M.	1 P.M.	3 P.M.	4 P.M.
1	30.00	30.25	29.75	29.90	30.05
2	30.50	31.05	29.80	29.00	29.60
3	29.95	30.00	30.05	29.95	29.90
4	30.60	29.70	29.80	29.65	29.85

Using Linear Regression, analyze the effect of the time of day on the truffle weights of samples.

5. Analyze the following data to calculate the parameters for the *x*-chart and draw the revised chart. Discuss the results. Using Linear Regression, analyze the effect of the time of day on the revised truffle weights of samples. Also calculate the *R*-chart parameters and discuss the results.

Subgroup →	I	II	III	IV	V
Time- →	10 A.M.	11 A.M.	1 P.M.	3 P.M.	4 P.M.
1	30.50	30.30	30.15	30.15	30.15
2	29.75	31.00	29.50	29.95	30.25
3	29.90	30.20	29.75	29.80	30.50
4	30.25	30.50	30.00	30.05	29.70

6. For the following tabulated data, draw the *p*-chart. NR = number of rejects and *n* = sample size. Also analyze the effect of the day on the number of rejects in each sample using regression analysis.

Subgroup No.	NR	n
1 - Monday	1	9
2 - Tuesday	2	9
3 - Wednesday	1	9
4 - Thursday	3	16
5 - Friday	1	9

7. A food processor specified that the contents of a jar of salsa should weigh 14 ± 0.10 ounces net. A statistical QC operation is set up, and the following data are obtained for one week:

Sample No				
1 - Monday	14.10	14.06	14.25	14.06
2 - Tuesday	13.90	13.85	13.80	14.00
3 - Wednesday	14.40	14.30	14.10	14.20
4 - Thursday	13.95	14.10	14.00	14.15
5 - Friday	14.05	13.90	13.95	14.60

   Using regression analysis, analyze the effect of the day on the sample means for the weight of the contents of the salsa jars.
   a. Construct an x-chart based on these five samples.
   b. Construct an R-chart based on these five samples.
   c. What points, if any, have gone out of control?
   d. Discuss the results.
8. Use a data check sheet to track the Dow Jones average, regularly reported on the financial pages of most newspapers. Record the Dow Jones closing index value on a data check sheet every day for one week. Do an Ishikawa analysis, trying to develop hypotheses concerning what causes the Dow Jones index to move the way it does. Draw scatter diagrams to see if the hypothesized causal factors are related to the Dow Jones.
9. In developing control lines for a "p" chart the total number of defective items from all samples is 3000, the number of samples is 150, and the sample size is 50. What would be the standard deviation used in developing the control lines?
10. What are the values of UCL and LCL for the following data? Sample size = 100. The fraction defective is 0.06 and the standard deviation is 0.01. The desired confidence level is 95.00%. Give your answer and we will compare your answer with those of our friends and colleagues down under.
11. In a double sampling plan the two acceptance numbers are: $c1 = 4$ and $c2 = 7$. The number of defectives found in the first sample is 5. Therefore, a second sample is taken. What is the maximum number of defectives allowed in the second sample for the lot to be accepted?
12. A manufacturing process is being monitored using a sample of size 50. The UCL and LCL on percentage defectives $\bar{p}$ are 6.30% (0.063) and 4.80% (0.048), respectively. On a given day 6 defective items were found in the sample of 50. Is the process in control?
13. In a single sampling plan, the manufacturer will prefer to have a large acceptance number c. Explain why you agree or disagree.
14. Buyers prefer to have lower values of c. Is this statement True or False? Explain.

Readings and References

Apte, U.M., and C.C. Reynolds, Quality Management at Kentucky Fried Chicken, *Interfaces*, 25(3), 1995, p. 6.

Arditi, D.H., and M. Gunaydin, Total Quality Management in the Construction Process, *International Journal of Project Management*, 15(4), August 1997, pp. 235–243.

Barkir, S.T., A Quality Control Chart for Work Performance Appraisal, *Quality Engineering*, 17(3), 2005, p. 429.

Berstein, A., Quality is Becoming Job One in the Office, Too, *Business Week*, April 29, 1991, pp. 52–54.

Blakeslee, J.A. Jr., Implementing the Six Sigma Solution, *Quality Progress*, July 1999, pp. 77–85.

Choi, T.Y., and K. Eboch, The TQM Paradox: Relations among TQM Practices, Plant Performance, and Customer Satisfaction, *Journal of Operations Management*, 17(1), December 1998, pp. 59–76.

Crosby, P.B., *Quality Is Free: The Art of Making Quality Certain*, New York: McGraw-Hill, 1979.

Davis, T.R.V., Breakdowns in Total Quality Management: An Analysis with Recommendations, *International Journal of Management*, 14(1), March 1997, pp. 13–22.

Deming, E.W., On Some Statistical Aids Toward Economic Production, *Interfaces*, 5(5), August 1975, pp. 1–15.

Dobbins, J.G., and W.J. Padgett, SPC in Printed Circuit Board Assembly, *Quality Progress*, July 1993, pp. 65–67.

Effner, J., Defining Quality for Machined Parts, *Furniture Design & Manufacturing*, 69(13), December 1997, pp. 106–109.

Elg, M., J. Olsson, and J.J. Dahlgaard, Implementing Statistical Process Control, *The International Journal of Quality and Reliability Management*, 25(6), 2008, p. 545.

Feeney, A.-M., and M. Zairi, TQM in Healthcare, *Journal of General Management*, 22(1), Autumn 1996, pp. 35–47.

Feigenbaum, A.V., Raising the Bar, *Quality Progress*, 41(7), July 2008, pp. 22–28.

Garvin, D.A., What does "Product Quality" Really Mean? *Sloan Management Review*, Fall 1984, p. 25.

Garvin, D.A., Competing on the Eight Dimensions of Quality, *Harvard Business Review*, November–December 1987, p. 101.

Garvin, D.A., How the Baldrige Award Really Works, *Harvard Business Review*, November–December 1991, pp. 80–93.

Garvin, D.A., Quality on the Line, *Harvard Business Review*, September–October 1983, pp. 63–71.

George, M., *Lean Six-Sigma for Service: How to Use Lean Speed and Six-Sigma Quality to Improve Services and Transactions*. New York: McGraw-Hill Trade, Spiral Edition, June 2003.

Goh, T.N., and M. Xie, Prioritizing Processes in Initial Implementation of Statistical Process Control, *IEEE Transactions on Engineering Management*, 45(1), February 1998, pp. 66–72.

Gonzalez-Benito, J., and O. Gonzalez-Benito, Operations Management Practices Linked to the Adoption of ISO14001, *International Journal of Production Economics*, 113(1), May 2008, p. 60.

Gryna, D.S., and F.M. Gryna, Quality in Banking Starts with Four Assessments, *Quality Progress*, August 1999, pp. 27–34.

Harrington, D., R.M. Khanna, and G. Deltas, Striving to Be Green: The Adoption of Total Quality Environmental Management, *Economics*, 40(23), December 2008, p. 2995.

Hammer, M., Process Management and Future of Six Sigma, *MIT Sloan Management Review*, Winter 2002, p. 26.

Harold, F.D., and H.G. Romig, Sampling Inspection Tables, 2nd edition. New York: Wiley, 1959.

Harris, C.R., and Y. Walter, Successfully Implementing Statistical Process Control in Integrated Steel Companies, *Interfaces*, 24(5), 1994, p. 49.

Hauser, J.R., and D. Clausing, The House of Quality, *Harvard Business Review*, May–June 1988, p. 3.

Harvey, J., Service Quality: A Tutorial, *Journal of Operations Management*, 16(1), 1998, pp. 583–597.

Hendricks, K.B., and V.R. Singhal, Does Implementing an Effective TQM Program Actually Improve Operating Performance? Empirical Evidence from Firms that Have Won Quality Awards, *Management Science*, 43(9), September 1997, pp. 1258–1274.

Hostage, G.M., Quality Control in a Service Business, *Harvard Business Review*, 53(4), July–August 1975, pp. 98–106.

Hosseini, J.F., and S. Nasser, A System for Analyzing Information to Manage the Quality-Control Process, *Interfaces*, 21(2), 1991, p. 48.

Iacobucci, D., The Quality Improvement Customers didn't Wait, *Harvard Business Review*, January–February 1996, p. 20.

Ireland, S., Quality and Nonprofit Organizations, *Quality Progress*, March 1999, pp. 96–99.

Ishikawa, K., *What is Total Quality Control? The Japanese Way*. Englewood Cliffs, NJ: Prentice-Hall, 1987.

Jones, T.O., and W.E. Sasser, Why Satisfied Customer Defect, *Harvard Business Review*, November–December 1995, pp. 89–99.

Kackar, R.N., Taguchi's Quality Philosophy: Analysis and Corninentary, *Quality Progress*, April 1986, pp. 18–23.

Kalinosky, I.S., The Total Quality System—Going Beyond ISO 9000, *Quality Progress*, June 1990, pp. 50–53.

Kerwin, K., When Flawless Isn't Enough, *Business Week*, December 8, 2003, pp. 80–82.

Khatri, N., et al., Medical Errors and Quality of Care: From Control to Commitment, *California Management Review*, 46(3), 2006, p. 115.

Knowledge@wharton.upenn.edu, 'Quality Fade': China's Great Business Challenge. Knowledge at Wharton, July 25–August 7, 2007.

Kumar, S., and Y.P. Gupta, Statistical Process Control at Motorola's Austin Assembly Plant; *Interfaces*, 23(2), 1993, p. 84.

Lawler, E.E., and S. Mohrman, Quality Circles After the Fad, *Harvard Business Review*, January–February 1985, pp. 65–71.

Lazarus, I.R., and K. Butler, The Promise of Six Sigma, *Management Healthcare Executive*, October 2001, pp. 22–26.

Liswood, L.A., New System for Rating Service Quality, *Journal of Business Strategy*, July–August 1989.

Lucier, G.T., and S. Seshadri, GE Takes Six Sigma Beyond the Bottom Line, *Strategic Finance*, May 2001, pp. 41–46.

Maki, R.G., and M.R. Milota, Statistical Quality Control Applied to Lumber Drying, *Quality Progress*, December 1993, pp. 75–79.

Manley, R., and J. Manley, Sharing the Wealth: TQM Spreads from Business to Education, *Quality Progress*, June 1996, pp. 51–55.

Mathes, N., et al., Statistical Process Control for Hospitals, *Quality Management in Health Care*, 16(3), July–September 2007, p. 205.

Nakhai, B., and J.S. Neves, The Deming, Baldrige and European Quality Awards, *Quality Progress*, April 1994, pp. 33–37.

Nick, C., 100 Years of Quality Control Innovation, *Beverage World*, 117(1651), January 1998, pp. 142–145.

Noori, H., The Taguchi Methods: Achieving Design and Output Quality, *The Executive*, 3, Fourth Quarter 1989, pp. 322–326.

Oppenheim, B.W., and Z.H. Przanyski, Total Quality Requires Serious Training, *Quality Progress*, October 1999, pp. 63–73.

Prokesch, S.E., Competing on Customer Service: An Interview with British Airways' Sir Colin Marshall, *Harvard Business Review*, November–December 1995, p. 101.

Reichheld, F.F., Learning from Customer Defections, *Harvard Business Review*, March–April, 1996, pp. 56–69.

Roth, H.P., How SPC Can Help Cut Costs, *Journal of Corporate Accounting and Finance*, 16(3), March–April 2005, pp. 21–30.

Roethlein, C.J., and P.M. Mangiameli, The Realities of Becoming a Long-term Supplier to a Large TQM Customer, *Interfaces*, 29(4), 1999, p. 71.

Rust, R.T., T. Keiningham, S. Clemens, and A.Z. Zahorik, Return on Quality at Chase Manhattan Bank, *Interfaces*, 29(2), 1999, p. 62.

Salegna, G., and F. Fazel, Obstacles to Implementing Quality, *Quality Progress*, July 2000, pp. 53–57.

Seawright, K.W., and S.T. Young, A Quality Definition Continuum, *Interfaces*, 26(3), 1996, p. 107.

Schroeder, R.G., et al., Six Sigma: Definition and Underlying Theory, *Journal of Operations Management*, 26(4), 2008, pp. 536–554.

Shewart, W.A., *Economic Control of Quality of Manufactured Product*, Princeton: D. Van Nostrand, 1931.

Sirkin, H., and G. Stalk, Jr., Fix the Process, Not the Problem, *Harvard Business Review*, July–August 1990, p. 26.

Soltani, E., P. Lai, and P. Phillips, A New Look at Factors Influencing Total Quality Management Failure, *New Technology, Work, and Employment*, 23(1–2), March 2008, p. 125.

Steeples, M.M., The Quality—Ethics Connection, *Quality Progress*, June 1994, pp. 73–75.

Stevenson, W.J., Supercharging Your Pareto Analysis, *Quality Progress*, October 2000, pp. 51–55.

Stewart, D.M., Piecing Together Service Quality: A Framework for Robust Services, *Production and Operations Management*, Summer 2003, pp. 246–265.

Sullivan, L.P., The Power of Taguchi Methods, *Quality Progress*, 20(6), 1987, pp. 76–79.

Taguchi, G., and D. Clausing, Robust Quality, *Harvard Business Review*, January–February 1990, p. 65.

Terziovski, M., D. Samson, and D. Dow, The Business Value of Quality Management Systems Certification Evidence from Australia and New Zealand, *Journal of Operation Management*, 15(1), February 1997, pp. 1–18.

Tribus, M., and G. Szonyi, An Alternative View of Taguchi Approach, *Quality Progress*, May 1989, pp. 46–52.

Wheeler, D.J., Use Individual Values to Compute Control Limits, *Quality*, 37(1), January 1998, pp. 26–28.

Quality: The Key to Growth for Small Companies and for America, *Business Week*, November 30, 1992.

The Qualitive Imperative, *Business Week*, October 25, 1991.

Chapter 9

Supply Chain Management

Readers' Choice—Supply chains function because demand chains drive them

Amini, M., R.F. Otondo, B.D. Janz, M.G. Pitts, Simulation Modeling and Analysis: A Collateral Application and Exposition of RFID Technology. *Production and Operations Management,* 2007, 16(5), pp. 586–598. The authors used RFID to track trauma patients in a hospital. RFID improved the tracking of patients' time spent in the trauma center.

Arruñada, B. and X.H. Vázquez, When Your Contract Manufacturer Becomes Your Competitor, *Harvard Business Review,* September 2006. The authors point out that outsourcing of the entire manufacturing of a product may pave the grounds for the contract manufacturer to become a competitor.

Butter, F.A.G. den and K.A. Linse, Rethinking Procurement in the Era of Globalization, *Sloan Management Review,* Fall 2008. The authors propose a framework that integrates cost and non-cost factors (related to cultural, institutional, and political differences) in a globalized market for procurement and sourcing.

Choi, T., and T. Linton, Don't Let Your Supply Chain Control Your Business, *Harvard Business Review,* December 2011, pp. 112–117. The authors mention that a heavy reliance on first-tier suppliers may lead to a weakened control over supply chain.

Delen, D., B.C. Hardgrave, and R. Sharda, RFID for Better Supply-Chain Management through Enhanced Information Visibility, *Production and Operations Management*, 2007, 16(5), pp. 613–624. The authors establish the business value of RFID for a retailer by analyzing the movement of RFID-tagged cases between distribution centers and retail stores.

Fisher, M.L., What Is the Right Supply Chain for Your Product? *Harvard Business Review*, March–April 1997, pp. 105–116. According to Fisher, product characteristics like product life cycle, demand predictability, product variety, and market standards for lead times and service must be considered while designing supply chains.

Gupta, S., C. Koulamas, and G.J. Kyparisis, E-Business: A Review of Research Published in *Production and Operations Management* (1992–2008), *Production and Operations Management*, 18(6), November–December 2009, pp. 604–620. This paper presents the developments in Internet-enabled technologies that are changing the business functions, the business processes, and the structures of business organizations.

Lee, H.L., The Triple-A Supply Chain, *Harvard Business Review*, October 2004, pp. 102–112. Lee advocate to design agile, adaptable, and aligned supply chains to be competitive.

Lee, H.L., Aligning Supply Chain Strategies with Product Uncertainties, *California Management Review*, 44(3), Spring 2002, pp. 105–119. Lee has proposed to integrate product uncertainty with supply uncertainty in designing supply chains.

Starr, M., Application of POM to e-business: B2C e-shopping. *International Journal of Operations and Production Management*, 23(1), 2003, pp. 105–124. According to Starr, P/OM can make significant contributions to the profitability of the Internet-based businesses.

Trent, R.J., and R.M. Monczka, Achieving Excellence in Global Sourcing, *Sloan Management Review*, Fall 2005. The authors have identified seven features that characterize organizations which are effective in global sourcing.

A supply chain is an interlinked system of numerous organizations that collaborate and cooperate in delivering the product and service to the final consumer. These organizations include manufacturers and service providers, suppliers of the manufacturers and service providers, suppliers of the suppliers, distributors, wholesalers and retailers, transporters, and financial institutions. These organizations could be independent of each other or part of a one larger organizational system. These organizations are called "partners" in the supply chain.

After reading this chapter, you should be able to:

- Define supply chain capacity and explain how it is measured.
- Identify upstream and downstream partners for a chosen product.
- Identify activities of acquisition chain management.
- Discuss the importance of the purchasing function.
- Explain the role of purchasing agents.
- Highlight the importance of ethics in purchasing.
- Describe the process of receiving, inspection, and storage.
- Describe the use of bids in purchasing.
- Identify steps for supplier certification.
- Discuss systems issues in global sourcing.
- Explain how distribution chains involve many partners.
- Discuss the role of e-business.
- Identify competition, conflict, collaboration, and coordination (C^4) issues.
- Discuss the importance of radio frequency identification (RFID).
- Explain how logistics and distribution planning can minimize costs.
- Make coordinated forecasting and inventory decisions.
- Describe the bullwhip effect and its implications for supply chains.

9.1 Introduction

The primary focus in supply chain management is managing the movement of material. Raw material and component parts flow from suppliers to manufacturer and are converted into finished products. Finished goods are then transported to the final consumer through several intermediate organizations.

For example, Figure 9.1 depicts the wine supply chain determined by GS1, an international not-for-profit association. GS1 is dedicated to the design and implementation of global standards and solutions to improve the efficiency and visibility of supply and demand chains globally and across multiple sectors. It co-established the Wine Traceability Working Group in 2003. The objective was to adapt the GS1 System for its implementation by the wine industry to facilitate compliance with the traceability-related provisions of the General Food Law—Council Regulation (EC) No. 178/2002. GS1 Global Office is located in Brussels, Belgium. The Working

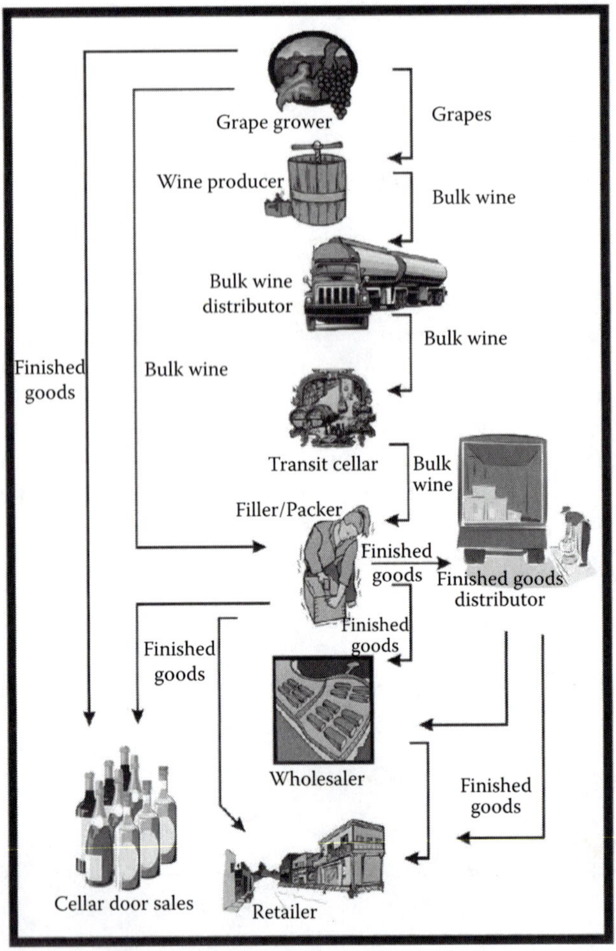

Figure 9.1 The wine supply chain. (Adapted from Wine Supply Chain Traceability, GS1 Application Guide, http://www.gs1.org/docs/traceability/GS1_wine_traceability.pdf. Copyright GS1, January 2005. All rights reserved.)

Group determined that the wine supply chain could be broken down into the following key areas:

- Grape grower
- Wine producer
- Bulk distributor
- Transit cellar
- Filler/packer

- Finished goods distributor
- Retailer

Each area was examined with a view to explaining traceability within that business process and to determine the relevant GS1 standards to be deployed.

A service supply chain may not involve the movement of materials but involves the design of interlinked operations. For example, the linkages between travel agents, airlines, hotels, and cruise lines to provide an overall pleasurable experience to the clients going on a cruise forms the supply chain. Figure 9.2 depicts a simplified tourism supply chain (TSC).

The customers, downstream end of the supply chain partners, are tourists; the retailers for the tourism products are travel agents, travel agencies, and online websites. Other partners in the TSC include resorts, hotels, airlines, cruise ships, etc. Tour operators influence the TSC activities to a large extent. Tour operators create TSC products—the complete holiday packages which are sold to the end customers—the tourists (Ujma, 2001). Zhang et al. (2009) describe a TSC structure in more detail.

In any supply chain, competition and conflict are inevitable between partners at different stages of the supply chain and among multiple partners at a given stage. For example, suppliers compete for winning the manufacturer's supply orders and retailers compete among themselves for increasing their market share. With the advent of e-business, the manufacturers have started competing with their own retailers by opening parallel Internet channels to sell their products. Competition, conflict, collaboration, and coordination (C^4) issues span across all stages of a supply chain (see, e.g., Gupta et al., 2009). Design of supply chains, therefore, involves not only minimizing the cost of moving material, but also managing the intricate behavioral relationships among supply chain partners.

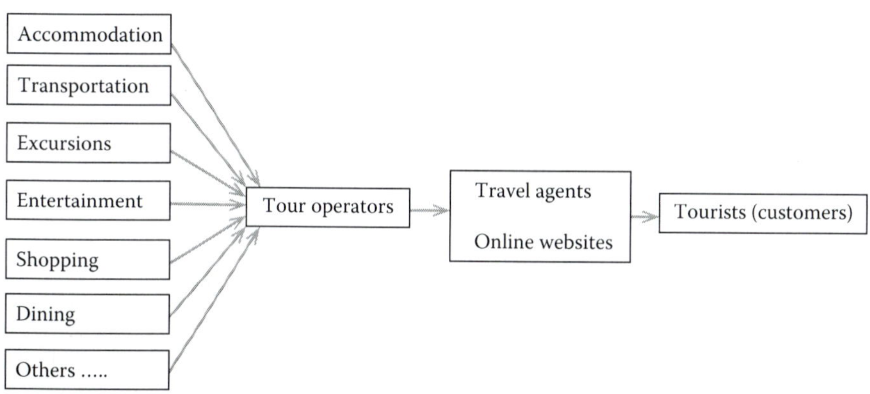

Figure 9.2 A simplified tourism supply chain.

Behavioral relationships among partners exist in all walks of life. When you live alone, you perform various functions and make choices to maximize your satisfaction and happiness—what to eat, which TV shows to watch, when to get up in the morning, what to wear, where to work, who are your friends, which parties to attend, location of your house (on the lake, big backyard or both), and so on. The scenario changes after you get married. Your spouse will have his/her own preferences for the same items. Do the choices of the spouses match? If not, how do you reconcile the differences as a team to optimize your satisfaction and happiness so that the individual choices do not lead to separation and divorce? Supply chains face similar dilemmas!

Supply chain management consists of three distinct phases:

- Acquisition chain—purchasing of materials
- Transformation process—production of goods
- Distribution chain management—distribution of finished products

The transformation process (production of goods) has been the main focus of other chapters in this book, and the functions that are essential in production of goods have already been discussed in other chapters.

Therefore, in this chapter, we will focus on the remaining two phases: acquisition and distribution chains. There are some functions that are common to both phases. For example, planning for transportation is to be performed both in the acquisition and distribution chains. On the other hand, there are functions that are unique to each phase. For example, locating suppliers is performed during the acquisition phase, and deciding on wholesalers and distributors is a function carried out in the distribution phase. The functions and activities included in the acquisition chain are required by the manufacturer and end at the manufacturer; the activities included in the distribution chain start from the manufacturer and end with the final consumer. Let us begin by studying activities that are involved in acquisition and later focus on distribution activities.

9.2 Acquisition Chain Management

We use the term acquisition chain to represent all partners that are linked to provide raw materials, component parts, and special services to a manufacturing company. *Traditionally this topic is known as purchasing and materials management.*

Materials management (MM) is the bulk of acquisition chain activities. Materials management is a tactical system of planning and control. Strategic planning came into play during start-up, when deciding what goods were to be made and what services were to be offered. However, strategies must always be reevaluated. If the system is not performing as was intended, then changes in strategies may need to be made. In this regard, two systems measures—"turnover" (T) and "days of inventory" (DOI)—are very helpful for monitoring how well purchasing and inventory managers do their job. They assist in assessing how well the strategic plans are faring. Strategies can be tweaked or changed significantly if they are not delivering according to plans. T and DOI measures can be useful in evaluating a system's performance under stressful conditions. In hard times, keeping inventories low is critical to cash-flow control. Insufficient cash flow can jeopardize the survival of the firm. Airline strategies with respect to materials management (e.g., cost of meals served, fuel consumed per passenger mile, airport servicing costs, etc.) were reevaluated after the bankruptcies of several major carriers. The main classes of materials that have to be purchased and managed in the acquisition chain include raw materials, component parts, and subassemblies. However, even the finished product can be procured as part of the overall make or buy strategy of a company. For example, food on board is catered; it must only be unpacked and heated before serving to passengers.

We have explored materials management in great depth in Chapter 5. Therefore, we will focus on the purchase activities in the following sections. The purchasing function is primarily concerned with managing vendors and suppliers.

The distinction between vendors and suppliers is a matter of local usage. One or the other and often both terms are used by various companies and/or industries in different regions of the United States and the world. The terminology describing outsourcing has been fragmented to describe off-shore, near-shore, best-shore, on-shore, and even all-shore sourcing. Canada has been called a secure near-shore location for US companies. The taxonomy of names for various types of external supplier relationships will continue to reflect economic and political issues such as the relative costs of labor, transportation costs, and the difficulty of getting a green card to enter the United States.

Acquisition chain management (ACM) connects the external sourcing of supplies to the internal scheduling of product to be delivered to the customer. Mismanage any part of the interlinked system and the adage that "a chain is as strong as its weakest link" applies. The internal MM system requires control of production materials, which are process flows. Further, there is control of work in process (WIP) on the plant floor and finished goods in the warehouse.

Some companies have external control over shipments of finished goods to distributors and customers as part of MM. Other companies limit MM to the regulation and use of incoming stock. The materials management system must be synchronized and coordinated in order to be effective. The key to synchronization is in knowing "when" materials are needed and "where." Coordination results in meeting the deadlines. It is worth noting in Figure 9.3 how many functions must be coordinated. Each deals with different aspects of materials management as part of the internal supply chain of the company. The incoming logistics system of the company is the outgoing logistics system of the supplier. Logistics, which are the tactics of materials management and product distribution, should always be in harmony with the company's strategies.

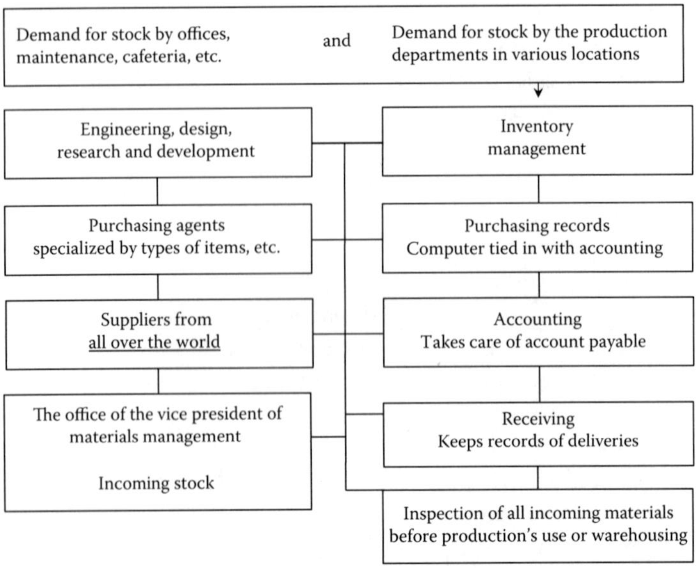

Figure 9.3 Systems perspective of materials management for incoming stock requirements.

Clearly defined quality standards for the raw materials, components, and subassemblies must be set by the materials managers for the producer. As some materials are transformed from raw materials to finished goods, they become less bulky. This characterizes analytic products, which start with tons of raw materials that are reduced to ever-smaller amounts of work in process based on the condition that less bulky goods can be transported at lower cost and require less storage space. Tons of uranium ore are reduced to pounds of uranium oxide which is concentrated into yellowcake and then further refined to extract pure uranium. This typifies an analytic process.

The opposite effect occurs for many manufactured products, which employ synthetic processes. These assemble components into bigger and heavier subassemblies and eventually finished products like farm tractors, diesel locomotives, automobiles, and commercial airliners. There is clearly an advantage in having such large and heavy finished goods near to their marketplace, so they do not have to be transported great distances to the customers.

Materials managers choose suppliers based on their locations. There has to be a balance between where to buy bulky raw materials and where to assemble big products. Decisions about where to buy materials and where to locate assembly for shipment to the marketplace are based on analyzing the costs of transport, handling, and storage. Labor-intensive operations try to be as close as possible to ample supplies of low-priced labor. Being well informed always represents an advantage because some alternatives are much better than the others.

To illustrate, a variety of organizations purchase all, or part, of the finished goods that they sell without acknowledging their use of outsourcing. The customer believes that the corporate brand name identifies the maker. This ploy may be because the organization does not have the capacity to meet demand and, therefore, contracts for the services of co-producers to augment their own output. Over the past decade, many companies have shut down their production facilities in the developed countries to gain labor cost advantages in developing countries. Finished goods may be shipped directly to customers or put into finished goods inventory in domestic distribution centers. From both a production and a transportation standpoint, there have to be tight quality controls with respect to meeting the established standards.

Co-produced (and co-packed) products have much of their value added by the supplier. The purchaser's profits are related to marketing, selling, and shipping the product. There are instances of where firms buy finished goods for sale in certain countries while making them in others. Analysis is based on running the numbers for alternative taxes and tariffs in the countries where finished goods are purchased and sold.

9.3 The Purchasing Function

Purchasing agents (PAs) and their buying organizations have the traditional role of bringing into the organization the needed supplies. Although this is very important, the role is changing, becoming more integrated within the organization and based on having a wide and informed net of important information. Sourcing (which is the generic name for purchasing every kind of supply needed by the organization) has become a global system.

The purchasing department in the twenty-first century is an information-gathering agency. Totally on the up-and-up, by being everywhere and able to listen, it is able to learn about new technologies being used by suppliers (as well as the organizations they supply) worldwide. It is on top of new materials, new suppliers, new distribution channels, new prices, and new processes that produce quality levels previously not attainable. Up-to-date purchasing departments have a global reach via satellites and telecommunications capabilities that constantly expand horizons.

This purchasing department is responsive to the marketing strategies of the various suppliers from whom it obtains required materials. The old and waning role of purchasing is to push suppliers on price. Shopping around for the best prices is no longer done by Stage II and IV companies (see Chapter 1). It is far from "best practice." The price-tag approach has been thrown out in favor of a long-term relationship with special trusted suppliers. In some cases, there are many vendors and in other cases just a few.

The importance of the buying function depends on the extent to which the company requires outside suppliers. The determination of what to make and what to buy is a P/OM decision; however, purchasing department information can be crucial. It is evident that the decision often depends on the terms to buy, including price, quality, delivery, and innovations, among others, which purchasing learns about and communicates to the P/OM team.

When the production department cannot make the product, or deliver the service, then the importance of the buying function increases. Few mail-order companies and/or department stores produce any of the materials they offer for sale. Supermarkets and membership warehouses (like Costco) use the capacity of co-packers to offer products with their own brand's name (Kirkland Signature is Costco's store brand). Purchasing provides the leverage for such products.

9.3.1 Purchasing Agents

Purchasing records provide a history of what has been done in the past. Documenting history is useful, including what the costs were, who the major suppliers were, what discounts were obtained, what quality levels were achieved, and delivery periods for specific items. Without documentation, the supplier history of a company will eventually be lost.

The skills and experience of PAs are not readily transferable between different industries. They may vary between companies even in the same industry. Differentiation exists by types of materials, buying and shipping terms, and supplier purchasing traditions.

It is not possible to discuss all of the intricate relationships that have been developed by buyers and suppliers in order to achieve maximum satisfaction for both parties. A number of important procedures are covered, but new ones are being developed all of the time that take advantage of changes in the information, storage, and transportation technologies.

The purchasing function is responsible for bringing the exact materials that production needs before they are needed, or just in time. Purchasing interfaces P/OM and its suppliers. P/OM may have some decisive requirements and some exceptional information for suppliers. It is not surprising, then, that this part of the supply chain is tightly coordinated with P/OM. Whatever the organization's structure, purchasing must be strongly linked to the P/OM team.

When the purchasing process is technical, the PA may be an engineer or a person who has worked with the production department. Purchasing is usually responsible for the following functions, which might be called the purchasing mission:

1. Ordering what is needed in the right quantities and then meeting all quality standards at the best possible prices—always achieving delivery reliability. Inventory Management models, discussed in Chapter 5, are useful for this function. This mission must be coordinated with P/OM and marketing with

respect to what is going to be needed and when it will be used. Purchasing must be assisted by P/OM in predicting the amount of scrap. By increasing order sizes to compensate, costly reorders can be avoided for a small number of units needed to complete orders.

2. Receiving inventories is part of the materials management function. Often, but not always, "receiving" is the responsibility of purchasing. Some organization must determine that deliveries are on time and that P/OM will have what it needs for its production schedules. Will it be just in time, or will extra stock be carried just in case? Such decisions require communication and coordination between P/OM and the PAs.
3. Inspecting the incoming goods to make sure that their qualities meet specifications and that the right quantities have been delivered. Statistical quality control (SQC) techniques, discussed in Chapter 8, are useful for performing this function.
4. PAs are specialists in knowing which suppliers to use. Sometimes, the PAs are in charge of certification of suppliers. Keeping informed about supplier changes is essential.
5. Purchasing is the materials management function to consult when (internal) engineering design changes (EDCs) occur that demand alterations of the specifications of purchased materials. What happens to the stock on-hand that is outmoded? How fast can the new specifications be made and shipped? If the company is constantly changing designs (which many are doing), then knowing which suppliers can cope with shifting demands is crucial. This function has to be coordinated with teams responsible for new product development.
6. Stability of supply relationships can have inestimable value; lack of stability can lead to serious crises.
7. Purchasing must be adept at coordinating the materials that are needed for start-ups. The management dynamics that are associated with start-ups are entirely different from those that operate successfully for mature products.

Coordinating the goals of materials management with those of process management is one of P/OM's greatest responsibilities. Scenarios applicable to Functions 1 and 4 are common where a variety of suppliers are dealt with, and about 1 out of 10 has serious failures. Consequently, there is regular shifting of suppliers over time, requiring P/OM adjustments.

A scenario applicable to Function 2 has the receiving dock forgetting to log in a shipment, which results in a "false" crisis when that item seems to have run out of stock. In some instances, the mistake is not traced and even though there is the necessary stock on-hand, it is lost in the warehouse.

The following story for Function 3 occurs repeatedly. Inspectors fail to check all of the quality standards and put defective received goods into inventory. The production line may be forced to shut down.

The situation that is involved with Function 5 is extremely serious in companies where technological change is moving at a rapid rate. Product malfunctions in the field lead to parts being redesigned and EDCs being issued, sometimes in bunches, in the hope that the problems will be corrected. Even with mature aircraft such as the Boeing 737, design changes in the reverse thruster were called for after it had been flying for many years.

9.3.2 *The Ethics of Purchasing*

PAs make buying decisions that involve enormous amounts of money. As labor costs decrease and material costs increase, the value of the job to buy materials becomes higher which results in PAs getting paid more. The fact that suppliers influence purchasing decisions through gifts is not considered illegal or unethical in some parts of the world. This disparity is an ethical conundrum. In the United States, it is neither ethical nor legal to "bribe" PAs. The fact that it is okay elsewhere poses irreconcilable problems until all parties sit down at the same table to discuss the quandaries and their solutions (e.g., bidding). Section 9.5 discusses bidding.

After years of dealing with a supplier, a friendly relationship can develop that allows an entirely ethical alliance. Both buyer and supplier value the long-term stability and goodwill of their relationship. In the business environment of the United States, personal relationships are not considered to be a reasonable basis for enterprise decisions. They exist, nevertheless, in less blatant form than in other cultures. For example, in Latin America and the Middle East, personal friendships are held to be business assets that reduce risk and have monetary value. Part of this cultural difference can be traced to the importance placed upon legal contracts in the United States that does not exist elsewhere. With the growth of global business, cultural and legal factors can play major roles in determining P/OM's success in handling the affairs of subsidiaries outside the United States (see the classic article by Hall, 1960).

9.4 Receiving, Inspection, and Storage

An important part of the materials management job is receiving shipments from suppliers. There is a need for a (receiving) unloading facility designed to take the supplies out of the shippers' conveyance. After unloading the supplies, there is usually a storage area in which to put them.

The design of this facility differs depending upon what is to be unloaded (type of supplies), what the supplies are unloaded from (trucks, freight cars, hopper cars, ships, planes, etc.), and where they are to be unloaded. Smart warehouses have been developed which permit optimal use of storage space and minimal time for retrieval. Bar codes and RFID are used to define unequivocally the location of every unit in storage.

The receiving facility is often called the receiving dock, and there is another location for shipping called a shipping dock. In many instances, these are the same place. In some companies, in the morning they are receiving docks and in the afternoon they are shipping docks. Most often, they are completely separate facilities. Wal-Mart uses cross-docking to transfer goods from incoming trucks at the receiving dock to outgoing trucks at the shipping dock. This means that a large percentage (of goods) never enters the warehouse but crosses from one dock to the other. Such great savings are made that many organizations have tried to emulate Wal-Mart but they have not been very successful because on-time scheduling must be synchronized. Computers are tracking shipping throughout the supply chain and correspondent computers are monitoring coordinated receipt of goods. Synchronization must be nearly perfect because the penalty for failure is disorganized chaos.

Cross-docking brings to mind Volkswagen's "modular consortium" plant in Resende, Brazil. Eight different subcontractors operate individual mini-assembly shops alongside of the main production line. In this case, receiving and shipping take place on the plant floor directly to the truck and bus production lines. As with cross-docking, great control systems are essential to achieve near-perfect synchronization. There is no leeway for deviations. The cross-docking model and the modular consortium model are credited with saving substantial amounts of money and time. Both are often cited as examples of how P/OM's creativity can improve the logistics of acquisition and distribution operations.

It is not unusual for the freight cars or hopper cars to be used as storage facilities with materials being unloaded as needed. Instead of moving chemicals and plastics from the hopper car to the warehouse, the factory draws directly on the reserves in the hopper cars. The DuPont Corporation has successfully cut down on the number of hopper cars with (non-value adding) inventories (sitting around on sidings) by coordinating its customers' needs with shipping schedules.

Supplies must be inspected to control the quality and quantity of what was ordered. Is the shipment exactly correct and has it been received undamaged? Specific quality checks are made using acceptance sampling methods (see Chapter 8). Accepted materials are moved to storage facilities—often the company warehouse. Many materials deteriorate and must be monitored for age. Accordingly, order size must be adjusted for usage rates and time elapsed since receipt and storage.

9.5 Requiring Bids before Purchase

Bidding is a process by which the buyer requests competing companies to specify how much they will charge for their product. Competitive bids can involve more than price. Sometimes, purchasing requests suppliers to submit competitive bids for both cost and delivery time. The main focus in most bidding cases is to satisfy a constituency (such as the US Congress) that purchases made under its watch have been made at the lowest reasonable cost. The Office of the Inspector General (OIG) in the US Department of Justice is a statutorily created independent watchdog whose mission is to detect waste and fraud. The OIG is charged with correcting purchasing abuses.

That explains why companies use bids to provide the appearance of purchasing decisions that are not influenced by gifts of any kind. Regarding the use of bids, it can be shown statistically that as the number of bidders increases, the expected profit that a company can make with a winning bid is reduced. The implications for purchasing (where the objective is cost control) are that costs can be reduced by inviting more companies to participate in the bidding process. However, there is a cost of reviewing each company's bid which can be time-consuming and expensive. Also, the credentials of bidding companies must be scrutinized and a basis for establishing trust needs to be forged. With trust and transparency, the need for bidding (in the first place) is diminished. However, governmental projects such as those sponsored by the US Air Force legally require bidding procedures that follow formal specifications and standardized procedures. The same also applies to certain industries and specific organizations that consider bidding a routine method for buying items that cost more than a given (significantly expensive) amount.

There are *always* two points of view with respect to bidding—the buyer's and the seller's. In a company, materials management has the buyer's point of view of bidding, which is to get the lowest costs, consistent quality, fastest delivery, vendor continuity and reliability, etc.

PAs might decide to use discretionary bidding to prevent charges of favoritism. Bidding controls expenditures when there can be significant variation in supplier charges. Bid requests state specifically all of the conditions that must be met and ask for details of what the supplier intends, including prices, delivery dates, quality specifications, checks, and assurances.

Bidding can be a costly process for the materials management buyer. This is especially the case when there are many criteria upon which competing suppliers will be rated. For example, the credit rating of the vendor may play a major role. Cost also rises when there are many firms that are bidders, but advantages of having many participants have been described above. On the other hand, for bidding to work, there must be at least two suppliers willing to bid for the job.

Oligopolies (where the market is dominated by only a few vendors) can present serious restrictive trade practices. Cartels exist (such as OPEC) which have formal agreements for collusion concerning volume of production and prices. The airline industry has been coping with such supply chain problems for one of its most important input materials for many years. In 1982, AT&T was found guilty of being a monopoly leading to divestiture of the Bell System. Many fragments of that breakup coalesced over the years so that at present there are only a few telecommunication companies in the entire USA. *Information flows* throughout the supply chain are subject to oligopolistic conditions. Supply chains cannot function without free-flowing but private information. For example, in many cases, the prices to be offered in bidding cannot be made public until the bidding is concluded.

Bidding by government agencies is usually standardized. Thus, the IRS sought and reviewed bids for updating its computer systems. The armed services utilize bids for military acquisitions. Bids are familiar in situations where industrial firms have no prior supplier arrangements and in which costly purchases (including engineering and construction jobs) are to be made. The federal government usually awards contracts to the lowest bidder subject to a set of external standards. Qualitative concerns that are taken into account by private industry are seldom permitted with government awards. Properly, this will change when P/OM is able to present the big systems picture to the watchdog agencies.

Bids can be requested where the price is fixed and the creativity and quality of the solution is at stake. Advertising agencies bid for accounts that have a set budget. Alternative bids are based on campaign creativity. The same applies to a P/OM request for proposals from a consulting organization where the budget allocation is fixed. Competing bids for computerized materials management systems are common.

With so much bidding action, it is not surprising that bidding (decision) models have been developed. Experienced e-Bay bidders use decision model software to guide them. *Sniping* is entering a bid at the last minute (or better yet, the last 10 seconds) to prevail at the moment further bidding ceases. The key is to prevent others from having further time to react. There are sniping services (such as eSnipe) available for a price on the web. There are selling strategies as well. Various small companies are auction-oriented participants in the supply chain. DealDash has developed a seller's strategy on TV that promises "easy to win, fair, and honest auctions." Many variations exist such as Dutch auctions (named after the method used for Dutch tulip auctions) start with a high price and gradually lower it until someone accepts a price which becomes the winning bid. Google used the Dutch auction technique to sell its stock to the public. The US Treasury uses Dutch auctions to sell securities as do many other countries.

Developments in information technology have facilitated an increasing use of the Internet for e-auctions. See Gupta et al. (2009) for a detailed discussion of e-auctions. e-Auctions can be used for consumer-to-consumer (C2C) and business-to consumer (B2C) auctions through commercial vendors like eBay and Yahoo! The e-commerce vendors such as FreeMarkets have focused on business-to-business (B2B) and e-procurement auctions. e-Auctions have become popular for both forward auctions and reverse auctions. Several buyers compete and bid for one seller's good(s) or service(s) in forward auctions, whereas several sellers compete and bid to fulfill the order of one buyer in the reverse auction systems.

There are many varieties of auctions that exist. Buying and selling in this way requires knowledge of the value of items being purchased as well as the auction system. Auctions and games of chance have similar characteristics. That is why bidding models rely on probabilistic analyses. To be proficient with auctions and bidding requires skills not normally associated with the proper education of an operations manager.

Hedging is another form of bidding and a game of chance. When the hedge works out, heroes are made. Here is how the hedge works: Delta Airlines buys "kerosene futures" which means paying today's prices for delivery in, say, 6 months. If the price of kerosene has gone up in 6 months, then Delta saves money. The reverse is true if the price in 6 months is less than it is at time the hedge is completed. In fact, Delta Airlines did a successful hedge because they predicted that kerosene costs would escalate when they did zoom. In fact, Delta also stopped hedging at the proper time. Hedging is viewed by many as necessary and intelligent gambling. That has never been part of the profile of a successful operations manager but it has been associated with the best PAs. It is vital that operations managers and PAs work together.

Bidding models show that as the number of bidders competing for a specific job increases, the size of the winning bid decreases. That is both apparent and logical. With more bidders, variability increases and so does competition. Perhaps this is a good reason for buyers to include many bidders. On the other hand, each extra bid adds to the ordering costs. There also is the fear that a low bid will be made by an organization that is less likely to produce quality work. Steps must be taken to ensure that quality is not compromised by price. Also, too many bidders may drive the expected profit so low that qualified suppliers refuse to join the bidding, leaving the field open to the less qualified.

A company may not choose to buy from the organization presenting the lowest bid. Price is almost never the only factor that needs to be taken into consideration when awarding a contract. Among other things, it is essential to consider quality and guarantees of quality, the experience of the supplier, the uncertainties of delivery, and the kind of long-term, supplier–producer relationship that is likely to develop. Certification may be required before a supplier is allowed to participate in a bidding contest.

Bidding is applicable for project procurement policies and for start-up purchase arrangements for the flow shop where the volume of materials is large. It is less relevant for the job shop, although it may make sense when costly components and/or relatively big batches are involved. Bidding is a useful protection when there is suspicion that special purchasing deals are being made between suppliers and company personnel.

9.6 Certification of Suppliers

Certification of suppliers is a process for grading suppliers to ensure that suppliers' organizations conform to standards that are essential for meeting the buyer's needs.

It is expensive and time-consuming to keep up-to-date. The certification process should be reserved for suppliers of A-type items with respect to both criticality and dollar volume. See Chapter 5 for a discussion of ABC classification.

In addition to establishing minimum standards that every supplier must meet, certification is like a bidding process for long-term relationships. Companies use the certification process to choose the best in the class—just like they use various criteria to hire students based on grades, dean's list, and personal evaluations. Most often, the requirements for suppliers are equivalent to the company's internal standards for itself with respect to excellence in quality and reliability.

The number of suppliers chosen can vary from one to several. Often, supplier organizations that do not make the grade are encouraged to improve. Many companies help potential suppliers upgrade those capabilities on which they are rated as deficient. Accepted suppliers are regularly reviewed to make certain that they maintain their "winning" status. Thus, although certification aims at long-term relationships, it is subject to reassessment. Often, smart buyers raise acceptance standards while assisting certified suppliers to meet the new and more stringent standards. A comprehensive dynamic systems plan alters conditions for certification overtime in line with the company's strategies. For example, Ford Motor Company has been pursuing the "Global Car Strategy based on five common platforms worldwide to leverage economies of scale." The implications for Ford suppliers everywhere are staggering.

Rating procedures include formal evaluations of price, quality, delivery time, and the ability to improve all three, and more. Suppliers' productivity improvement programs are expected to result in lower prices. Suppliers' total quality management (TQM) programs are monitored for expected improvements. ISO 9000 standards and the Baldrige Award criteria provide the foundation. Lead-time management programs track delivery time reduction which is a major concern. Time-based management concepts applied to the acquisition chain, the transformation process, and the distribution chain provide powerful new bases for evaluation.

The buyer's materials management information system (MMIS) has to be able to handle many suppliers and potential suppliers for hundreds and even thousands of A-type items. Who does this successfully? The list is impressive. Amgen, Apple, AT&T, Chrysler, Dell, FedEx, Ford, General Motors, Hewlett-Packard, Honda, IBM, Motorola, Nokia, Target, Texas Instruments, Toshiba, Toyota, UPS, Wal-Mart are only a few of the companies that have made public their use of certification programs. Certification procedures blossom with the use of the systems approach.

Problems 6–8 in the Problem section require using a Scoring Model (see Chapter 10) to evaluate and certify a group of vendors. Instead of using text material to repeat what has been said before, but in a different context, we have put the challenge of learning to apply the scoring method for certification on our students. All of the issues related to scoring models should be reviewed at this time in order to comprehend the issues involved in supplier certification procedures. We can talk in words about certification but when we deal in numbers the power of measurement in conjunction with discussion of qualitative factors becomes apparent.

9.7 Global Sourcing

There is a difference between *international purchasing* and *global sourcing*. According to Trent and Monczka (2005), "International purchasing involves a commercial transaction between a buyer and a supplier located in different countries. Global sourcing, on the other hand, involves integrating and coordinating common items, materials, processes, technologies, designs and suppliers across worldwide buying, design and operating locations." Based on their research, the authors have identified the following seven features that characterize organizations which are effective in global sourcing:

- Executive commitment to global sourcing
- Rigorous and well-defined processes
- Availability of needed resources
- Integration through information technology
- Supportive organizational design
- Structured approaches to communication
- Methodologies for measuring savings

Butter and Linse (2008) mention that procurement decisions in the era of globalization, "... are no longer based entirely on an understanding of direct purchase costs or on easily observable transaction costs, such as transport expenses and import duties, but on many other types of transaction costs as well, including those related to cultural, institutional and political differences." The authors propose a framework that integrates all these costs in a globalized market for procurement and sourcing.

The term OEM which we will be using below is acknowledged to be confusing because it has been applied in three ways. As an original equipment manufacturer, the OEM makes component parts that are purchased by another company for inclusion in their product. It is also used to describe the company that buys the component parts. To further complicate the picture, resellers who purchase the component and then brand and sell it with some added service under their own name are often referred to as OEMs. For this text, we use the second definition above. For example, Apple is a famously successful OEM buying component parts from many suppliers including some of its competitors.

Choi and Linton (2011), in the article, *Don't Let Your Supply Chain Control Your Business*, mention that "a heavy reliance on first-tier suppliers is dangerous for OEMs. It weakens their control over costs, reduces their ability to stay on top of technology developments and shifts in demand, and makes it difficult to ensure that their suppliers are operating in a socially and environmentally sustainable fashion."

In the same vein, Arrunada and Vazques (2006) mention that a manufacturer may be in a difficult situation if its contract manufacturer becomes its competitor. An OEM may outsource the entire manufacturing of a product to a contact manufacturer (CM). It is not unlikely, observe the authors, "Having manufactured an OEM's product in its entirety, the CM may decide to build its own brand and forge its own relationships with retailers and distributors—including those of the OEM."

To avoid such a situation, the authors suggest that the OEM should note: "Doing so requires a few things: modesty about revealing one's secrets; caution about whom one consorts with; and a judicious degree of intimacy, loyalty, and generosity toward one's partners and customers. OEMs can also elude CMs' backbiting tendencies by using their surplus intellectual property to enter markets beyond those for their core products. Ironically, CMs' barrier-breaking abilities, otherwise used to invade OEMs' markets, can offer OEMs access to new markets—and sometimes a way out of the dilemma."

To summarize this material, global sourcing requires the systems approach to be efficient and effective. There are many facets to understand, some of which are involved with the specific geography and economic conditions of regions of the world and individual countries.

9.8 Distribution Chain Management

As discussed earlier, the distribution chain starts once an item has been manufactured. The objective is to deliver the product to the consumer at the right time and at minimum cost. However, Hau Lee (2004) states, "Evidently, it isn't by becoming more efficient that the supply chains of Wal-Mart, Dell, and Amazon have given those companies an edge over their competitors. According to my research, top-performing supply chains possess three very different qualities. First, great supply chains are agile. They react speedily to sudden changes in demand or supply. Second, they adapt over time as market structures and strategies evolve. Third, they align the interests of all the firms in the supply network so that companies optimize the chain's performance when they maximize their interests. Only supply chains that are agile, adaptable, and aligned provide companies with sustainable competitive advantage."

Product characteristics are important in designing a supply chain. According to Fisher (1997), these characteristics include "product life cycle, demand predictability, product variety, and market standards for lead times and service (the percentage of demand filled from in-stock goods)." Fisher has found that, "if one classifies products on the basis of their demand patterns, they fall into one of two categories: they are either primarily functional or primarily innovative. And each category requires a distinctly different kind of supply chain. The root cause of the problems plaguing many supply chains is a mismatch between the type of product

		Demand Uncertainty	
		Low (Functional Products)	*High (Innovative Products)*
Supply Uncertainty	*Low (Stable Process)*	Efficient supply chains	Responsive supply chains
	High (Evolving Process)	Risk hedging supply chains	Agile supply chains

Figure 9.4 Matched Strategies. (Adapted from Lee, H. L., Aligning Supply Chain Strategies with Product Uncertainties, *California Management Review*, 44 (3), Spring 2002, pp. 105–119.)

and the type of supply chain." Functional products, for example, food and gas, satisfy primary basic needs. These products have predictable and generally stable demand pattern and enjoy long product life cycles. Fashion apparel and personal computers fall in the category of innovative products. According to the author, the ideal supply-chain strategy is to be *more efficient* for functional products and *more responsive* for innovative products

Lee (2002) has integrated product uncertainty with supply uncertainty and has proposed a 2D matrix for designing supply chains. He has categorized supply chains as: efficient, responsive, risk hedging, and agile. Figure 9.4 shows which type of supply chain is suitable for which combination of product and process characteristics.

Once a strategic framework, based on product characteristics, has been established, the manufacturer needs to establish the distribution chain to make the product available to the end customer at the right time.

Consider the supply chain shown in Figure 9.5. This supply chain consists of a manufacturer, distributor(s), wholesaler(s), and retailers(s). A partner to the right is called a downstream partner and a partner to the left is called an upstream partner.

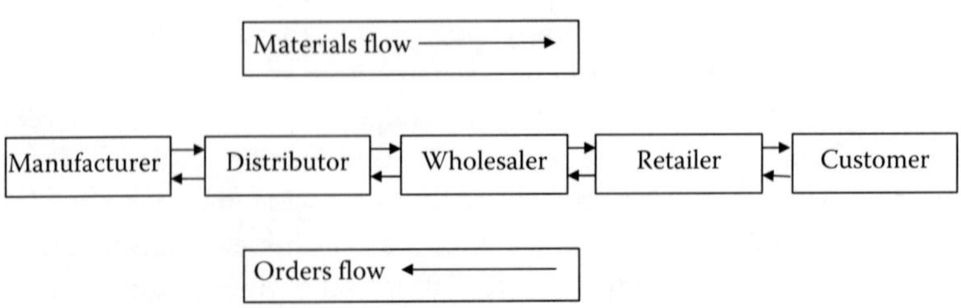

Figure 9.5 Supply chain with four levels.

Thus, retailer is a downstream partner of wholesaler and distributor is an upstream partner. Materials flow downstream and orders flow upstream. Information flows in both directions.

There are several decisions that have to be made in designing this supply chain. The manufacturer can ship the product directly to the retailer or have intermediate partners like wholesalers and distributors. The manufacturer may even choose to use an e-channel in which the customers place orders online and the product is delivered directly to the customers without any intermediaries. Many combinations are used. The "product" dictates the design to a large extent at this stage. For example, orders for products like TVs, computers, furniture, books, etc. can be placed online (e-channel or e-tailing) and the product is delivered directly to the customer. Whereas, for alcoholic beverages like wine, beer, whisky, etc., a consumer may not be able to buy directly from a manufacturer due to government regulations (the rules vary from state to state in the United States). The e-channel is becoming popular with advances in Internet technology. Amazon is an example of a highly successful e-tailer. This alternative is discussed in more detail in the section on e-business.

The manufacturer also has to decide the numbers of wholesalers, distributors, and the retailers. The network for the flow of goods has to be designed. A decision is to be made about the location of the manufacturing plant or multiple plants. The locations of the supply chain partners have to be determined. The chapter on location has discussed various techniques for making location decisions. In a supply chain, the relative location of these partners is important because that establishes the distribution network and affects the cost of goods transported.

The manufacturer also has to decide the means of transportation: trucks, trains, airplanes, ships, etc. The choice depends on the product and the cost of transportation. For example, perishable products like fresh food items may have to be transported by air. Refrigerated trucking is another popular option.

In case trucks are used for transportation, should the manufacturer own the trucks or use a trucking company? This is an important decision and affects cost. Owning the trucks increases the fixed costs, whereas using a trucking company increases the operational (variable) cost. The trucking company has to achieve a profit and this must also influence the financial decisions.

See the breakeven analysis in Appendix A. A manufacturing company is considered more vertically integrated if it owns the intermediaries. A company that owns trucks is more vertically integrated as compared to the one that uses a trucking company.

We shall study the distribution chain under the following topics:

- e-Business
- Logistics
- Forecasting and inventory

9.9 e-Business

Developments in Internet-enabled technologies are changing the business functions, the business processes, and the structures of business organizations. See Gupta et al. (2009) for a detailed discussion of e-business developments that are presented in this section. The text in this section has been reproduced with the permission of the authors.

Today, web-based functions span across product design, e-auction and procurement, vendor development, customer relations management, logistics and distribution, and pricing. The enabling web-based technology integrates various business functions and improves communication among business partners in a supply chain. Overall, the Internet has posed many challenges and has provided many opportunities to supply chain managers.

e-Business is a multidimensional discipline involving the application of technology, the study of customers' attitudes, expectations, and satisfaction, the identification of internal organizational environment, the study of the relationships among partners in the supply chain, the development of collaborative strategies and coordination mechanisms, and the development of analytical models for operating (e.g., inventory and pricing) decisions. The e-business area has been influenced by the developments in many academic fields that include but are not limited to the following: behavioral sciences, computer science, economics, information systems, marketing, operations management, operations research/management science, and technology management.

We discuss the developments in this nascent yet expanding field in the following three subsections: e-business system design and competition, conflict, collaboration and coordination (C^4), and radio frequency identification (RFID).

9.9.1 e-Business System Design

The design of e-business systems has become an important and major organizational endeavor. P/OM can make significant contributions to the profitability of the Internet-based businesses (Starr, 2003). Designing a user-friendly web interface has become crucial in order to improve customer satisfaction and ensure the ultimate success of e-business activities. The research on e-business system design shows that system flexibility, quality of service, product attributes, and perceived ease of using the e-business systems are important factors that influence customer satisfaction and loyalty. The design of e-business system should also take into account the customer characteristics in the case of heterogeneous customers. It was also observed by the researchers that e-process adoption is easier if the internal organizational environment supports the e-process and the e-process leads to improved organizational performance. Some of the relevant research studies include: Ba and Johnson (2008), Boyer and Olson (2002), Field et al. (2004), Heim and Sinha (2002), Rabinovich et al. (2008), Tatsiopoulos et al. (2002), and Tsikriktsis et al. (2004).

9.9.2 Competition, Conflict, Collaboration, and Coordination (C^4)

In any business environment, more so in e-business, competition and conflict are inevitable between partners at different stages of the supply chain and among multiple partners at a given stage. For example, suppliers compete for winning the manufacturer's supply orders (the case of e-procurement) and retailers compete among themselves for increasing their market share at the retailing stage. With the advent of e-business, the manufacturers have started competing with their own retailers by opening parallel Internet channels to sell their products—a situation most prevalent at the distribution stage

The advent of the Internet has paved the way for the emergence of mixed-channel supply chains in which a manufacturer, through a direct (Internet) channel, competes with his/her own traditional brick-and-mortar retailer for the same consumer market. This creates a situation of conflict. Some of the strategies to reduce conflict include: revisions in the wholesale prices, diversion of customers to the direct channel by the reseller for a commission, and fulfilling the demand only through the reseller (see Tsay and Agrawal, 2004); the wholesale prices remain unchanged and the retailer sets his/her prices, the wholesale prices are adjusted to maintain the retail prices at the previous level and the wholesale price is determined to maximize the wholesaler's profit with no guarantee of maintaining the base wholesale or retail prices (see Cattani et al., 2006); the retailer may be allowed to add other features and value to differentiate his/her product offering from the one sold through a direct channel by the manufacturer, the retailer sets the price of the modified product, and the manufacturer may also make a side payment to the retailer to reduce conflict (see Mukhopadhyay et al., 2008).

Competition results in conflict which in turn leads the competing entities to collaborate and coordinate to arrive at a win-win situation for everyone.

9.9.3 Radio Frequency Identification

RFID is an enabling technology for real time data collection and has a great potential to support and promote e-business activities. RFID uses wireless non-contact radio-frequency electromagnetic fields for data transfer, to automatically identify and track tags attached to objects (http://en.wikipedia.org/wiki/Radio-frequency_identification). RFID tracks the movement and flow of items in a supply chain and provides visibility to managers about the location and condition of the tracked items. The real-time information is valuable because it helps one to increase asset utilization and to minimize inventory- and logistics-related costs. RFID also minimizes delays in information transmission, leading to improved information sharing among the partners in a supply chain. The design of RFID systems is critical because installing RFID technology infrastructure entails a large initial investment and significant potential risks in technology adoption. In this section, we study the business value that an RFID system provides and the issues in adoption and implementation of RFID.

9.9.3.1 Business Value of RFID

The business value of RFID comes primarily from the visibility it provides to managers about the items tracked. RFID also provides the visibility across partners in a supply chain and has an impact on reducing information asymmetries.

Dutta et al. (2007) identify the following three stages in the evolution of RFID business value: technology deployment and integration, integration with business processes, and development of new business architectures for employees, policies, and organizational structures. They propose three dimensions for an RFID value proposition that include RFID technology, the quantification of the RFID business value, and the incentives for RFID adoption and implementation. The business value of RFID emanates from labor cost savings, shrinkage reduction, and inventory visibility. According to Lee and Ozer (2007), the estimates of the value of RFID due to labor cost savings are relatively more reliable when compared to other claims because they are derived from detailed time-and-motion studies. However, the estimates of the value of RFID resulting from inventory savings, shrinkage reduction, out-of-stock reduction, and/or sales increases lack academic rigor because these estimates are invariably based on the unreliable guesses of technology consultants and vendors.

Amini et al. (2007) and Delen et al. (2007) analyze the business value of RFID in two different industrial contexts. Amini et al. (2007) used RFID to track trauma patients in a hospital and used the data collected in a simulation model. RFID improved the tracking of patients' time from about 25% to 80% in the trauma center where patients spent about 10–12 h for their treatment. RFID technology's ability to collect data in a passive manner avoids interference with medical procedures. RFID-based simulation models help in analyzing healthcare processes more thoroughly based on the data collected for process cycle time, patient throughput rate, and equipment and personnel utilization.

Delen et al. (2007) establish the business value of RFID for a retailer by analyzing the movement of RFID-tagged cases between distribution centers and retail stores. The analysis provides insights about the distribution of lead times among different products and different combinations of distribution centers and retail stores. The RFID-generated information also helps in tracking recalls, delivering products to the stores as per schedule, and in studying the backroom process that involves moving the products to the sales floor. In addition to providing immediate visibility, RFID data utilization leads to gains due to limited process changes first and then to the major modifications in the logistics system.

9.9.3.2 Adoption and Implementation of RFID

The return on investment, the business value, and the selection of partners are important considerations at the strategic level in RFID investment projects (Ngai et al., 2007). The authors present a case study of the design, development, and implementation of an RFID-based traceability system to increase inventory visibility during the maintenance cycle in order to minimize delays and item misplacements in the maintenance department of an aircraft engineering company. The critical success factors for RFID implementation include: (1) high organizational motivation, (2) implementation process efficiency, (3) effective cost control, and (4) RFID skills and knowledge transfer. Several deployment issues that impede RFID implementation include: (1) lack of in-house RFID expertise, (2) inadequate technology support from local RFID vendors, (3) existence of different sets of industry standards, (4) unreliable hardware performance, and (5) underdeveloped RFID middleware. Overall, the study reports that the RFID-based traceability system has resulted in improved lead times, competitive differentiation, savings from reusing RFID tags, breakthrough productivity by automation, reduction of human errors in handling the repairable parts, improved inventory management, reduced manpower, and manual data recording, real-time monitoring and access to detailed information, reduction of repairable parts loss, and improved customer relationships.

Barratt and Choi (2007) investigate organizational responses to RFID mandates through a case study of a large defense contractor and conclude that the responses of four different business units of the contractor ranged from full compliance to non-compliance or unwillingness to comply. The drivers for the different levels of responses included each unit's perception of institutional rationalization, technical rationalization, perceived uncertainty, and internal coupling within the organization.

Whitaker et al. (2007) study the relationships between RFID adoption and factors involving financial and IT resources of an organization, the partner mandates, and the data standards. According to the authors, an organization can enjoy more RFID benefits by partner mandate, increased RFID spending, and IT application upgrade.

9.10 Logistics

Logistics systems outline the distribution strategy. We use the example of Rukna Auto Parts Company to illustrate how distribution strategies are determined.

Rukna Auto Parts Company has three plants located in Miami, FL (20,000), Tempe, AZ (40,000), and Columbus, OH (30,000). References to the states are omitted henceforth. The numbers in parentheses show the manufacturing capacity of each plant. Rukna supplies auto parts to four distributors MKG, Inc. (25,000), ASN, Inc. (13,500), GMZ, Inc. (16,800), and AKLA, Inc. (34,700). The numbers in parentheses show the demand of each distributor. Therefore, the totals of supply (90,000) and demand (90,000) match. In a more general and complex problem, the supply and demand may not match.

9.10.1 Transportation Cost

Table 9.1 shows the cost of transportation per unit from each plant to each distributor. For example, it costs $3.00 per unit to ship one unit from Miami to ASN, Inc. This table also shows the capacity of each plant and the demand of each distributor.

EXAMPLE

Rukna wants to develop a distribution strategy to minimize the cost of transportation (distribution). Specifically, Rukna wants to know the number of units to be shipped from each plant to each distributor.

Solution

The best solution to this problem is given in Figure 9.6. The transportation method of linear programming has been used to solve this problem. See Appendix A for more details. Cells C4–F6 (highlighted cells) give the number of units shipped from a plant to a distributor. For example, cell C6 gives the number of units shipped from Columbus to MKG (13,200 units). All costs are given in cells C13–F15. For example, cell C15 ($2.50) gives the shipping cost per unit from Columbus to MKG. The total transportation costs for each combination of the plant and distributor are given in cells C19–F21. The values in these cells (C19–F21) are obtained by multiplying the number of units shipped by the shipping cost per unit in the respective cells for each combination of the plant and distributor. For example, value in cell C21 ($33,000) is obtained by multiplying the value in cell C6 (13,200) and C15 ($2.50).

The total transportation cost is given in cell G22 ($203,525). This is the minimum value of cost of transportation for this problem.

Table 9.1 Distribution Cost per Unit for Rukna Auto Parts Company

		Distributors				
		MKG, Inc.	*ASN, Inc.*	*GMZ, Inc.*	*AKLA, Inc.*	*Capacity*
Plants	Miami	$1.00	$3.00	$3.50	$1.50	20,000
	Tempe	.00	$1.75	$2.25	$4.00	40,000
	Columbus	$2.50	$2.50	$1.00	$3.00	30,000
	Demand	25,000	13,500	16,800	34,700	

	Rukna Auto Parts Company							
		Distributors						
		MKG	ASN	GMZ	AKLA	Total Shipped	Constraint Type	Capacity
Plants	Miami	11,800	0	0	8,200	20,000	=	20,000
	Tempe	0	13,500	0	26,500	40,000	=	40,000
	Columbus	13,200	0	16,800	0	30,000	=	30,000
	Total Received	25,000	13,500	16,800	34,700	90,000		90,000
	Constraint Type	=	=	=	=			
	Demand	25,000	13,500	16,800	34,700	90,000		

	Unit Cost of Transportation				
		MKG	ASN	GMZ	AKLA
Plants	Miami	$1.00	$3.00	$3.50	$1.50
	Tempe	$5.00	$1.75	$2.25	$4.00
	Columbus	$2.50	$2.50	$1.00	$3.00

	Total Transportation Cost					
		MKG	ASN	GMZ	AKLA	Total Cost
Plants	Miami	$11,800	$0	$0	$12,300	$24,100
	Tempe	$0	$23,625	$0	$106,000	$129,625
	Columbus	$33,000	$0	$16,800	$0	$49,800
	Total Cost	$44,800	$23,625	$16,800	$118,300	$203,525

Figure 9.6 Solution to Rukna Auto Parts Company.

9.11 Forecasting and Inventory Decisions in Supply Chain

We have discussed forecasting in Chapter 3 and inventory control in Chapter 5. In this chapter, we highlight the importance of integrated decisions in these two areas. Bottlenecks may be created if information is not shared and the decisions are not coordinated.

Consider the supply chain shown in Figure 9.7 that consists of retailers, distributors, producer, and suppliers. The orders flow from customers to retailer to distributors to producer to suppliers. Materials (supplies and final product) flow from suppliers to producer to distributors to retailers to customers.

Any one of the components in the supply chain can be a bottleneck. The capacity of a supplier can be a bottleneck, for instance. Such a bottleneck will cause the producer to cut production runs. Simultaneously, the producer will notify all other suppliers to ship only a proportion of the regular order. Meanwhile, the manufacturer's warehouse will receive reduced shipments of the finished goods.

The distributor also will be shortchanged. Because the distributor does not have a sufficient inventory to meet all of the demands, retailers' orders will be cut. Finally, customers in the stores will be informed that they have to wait until more goods arrive. This situation, if not quickly remedied, can get progressively worse. Competitors, if they can, will surely take advantage of the failure to "hear the customer's voice."

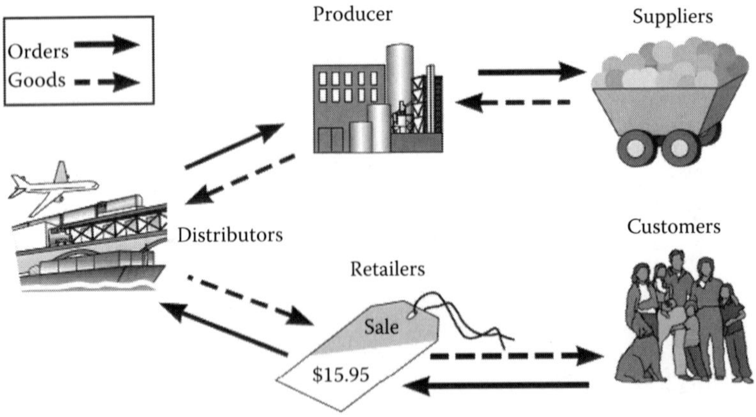

Figure 9.7 Chain with producers, distributors, and retailers.

In this scenario, every supply chain player is suffering because of the problems of one supplier. The systems approach deals with such interconnectedness of the supply chain. It focuses on finding ways to reveal what kinds of problems each participant can impose on others and on how to remedy failures to meet demand with quality products.

9.11.1 Contingency Planning for Supply Chain Capacity Crises

The producer may look around for a new supplier, or at least one that can make up the shortfall. A potential backup, even though more costly while the emergency exists, should have been known on a contingency basis. Before the supplier failed to deliver, were there any pre-warnings? Did the supplier provide all necessary information about the pending problem? Could the supplier and the producer have worked together to eliminate or reduce the severity of the problem? After-the-fact is too late. Contingency planning for supply chain capacity crises *must be done* to avoid crippling damage. At the same time, the other suppliers will be looking for new customers. Their contingency planning should have taken into account the fact that, through no fault of their own, this crisis has arisen for each of them. When these suppliers find other customers, they may have reduced ability to supply the producer. The distributor may start stocking a competing brand, with serious long-term consequences for the producer's market share. This is a good example of the need to use a broad systems perspective to connect all of the critical functions participating in the supply chain.

Retailers may find that the new competing brand is a hit with their customers. The loss of loyal customers is a serious blow stemming from this supply problem. One supplier's failure can destabilize an entire system, causing reduction in profit and losses in competitive leverage that will be hard to regain.

9.11.2 A Supply Chain System—The Better Beer Company Game

We study the operations of a supply chain system through the example of what is popularly known as "beer game." The supply chain is depicted in Figure 9.7. Forecasting and inventory decisions are made by the partners in the supply chain in this game. The game simulates the situation in such a way that the players recognize the factors that are important. If the game is properly designed, it should lead to players improving their real-world performance. That results when players understand the drivers of outcomes and throughput.

Retailers, distributors, and producers (manufacturers) can order too much (beer) or too little (beer) when information about actual demand for the brand is delayed along the linkages of a supply chain. The effects of ordering too much or too little as a result of information delays can be costly. The resulting imbalance of demand with existing capacity is important to understand. The game shows how the performance of a linked system is tied to the adequacy of forecasts about future demand.

There are measures that can be taken to improve communications and reduce information delays. There are methods for improving forecasts as well (collaborative forecasting). For example, it may not be possible to improve the national estimates of demand unless they are looked at as the sum of regional demands. Once improvement has been made in timeliness of information and better forecasts, the game players can achieve decreased costs.

Forrester (1961) used simulation to demonstrate many of these effects in 1961 with his seminal work on systems dynamics. This was at the Sloan School at MIT, where the "beer game" was developed in the 1960's.

Tables in Figures 9.8 and 9.9 show the data for simulations that reflect the effects of delays on supply chain decisions made by producers, wholesalers, retailers, suppliers, and customers. These simulations are identified with the *Better Beer Company* whose product, called *Woodstock*, is at the heart of the simulation.

Begin week 1 with 12 cases of stock on-hand (SOH). Supply at the start of the week is 4 cases, so net SOH is 16 cases. The demand is for 4 cases, so end of week 1 SOH is 12 cases. The retailer orders four cases (see column "Order Quantity"). Assume lead time to be 3 weeks. Therefore, the four cases will be delivered at the beginning of week 5. Continue to read Figure 9.8 in this fashion. Note that in weeks 10, 11, and 12, the end SOH rose to 24 cases, double the normal amount.

Figure 9.9, the next link in the supply chain, connects the producer (manufacturer) and the distributor. To read the table, follow the same procedure as with Figure 9.8. Thus, week 1 begins with 64 cases of beer at the distributor and end SOH is 64. Lead time is 4 weeks, so the order placed at the end of week 1 is received at the beginning of week 6.

These tables show how the retailer, spotting what seems like a major increase in demand (from 4 in week 1 to 8 in week 2) orders substantially more stock to avoid

Week	Begin SOH	Supply	Net SOH	Demand	End SOH	Order quantity	Delivery week
1	12	4	16	−4	12	4	5
2	12	4	16	−8	8	8	6
3	8	4	12	−8	4	16	7
4	4	4	8	−8	0	12	8
5	0	4	4	−7	−3	12	9
6	−3	8	5	−6	−1	8	10
7	−1	16	15	−5	10	4	11
8	10	12	22	−8	14	4	12
9	14	12	26	−6	20	0	13
10	20	8	28	−4	24	0	14
11	24	4	28	−4	24	0	15
12	24	4	28	−4	24	0	16
13	24	0	24	−4	20	0	17
14	20	0	20	−6	14	4	18
15	14	0	14	−8	6	8	19
16	6	0	6	−8	−2	10	20
17	−2	0	−2	−4	−6	4	21
18	−6	4	−2	−4	−6	4	22
19	−6	8	2	−4	−2	4	23
20	−2	10	8	−4	4	4	24

Figure 9.8 Supply chain simulation of retailers ordering from distributors. *Note:* Demand is shown as a negative number because demand depletes the inventory level.

outages. The retailer's order for more products takes 3 weeks lead time between placing the order and receiving the shipment. If the increase was just a random pulse in demand, then other retailers might counterbalance the effect by ordering less, because, by chance, they had less than normal demand. The distributor would hardly notice such effects, which tend to cancel each other. The retailer would gradually correct for the extra SOH that resulted.

On the other hand, if there is a *common cause* for many retailers to experience increased demand, they would all increase their order sizes, perhaps even overreacting with an extra large order to avoid going out of stock. The escalation in order sizes from many retailers would be a red flag to the distributor, who would greatly increase orders placed with the producer.

The 4-week lead time that the producer takes to fill the distributor's orders adds to the delay and increases the chances of large oscillations occurring for all supply

Week	Begin SOH	Supply	Net SOH	Demand	End SOH	Order quantity	Delivery week
1	64	16	80	−16	64	16	6
2	64	16	80	−32	48	32	7
3	48	16	64	−32	32	64	8
4	32	16	48	−32	16	64	9
5	16	16	32	−28	4	48	10
6	4	16	20	−24	−4	32	11
7	−4	32	28	−20	8	16	12
8	8	64	72	−16	56	16	13
9	56	64	120	−16	104	0	14
10	104	48	152	−16	136	0	15
11	136	32	168	−16	152	0	16
12	152	16	168	−16	152	0	17
13	152	16	168	−16	152	0	18
14	152	0	152	−24	128	16	19
15	128	0	128	−32	96	32	20
16	96	0	96	−32	64	48	21
17	64	0	64	−64	0	80	18*
18	0	80	80	−94	−16	128	23
19	−16	16	0	−64	−64	80	24
20	−64	32	−32	−32	−64	16	25

*FedEx delivery

*Note that the order for 80 cases made in week 17 is expedited via FedEx at extra cost to be delivered at the beginning of week 18.

Figure 9.9 Supply chain simulation of distributors ordering from producers (manufacturers).

chain participants. These oscillations are known as bullwhip effect and described in the sections on bullwhip effect later in the chapter.

Figures 9.10 and 9.11 show the SOH and orders placed by the retailer and distributor, respectively. It is evident that large oscillations are costing all participants a great deal. This is in spite of the fact that a review of the orders made by both the retailer and the distributor leads to the conclusion that the ordering policies followed were sensible.

Figure 9.12 compares the end SOH results for the retailer and the distributor. The effect had seemed enormous to the retailer. However, when the comparison is made with the distributor, the retailer's swings were gentle. The effect is going to be even worse at the producer's level.

If the increased demand seems to be sustained over a reasonable period of time, the producer might invest in more capacity (equipment and people) for what seems

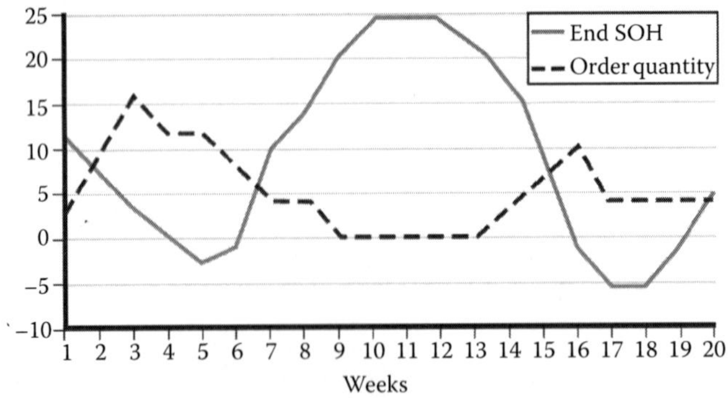

Figure 9.10 Supply chain simulation shows how the retailer's stock on-hand (SOH) oscillates.

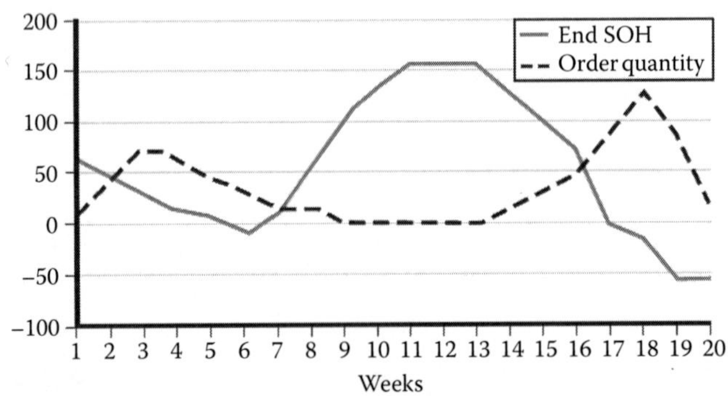

Figure 9.11 Supply chain simulation shows how the distributor's stock on-hand (SOH) oscillates.

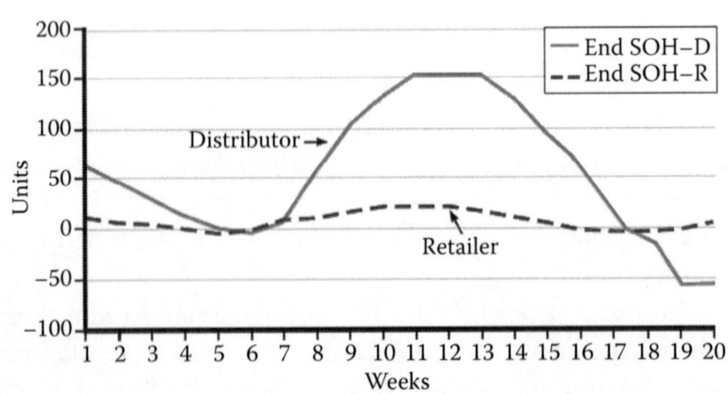

Figure 9.12 Supply chain simulation compares the SOH curves of the distributor and retailer.

like a reasonable change in demand levels. Patient evaluation of the permanency of common causes is easy to counsel, but hard to realize in the face of possible stock outages and unhappy customers, retailers, and distributors. Still, it seems reasonable to caution about the illusion of sustained common causes.

The producer could not resist and invested in more capacity. The final episode in this scenario is that the cause of the increased demand at the retail level was temporary. It was a common cause, such as might be created by a sensational TV program or a published newspaper article about the product. In reality, the glamour and appeal were gradually forgotten, and the special selling effects eventually disappeared.

The producer now has more capacity than is likely to be used for a long time, if ever. The producer drops the price to be able to sell everything that its new capacity enables it to make. The distributor orders more to take advantage of the probable increase in demand that follows a drop in price. The retailer also is working on lower margins and prefers to push the competitive products.

In the simulation game, the product can be beer or soft drinks, cosmetics, or food. Supply chains can be extended to industrial products for OEMs. The basic idea of linked supply and demand is valid as long as the relevant supply chain has been identified.

9.12 Bullwhip Effect

The *bullwhip effect* refers to amplified demand variability in a supply chain. It has been observed in many industries that even if the variability in demand (orders by customers) is low at the retailer level, more variability is observed in the orders placed by the upstream supply chain partners to replenish their inventories. In other words, the orders by a retailer to its wholesaler to replenish the stock are likely to fluctuate more than the demand at the retailer. This phenomenon continues up the supply chain. The bullwhip phenomenon was first identified by Proctor and Gamble (P&G). In the words of Lee et al. (1997), "Not long ago, logistics executives at Procter & Gamble (P&G) examined the order patterns for one of their best-selling products, Pampers. Its sales at retail stores were fluctuating, but the variabilities were certainly not excessive. However, as they examined the distributors' orders, the executives were surprised by the degree of variability. When they looked at P&G's orders of materials to their suppliers, such as 3M, they discovered that the swings were even greater. At first glance, the variabilities did not make sense. While the consumers, in this case, the babies, consumed diapers at a steady rate, the demand order variabilities in the supply chain were amplified as they moved up the supply chain. P&G called this phenomenon the 'bullwhip' effect. (In some industries, it is known as the 'whiplash' or the 'whipsaw' effect.)" Lee et al. (1997) studied the bullwhip phenomenon, identified its causes, and suggested some remedies.

The major cause for bullwhip effect is the lack of information about the actual demand by supply chain partners. Each partner in the supply chain does its own demand forecasting and places replenishment orders. This leads to inconsistency if the entire supply chain system is not well coordinated. The forecasting techniques have been discussed in Chapter 3. However, for effective management of supply chains, an integrated forecasting system needs to be developed. Such a system requires transparent information system, trust among supply chain partners, ability to make and adjust forecast at each level of supply chain. Collaborative forecasting has to be done on a periodic basis—monthly, weekly, etc.

de Ton et al. (2005) have reported how Philips Electronics synchronizes its supply chain to end the bullwhip effect. The authors report, "By applying stochastic multiechelon inventory theory, it developed an advanced planning and scheduling system that supports weekly collaborative planning of operations by Philips Semiconductors and one of its customers, Philips Optical Storage. The project has brought substantial savings. A conservative estimate shows minimum yearly savings of around US$5 million from $300 million yearly turnover. More important, Philips Optical Storage now has a more flexible and reliable supplier that can virtually guarantee quantities and delivery times. Philips Semiconductor is rolling out its new approach to other customers."

Summary

This chapter describes the supply chain management system designed, implemented, and monitored to deliver the proper products at the right times to the appropriate end customers. We have divided the SCM system into three parts: acquisition chain, transformation process, and distribution chain. The foci of this chapter are on the acquisition chain and the distribution chain.

The acquisition chain is concerned with acquiring materials, component parts, supplies, and services so that a manufacturer can produce the desired product. We spent a major portion of this discussion on sourcing and purchasing.

Purchasing is a crucial player on the MM team. It can take advantage of the global reach provided by expanding telecommunications technologies. The learning organization uses purchasing to gain opportunities for improving competitiveness. "Turnover" and "days of inventory" are important measures of the competitive performance of material managers. From another point of view, quality assurances and trusted suppliers can be essential. Turnover and DOI capture part of the firm's success with customer satisfaction with quality. To be organized, supplier certification procedures are valuable. Also covered are the MM functions of receiving, inspection, and storage. MM's use of bidding is discussed as well as the categorization of parts in terms of their criticality.

In the section on distribution management, we discussed e-business, logistics, forecasting, and inventory. We have included the description of the beer game (and its ability to generate the bullwhip effect) to facilitate students' learning of interactions between supply chain partners.

Review Questions

1. Explain analytic processes and give some examples.
   a. What characterizes analytic processes in terms of supply chain costs?
   b. Draw a chart for making a gold wedding ring.
2. Explain synthetic processes and give some examples.
   a. What characterizes synthetic processes in terms of supply chain costs?
   b. Draw a chart for building a swimming pool.
3. When Boeing's 737 reverse thruster was redesigned, what problems occurred for P/OM and the PAs in charge of materials involved?
4. How do oligopolies differ from monopolies and what are the relationships of both to the well-functioning supply chain?
5. What do cartels have to do with supply chain management? What strategies might be used to help when cartels seem to be raising the price on critical materials?

Problems

1. Use the weighted sum method to score the vendors in the table below.
2. Compare the results; select two vendors to certify and list issues that are not included in the numerical analysis that you would like to have considered before granting the certifications.

   Vendor Scoring Certification Model
   Weights add to 100; Best cell score is 10; Best total score is 1000.
   Weight and factor scales used: 1 to 10 for each cell entry; 10 being best.

   Calculations for Vendor A:

Factors	*Weights*	*Vendor A*	*Vendor B*	*Vendor C*	*Vendor D*
Price/unit (1)	30	9	5	5	7
Average quality level (2)	20	7	3	6	7
Quality consistency (3)	10	5	3	7	7
Delivery LT (4)	7	8	5	5	7

JIT ability (5)	2	4	3	6	7
Design flexibility (6)	1	4	4	7	7
Design change speed (7)	5	3	6	4	7
Product line diversity (8)	8	6	7	6	7
Services promised (9)	6	5	8	7	7
Attitude perceived (10)	11	6	10	4	7
Totals	100	687	529	548	700

The weighted sum measure is calculated by: $30 \times 9 + 20 \times 7 + 10 \times 5 + 7 \times 8 + 2 \times 4 + 1 \times 4 + 5 \times 3 + 8 \times 6 + 6 \times 5 + 11 \times 6 = 687$.

3. Consider the data given in Rukna Auto Parts Company (Section 9.10) and answer the following two questions.
   a. Find the optimal solution to the problem if the shipment from Miami to MKG is not feasible.
   b. What will be the distribution strategy if the capacity of Tempe plant was only 35,000 units? (Note: Total supply is less than the total demand).
4. *A Beer Game Simulation: The* beer game simulation is designed to give the participants a learning opportunity to design and operate supply chains. It puts them in a real life environment where they have to manage interactions among different partners that are involved in bringing a given product (beer in this case) from the manufacturer to the end consumer. The supply chain in the beer game consists of the following four partners: manufacturer, distributor, wholesaler and retailer. Available on the instructors's resource CD (IRCD) or at www.beergame.org/.

Reading and References

Amini, M., R.F. Otondo, B.D. Janz, and M.G. Pitts, Simulation Modeling and Analysis: A Collateral Application and Exposition of RFID Technology, *Production and Operations Management*, 16(5), 2007, pp. 586–598.

Aron, R., and J.V. Singh, Getting Offshoring Right, *Harvard Businesss Review*, December 2005, pp. 135–143.

Arruñada, B., and X.H. Vázquez, When Your Contract Manufacturer Becomes Your Competitor, *Harvard Business Review*, September 2006.

Ba, S., and W.C. Johansson, An Exploratory Study of the Impact of e-service Process on Online Customer Satisfaction, *Production and Operations Management*, 17(1), 2008, pp. 107–119.

Barratt, M., and T. Choi, Mandated RFID and Institutional Responses: Cases of Decentralized Business Units, *Production and Operations Management*, 16(5), 2007, pp. 569–585.

Blackburn, J.G., Supply Chain Strategies for Perishable Products, *Production and Operations Management*, 18(2), March–April 2009, pp. 129–137.

Boyer, K.K., and J.R. Olson, Drivers of Internet Purchasing Success, *Production and Operations Management*, 11(4), 2002, pp. 480–498.

Brienzi, M.S., How Kodak Transformed its Service Parts Supply Chain, *Supply Chain Management Review*, October 2005, pp. 25–32.

Butter, F.A.G. den, and K.A. Linse, Rethinking Procurement in the Era of Globalization, *Sloan Management Review*, Fall 2008.

Choi, T., and T. Linton, Don't Let Your Supply Chain Control Your Business, *Harvard Business Review*, December 2011, pp. 112–117.

Chopra, S., and M.S. Sodhi, Looking for the Bang from the RFID Buck, *Supply Chain Management Review*, May–June 2007, pp. 34–41.

Cook, R.L., B. Gibson, and D. MacCurdy, A Lean Approach to Cross-Docking, *Supply Chain Management Review*, March 2005, pp. 54–59.

Cross, G.J., How E-Business is Transforming Supply Chain Management, *Engineering Management Review*, 28, Third Quarter 2000, pp. 17–19.

Delen, D., B.C. Hardgrave, and R. Sharda, RFID for Better Supply-chain Management through Enhanced Information Visibility, *Production and Operations Management*, 16(5), 2007, pp. 613–624.

de Waart, D., and S. Kemper, 5 Steps to Service Supply Chain Excellence, *Supply Chain Management Review*, January–February 2004, pp. 28–35.

Dutta, A., H.L. Lee, and S. Whang, RFID and Operations Management: Technology, Value, and Incentives, *Production and Operations Management*, 16(5), 2007, pp. 646–655.

Field, J.M., G.R. Heim, and K.K. Sinha, Managing Quality in the e-service System: Development and Application of a Process Model, *Production and Operations Management*, 13(4), 2004, pp. 291–306.

Fisher, M.L., Making Supply Meet Demand. *Harvard Business Review*, May–June 1994, pp. 83–93.

Fisher, M.L., What Is the Right Supply Chain for Your Product? *Harvard Business Review*, March–April 1997, pp. 105–116.

Fleck, T., Supplier Collaboration in Action at IBM, *Supply Chain Management Review*, March 2008, pp. 30–37.

Forrester, J.W. *Industrial Dynamics*. New York: Wiley, 1961.

Fugate, B.S., and J.T. Mentzer, Dell's Supply Chain DNA, *Supply Chain Management Review*, October 2004, pp. 20–24.

Gaukler, G.M., R.W. Seifert, and W.H. Hausman, Item-level RFID in the Retail Supply Chain, *Production and Operations Management*, 16(1), 2007, pp. 65–76.

Gupta, S., R. Verma, and L. Victorino, Empirical Research Published in Production and Operations Management (1992–2005): Trends and Future Directions, *Production and Operations Management*, 15(3), 2006, pp. 432–448.

Gupta, S., C. Koulamas, and G.J. Kyparisis, E-Business: A Review of Research Published in *Production and Operations Management* (1992–2008), *Production and Operations Management*, 18 (6), November–December 2009, pp. 604–620.

Hall, E.T., The Silent Language in Overseas Business, *Harvard Business Review*, 38(3), May–June 1960, pp. 87–96.

Handfield, R.B., and K. McCormack, What you Need to Know About Sourcing From China, *Supply Chain Management Review*, September 2005, pp. 28–33.

Heese, H.S., Inventory Record Inaccuracy, Double Marginalization, and RFID Adoption, *Production and Operations Management*, 16(5), 2007, pp. 542–553.

Heim, G.R., and K.K. Sinha, Service Process Configurations in Electronic Retailing: A Taxonomic Analysis of Electronic Food Retailers, *Production and Operations Management*, 11(1), 2002, pp. 54–74.

Hu, J., and C.L. Munson, Speed versus Reliability Trade-offs in Supplier Selection, *International Journal Procurement Management*, 1(1–2), 2007, pp. 238–259.

Karaer, O., and H.L. Lee, Managing the Reverse Channel with RFID-enabled Negative Demand Information, *Production and Operations Management*, 16(5), 2007, pp. 625–645.

Kok, T. de, F. Janssen, J. van Doremalen, and E. van Wachem, Philips Electronics Synchronizes Its Supply Chain to End the Bullwhip Effect, *Interfaces*, 35(1), 2005, p. 37.

Lee, H.L., Aligning Supply Chain Strategies with Product Uncertainties, *California Management Review*, 44(3), Spring 2002, pp. 105–119.

Lee, H.L., The Triple-A Supply Chain, *Harvard Business Review*, October 2004, pp. 102–112.

Lee, H.L., and C. Billington, Managing Supply Chain Inventory: Pitfalls and Opportunities, *Sloan Management Review*, Spring 1992, pp. 65–73.

Lee, H.L., and C. Billington, The Evolution of Supply Chain-Management Models and Practice at Hewlett Packard, *Interfaces*, 25(5), September–October 1995, pp. 42–63.

Lee, H.L., and O. Ozer, Unlocking the Value of RFID, *Production and Operations Management*, 16(1), 2007, pp. 40–64.

Lee, H.L., V. Padmanabhan, and S. Whang, The Bullwhip Effect in Supply Chains, *Sloan Management Review*, 1997.

Liker, J.K., and T.Y. Choi, Building Deep Supplier Relationships, *Harvard Business Review*, December 2004, pp. 104–113.

Linton, J.D., R. Klassen, and V. Jayaraman, Sustainable Supply Chains: An Introduction, *Journal of Operations Management*, 25(6), November 2007, pp. 1075–1082.

Magretta, J., The Power of Virtual Integration: An Interview with Dell Computer's Michael Dell, *Harvard Business Review*, March–April 1998, pp. 72–84.

Marshall, F., What is the Right Supply Chain for Your Product? *Harvard Business Review*, March–April 1997, pp. 105–116.

Metters, R., Quantifying the Bullwhip Effect in Supply Chains, *Journal of Operations Management*, 15, 1997, pp. 89–100.

Mollenkopf, D.A., and D.J. Closs, The Hidden Value in Reverse Logistics, *Supply Chain Management Review*, July–August 2005, pp. 34–43.

Murphy-Hoye, M., H.L. Lee, and J.B. Rice, A Real World Look at RFID, *Supply Chain Management Review*, July–August 2005, pp. 18–26.

Ngai, E.W.T., T.C.E. Cheng, K.-H. Lai, P.Y.F. Chai, Y.S. Choi, and R.K.Y. Sin, Development of an RFID-based Traceability System: Experiences and Lessons Learned from an Aircraft Engineering Company, *Production and Operations Management*, 16(5), 2007, pp. 554–568.

Pisano, G.P., and R. Verganti, Which Kind of Collaboration Is Right for You? *Harvard Businesss Review*, 86(12), December 2008, pp. 78–86.

Plambeck, E.L., The Greening of Wal-Mart's Supply Chain, *Supply Chain Management Review*, July–August 2007, pp. 18–25.

Rabinovich, E., A. Maltz, and R.K. Sinha, Assessing Markups, Service Quality, and Product Attributes in Music CD's Internet Retailing, *Production and Operations Management*, 17(3), 2008, pp. 320–337.

Reeve, J.M., and M.M. Srinivasan, Which Supply Chain Design is Right for You? *Supply Chain Management Review*, May–June 2005, p. 57.

Sinha, K.K., and E.J. Kohnke, Health Care Supply Chain Design, *Decision Sciences*, 40(2), May 2009, pp. 197–212.

Slone, R.E., J.T. Mentzer, and J.P. Dittmann, Are you the Weakest Link in Your Supply Chain? *Harvard Business Review*, September 2007, pp. 116–127.

Starr, M., Application of POM to e-business: B2C e-shopping, *International Journal of Operations and Production Management*, 23(1), 2003, pp. 105–124.

Tatsiopoulos, I.P., S.T. Ponis, E.A. Hadzilias, and N.A. Panayiotou, Realization of the Virtual Enterprise Paradigm in the Clothing Industry Through e-business Technology, *Production and Operations Management*, 11(4), 2002, pp. 516–530.

Tiede, T., and K. ReeLee, What is an Optimal Distribution Network Strategy? *Supply Chain Management Review*, November 2005, pp. 32–39.

Trent, R.J., What Everyone Needs to Know About SCM, *Supply Chain Management Review*, March 2004, pp. 52–59.

Trent, R.J., and R.M. Monczka, Achieving Excellence in Global Sourcing, *Sloan Management Review*, Fall 2005.

Tsikriktsis, N., G. Lanzolla, and M. Frohlich, Adoption of e-processes by Service Firms: An Empirical Study of Antecedents, *Production and Operations Management*, 13(3), 2004, pp. 216–229.

Ujma, D., Distribution Channels for Tourism: Theory and Issues. In D. Buhalis & E. Laws (Eds.), *Tourism Distribution Channels: Practices, Issues and Transformations* (pp. 33–52). London: Continuum International Publishing Group, 2001.

Whitaker, J., S. Mithas, and M.S. Krishnan, A Field Study of RFID Deployment and Return Expectations, *Production and Operations Management*, 16(5), 2007, pp. 599–612.

Zhang, X., H. Song, and G.Q. Huang, Tourism Supply Chain Management: A New Research Agenda 2009, *Tourism Management*, 30, 2009, pp. 345–358.

Chapter 10

Long-Term Planning (Facilities, Location, and Layout)

Readers' Choice: Plans are important but the planning is more so

Benjaafar et al., Next Generation Factory Layouts: Research Challenges and Recent Progress, *Interfaces,* November–December 2002, pp. 58–78. This paper reports on the recent challenges and progress on developing layouts and discusses the work done by the Consortium on Next Generation Factory Layouts (NGFL).

DeForest, M.E., Thinking of a Plant in Mexico? *The Academi of Management Executive,* 8(1), 1994, pp. 33–40. This paper discusses the potential impact of the tripartite US/Canada/Mexico North American Free Trade Agreement (NAFTA) agreement on location decisions and points out that non-economic factors, for example, behavioral expectations may be important.

Mentzer, J.T., Seven Keys to Facility Location, *Supply Chain Management Review,* 12(5), May 2008, pp. 25–31. The paper highlights the importance of seven key factors while choosing locations based on computer models. These factors include: land, labor, capital, sources, production, markets, and logistics.

Partovi, F.Y., An Analytic Model for Locating Facilities Strategically, *Omega*, 34(1), January 2006, pp. 41–55. This paper incorporates external (customers, competitors, and the location characteristics) and internal (manufacturing processes) to develop alternatives for facility location. The applicability of the model is illustrated through a case study.

Roodberger, K.J., and I.F.A. Vis, A Model for Warehouse Layout, *IIE Transactions*, 38(10), October 2006, pp. 799–811. A method to minimize average travel distance for the order pickers in warehouses is proposed in this paper. The required storage space and the pick list size are important variables influencing the optimal decision.

Snyder, L.V., Facillity Location Under Uncertainty, *IIE Transactions*, 38(7), July 2006, pp. 537–554. The authors discuss the importance of developing facility location problems in a dynamic uncertain environment in which costs, demands, travel times and other input factors that affect location decisions continuously change. The paper summarizes and reviews optimization techniques for such scenarios.

Swamidass, P.M., A Comparison of the Plant Location Strategies of Foreign and Domestic Manufacturers in the US, *Journal of International Business Studies*, 21(2), Second Quarter 1990, pp. 301–317. This paper compares strategies of domestic and foreign manufacturers for locating plants in USA. The authors find that the location decisions of foreign manufacturers have a strong association with market size, and the distinction between foreign and domestic manufactures tend to diminish with the passage of time.

Zeng, A., et al., Design an Efficient Warehouse Layout to Facilitate the Order-Filling Process, *Production and Inventory Management Journal*, 43(0), 3rd-4th Qtr., 2002, pp. 83–88. This case study shows the application of various theoretical techniques to improve order-filling process in Brierly Lombard and Company Inc. (B&L) located in central Massachusetts.

Facilities are the plant and the office within which P/OM does its work. In addition to the buildings and the spaces that are built, bought, or rented, facilities also include equipment used in the plant and the office. There are four main components of facilities planning and they strongly interact with each other. These four include the following. *First*, where to locate the plant, or the branch, or the warehouse? This is a regional determination. The *second* issue is to find the specific structure and site to use. The *third* factor comes before moving in and the purpose is to design the layout. The *fourth* is to select furniture, lighting, decorative features, and equipment for doing the job. In fact, these four components interact, and none of them can be considered exclusive of the others. A great amount of expertise and management ability is required for facilities planning. A better job is done by broad-based visionaries and generalists who know how to work together as a team. The systems point of view has *never* been shown to be a weakness.

For facilities planning regarding location and layout, strategic thinking is essential to avoid suboptimization. The easiest way to define suboptimization is that it is less good than optimization. Suboptimization is often the result of optimizing a subsystem instead of the whole system. It is a result, or series of results, that strays from achieving the objectives that drive strategies. For example, say that a theme park assigns the manager of each of its attractions the goal of minimizing waiting time of customers. That objective for each ride will produce different effects than a single strategic objective to minimize the total waiting time of customers over the course of many rides and an entire day of being in the park. The sum of suboptimization results will not add up to an optimal solution. The sequence of rides taken, the timing of starts and completions, the encouragement of choosing certain intervals to eat and using the restrooms reflect the larger strategic framework.

When the real problem is too big to grasp because there are many options, the systems approach is indicated. Dividing the (facilities) problem into regional sub-problems to make it workable can jeopardize the quality of the solution. It is likely to yield a suboptimal (less than best) solution. Location, site, and building decisions are major components of the larger facilities planning problem. Strategies that yield suboptimal results can damage competitive capabilities. Thus, layout and job design are partners in the achievement of a good flow shop. They need to be considered together. In the job shop, production scheduling, shop-floor layout, and job design are interdependent and best approached as a system. The same kind of reasoning applies to the optimal facility design for a project. In all such situations, facilities planning is a multidimensional systems problem. The appropriate strategy must encompass all relevant elements of the minimum coherent system. To sum up: Suboptimization of the entire system is often more desirable than optimization of a large number of subsystems.

After reading this chapter, you should be able to:

- Explain the four distinct parts of facilities planning.
- Discuss who is responsible for doing facilities planning.
- Describe the nature of facilities planning models.
- Explain why planning for the design of facilities requires the systems approach.

- Describe the application of the transportation model for location decisions.
- Apply the transportation model to solve location decision problems.
- Determine the relative advantages of renting, buying, or building.
- Show how to use scoring models for facility selection decisions.
- Describe what the job of facility layout entails.
- Explain how job design and workplace layout interact.
- Evaluate the use of quantitative layout rules (algorithms).
- Use load-distance matrices to design and evaluate layouts.
- Discuss the use of heuristics to improve layouts of plants and offices.

10.1 Facilities Planning

1. *Location of facilities.* Where, in the geographical sense, should the various operations be located? Location can apply to only one facility but more generally includes a number of locations for doing different things. Thus, "where is it best to …" fabricate and assemble each product, locate the service center, situate the sales offices, and position the administration? These decisions impact the supply chain in many ways. Each facility location decision (and there are many decisions for normal supply chain systems) determines a set of times, costs, and risks that remain long after the data used for the decision have been discarded and the name or names of the decision makers can no longer be remembered. It is a good idea to *keep a permanent file for all location decisions.* In that way, pertinent history can be reconstructed and questions about reengineering the decision can be explored.

   Many location factors have to be dealt with in a qualitative fashion. Common sense prevails. Qualitative factors such as being near to supplies of raw materials and/or sources of skilled labor vie with being close to customers. Seldom can all desires be satisfied. The options may be to choose one over the other or to compromise with an in-between location. There are similar problems for distributors who want to be located close by all kinds of transportation facilities (road, rail, airports, and marine docks).

   Other location factors are better dealt with in a quantitative fashion because common sense does not work. The reason is that some real problems are too big to grasp all of the possible options. Because of this, optimal location assignments are often counterintuitive. For example, if a transportation analysis is made manageable by dividing the problem into regional sub-problems, the solution may be seriously suboptimal (far from being the best that is possible). The systems approach, which endeavors to include all relevant factors, is needed in this case.

The term "systems approach" has many different meanings depending upon the character of the field to which it is applied. We use it to solve problems. Therefore, we mean that all relevant factors which may enable solution of the quandary must be included in the analysis of the problem and the synthesis of its solution (see Section 1.5.1).

At the same time, there are strategic issues concerning centralization or decentralization of facilities. Should there be a factory in each country? Would a central warehouse be more desirable than regional warehouses or can they both be used to an advantage? Such decisions might be best reached with a combination of qualitative and quantitative considerations.

2. *Structure and specific site selection.* In what kind of facility should the process be located? How should the building or space be chosen? Is it best for the company to build the facility, buy, or rent it? Choice of a specific building is often decided after the location is chosen. However, there are circumstances where the location, site, and structure should be considered together. This makes structure and site decisions complex problems best resolved by using the systems approach.
3. *Equipment choice.* What kind of process technology is to be used? This decision often dominates structure and site options. In turn, there may be environmental factors that limit the choice of locations. Equipment choice can include transport systems and available routes. Interesting systems problems arise that combine all three: location, structure, and equipment aspects of facilities planning.
4. *Layout of the facility.* Where should machines and people be placed in the plant or office? There is interaction of equipment choice with layout, structure, site, and location. The size of the facility is determined by both current needs and future projections to allow for growth. Once the location, site, and structure are determined, layout details can be decided.

What path do conveyors, AGVs (automated guided vehicles), and other transport systems take? AGVs are computer-controlled along prescribed transport paths. Other transport systems include manual systems (like wheelbarrows) and forklift trucks, among others. Layout is an interior design problem that strongly interacts with structure and specific site selection and equipment choice.

Change in technology and even in purpose should lead to reexamination of layout decisions, which, in turn, may lead to site and structure reconsiderations. Relocation necessities (caused by political and economic factors) may trigger an entire change sequence. Since decisions about facilities are *key factors in supply chain systems planning*, openness to dynamic forces of technology and systems status is vital.

Of all the four issues discussed above, usually location is addressed first because its consequences may dominate other decisions. This is particularly true when transportation costs are a large part of the cost of goods sold. Location also may be crucial because of labor being available only in certain places. Closeness to market can affect demand. Closeness to suppliers can affect prices and the ability to deliver just in time. Tax considerations often play a part.

Once the location is known, a search for appropriate building structures can be done. However, when several locations are considered suitable and when buildings have been identified that qualify, then the various packages of locations and structures have to be considered together.

10.2 Who Does Facilities Planning?

The treatment of all four of the facilities planning issues is classic P/OM. Their importance is not in question, but the way that these four issues are treated and the role of P/OM in the planning process have been changing. In the global world of international production systems, international markets, and rapid technological transfers, facilities planning requires a team effort. It is no longer the sole responsibility of P/OM.

Team effort is required to deal with the issues properly. They deserve consideration by the entire strategic planning team of which P/OM is one important member. Beside the fact that all functional areas should be participants in location decisions, there are additional considerations.

Governmental regulations—from international to national to municipalities—have to be addressed. Legal issues need to be resolved. Lawyers from many countries often are consulted. There are negotiations with communities regarding such matters as financial incentives and tax breaks.

To rent, buy, or build requires specialists who know domestic and international real-estate markets. Trade rules and tariffs are special for trading blocks and their partners. They are constantly changing for General Agreement on Tariffs and Trade (GATT), North American Free Trade Agreement (NAFTA), European Union (EU), and other trade groups.

Currency conditions and banking restrictions must be interpreted by specialists. The list of special complications to consider is long and particular to each situation.

10.3 Models for Facility Decisions

P/OM's role and its contributions to facilities management are primarily of three kinds. *First*, P/OM's experience, with what succeeds and what fails for facilities planning, adds valuable insights into the planning team. *Second*, P/OM knows how to use facilities planning models developed by operations researchers, management scientists, and location analysts. The *third* is the input of international partners who understand the real estate issues as well as the governmental regulations.

Location decision models can use measures of costs and preferences to reach location decisions. Transportation models (TMs) use costs, and scoring models use combinations of costs and preferences. Other models—such as breakeven analysis (discussed in Appendix A) can help to select a facility and to choose equipment. Plant, equipment, and tooling decisions need close consultation between engineering, P/OM, and finance. These issues need to be guided and coordinated with top management.

Supermarket and department store locations are often chosen to be at the center of dense population zones. A center of gravity model can be used to find the population center, or the sales volume center—instead of the center of mass—which an engineer finds for structures or ships. Columbus, Ohio, is a popular distribution center because a circle around it within a 600-mile radius encloses a large percentage of U.S. retail sales. 61% of the U.S. population and 63% of U.S. manufacturing facilities lie within 600 miles of Columbus. Such a large market cannot be reached from any other state (www.conway.com/oh/distribution/ohiobody.htm).

Plant layout models for flow shops require detailed engineering in cooperation with P/OM process requirements. The technical aspects of the process are vital in determining layout for a flow shop. Layout for the job shop may be one of the few areas where P/OM is expected to do the job on its own. In actuality, good layout decisions may be based more on models that assist visualization than on models that measure layout-flow parameters. For example, computer-aided design software provides detailed 3D models or 2D drawings of floor plans.

Nevertheless, there is a need to know that there are layout models for minimizing material transport distances. This is the measured path that work in process must travel in the plant. These models are interesting for special applications, as well as for their quantitative properties.

How models are used is what counts. Some applications can be criticized as missing the complex issues. The study of P/OM provides the basis for exercising good judgment about choosing models and methods. Layout goals now encourage communication and teamwork instead of imitating the separate functions of a machine shop. Layout analysis must be based on good interactions between all interested parties.

10.4 Location Decisions—Qualitative Factors

Best location is related to the function of the facility and the characteristics of its products and services. Also, location decisions always are made relative to what other alternatives exist—just as building decisions are made relative to what else is available. When the building and location decision is taken jointly, it is possible to decide to wait for another alternative to appear. For example, the decision to consider building a factory or office structure at a specific location creates another alternative to consider. This enlarges the scope of the problem and provides another example of the need for a systems-oriented perspective to resolve some location issues.

10.4.1 Location to Enhance Service Contact

Service industries locate close to their customers to achieve the kind of contact that characterizes good service. Bank tellers and ATMs (automated teller machines) are such contact points. No one wants to travel many miles to make deposits or withdrawals. The closer bank will get the business. Of course, the closest bank is online. There are a number of banks that are strictly online.

Branches of banks, gas stations, and fast-food outlets, are scattered all around town, because distance traveled is one of the main choice criteria used by customers. Shopping malls are located so that many people find it convenient to drive to them. For retail business, best location is decided by the ability to generate high customer contact frequency.

An interesting exception to the advantage of proximity for contact is the services rendered to vacationers. Traveling many miles for sun and surf, or for snow and skiing, the service starts with the airline providing transportation. Then the hotel or resort offers food, shelter, sports, and entertainment.

Facilities planning and management are crucial to success in the hotel and resort business. Location may be at the top of the list. Services, in general, are strongly affected by location, structure, site, equipment, and layout because they all participate in making contact with customers successful.

Government institutions locate services close to the citizens who need them. Municipal governments provide police and fire protection to those who live within the municipality and pay the taxes. License plates for boats and cars are obtained at the state tag offices in many states. Hard liquor can only be obtained at state-run stores in many monopoly or control states. Effective federal service requires regional offices. For example, at the federal level there are offices for Veterans Affairs, Food & Drug Administration, the Agricultural Department, the Labor Department, and the ubiquitous U.S. Post Offices.

10.4.2 Just-in-Time Orientation

Extractors like to be close to their raw materials. Historically, at breakeven, gold mining reduces a ton of ore to 4.5 grams of gold. Profitable mines need to have ore that reduces to more than 4.5 g of gold. This assumes that competent P/OM controls the costs of extracting tons of earth and reducing these tons to grams of gold ore. It makes sense for the reduction process to be as near to the mines as possible. The price of gold on world markets plays a part. When the price of gold is high, the breakeven point is lower than 4.5 g. Mines can make a profit because each gram of gold fetches a higher price.

Fabricators like to be close to their raw materials and customers, but that cannot always be the setup. A balanced choice has to be made, which will depend on the specific product and the location of its raw materials, fabrication, and market places. Assembly plants try to keep their component suppliers close-by, and they encourage their suppliers to deliver on a just-in-time basis. The advantage to the supplier is that enough volume of business must be given by the producer to justify the nearby location. Similarly, for logical reasons, services usually profit by being close to the customer. Creative service solutions often depend upon reciprocal interactions.

The decision to be a nearby just-in-time supplier requires a mutual interchange of trust and loyalty. Just-in-time suppliers benefit from the proximity by being a single source (or one of very few sources) to their customer. Communication can be face-to-face and frequent. Customers benefit from reduction of inventory. Suppliers benefit from stability and continuity.

10.4.3 Location Factors

Six factors that can affect location decisions are:

1. *Process inputs.* Closeness to sources is often important. The transportation costs of bringing materials and components to the process from distant locations can harm profit margins.
2. *Process outputs.* Being close to customers can provide competitive advantages. Among these are lower transport costs for shipping finished goods and the ability to satisfy customers' needs and respond rapidly to competitive pricing—which can provide market advantages.
3. *Process requirements.* There can be needs for special resources that are not available in all locations (i.e., water, energy, and labor skills).
4. *Personal preferences.* Location decision makers, including top management, have personal biases for being in certain locations that can override economic advantages of other alternatives.
5. *Governmental issues.* Tax, tariff, trade, and legal factors usually matter. Trade agreements and country laws are increasingly important in reaching global location decisions.
6. *Site and plant availabilities.* The interaction between the location and the available facilities makes location and structure-site decisions interdependent.

Shipping costs are the primary concern for the first factor, viz., process inputs. These may be a function of shipping distances. However, there are more likely to be price advantages for specific materials in certain regions and countries.

For the second factor, viz., process outputs, being close to the customer have a variety of advantages including shipping distance for delivery and nearness for direct contact to remedy complaints and to consult on product changes. Being close to the customer facilitates design discussions and suggestions applicable to all phases of the supply chain channel linking producers and customers.

Factor three has alternatives such as processing bulk materials at the mining site to reduce their mass, and then further refining at a location close to the customer. In this case, process and transportation costs interact. For special medical procedures people have been known to travel far from home. For a great vacation, Hawaii and Tahiti beckon.

The fourth factor is individual, intangible, and often encountered. Often it is related to the manager's family preference. For the fifth factor, taxes and tariffs can add costs to either the production or the marketing elements. Legal costs are difficult to estimate and can be substantial. The sixth factor can require the decision to defer making a decision.

10.5 Structure and Site Selection

It would be highly unusual to choose a structure without having carefully considered location preferences. It is quite usual to choose a location and then search for a specific structure and site. Often, the list consists of combinations. Various locations, each with attractive sites and/or structures, are cataloged. Location and structure-site decisions eventually are considered simultaneously. If no structures are found in a chosen location, building is often an option.

Thus, North Carolina and Tennessee could be chosen as tax-advantage states in the United States. If market differentials exist, they also could play a part in the selection among regions of the country. After choosing location, the search for sites and structures becomes more specific. In comparing alternatives, all relevant elements of location, site, and structure should be included. If relocation is involved, the decision may be to select a new site or stay with the current one, and redesign and rebuild the facility.

Work configuration influences structure decisions. Flow shop structures permit serialized, sequential assembly with materials being received and introduced to the line as close as possible to the point of use. Access by suppliers at many points along the walls of the building is needed. Gravity-feed conveyors can be used instead of mechanized conveyors when the structure is multistoried. There are many similar considerations that arise that relate the building type with the work configuration. Part of the design of a good flow shop is the design of the structure.

Job shops do not entail large investments in the design of the process. In general, good flow shops do require large capital investment. It follows that the kind of structure that will be adequate to house job shop processes is less restricted than for flow shops. More real-estate choices are suitable for job shops than for flow shops. Rentals are more likely to be feasible for job shops than for flow shops or for flexible manufacturing systems (FMSs).

Service industries are often associated with particular kinds and shapes of structures. Airports, hospitals, theaters, and educational institutions typify the site-structure demands for service specifics. Technological information and knowledge of real-process details are required to reach good decisions.

Companies that build their own facilities to match work configuration requirements make fewer concessions. Continuous-process industries—like petrochemicals—have to build to process specifications. Even for the job shop, special requirements for space and strong floor supports—say for a large mixing vat—can influence the choice of structure. When renting, building, or buying, expert help from real-estate specialists, architects, and building engineers should be obtained to ensure proper evaluation of an existing facility or to plan a new structure. Among the facility elements to be considered are

- Is there enough floor space?
- Are the aisles wide enough?
- How many stories are desirable?
- Is the ceiling high enough?
- Are skylights in the roof useful?
- Roof shapes permit a degree of control over illumination, temperature, and ventilation. What are the maintenance requirements for roofs?

For new construction, in addition to costs, speed counts. Building codes may be too restrictive. Industrial parks may be appealing. Special-purpose facilities usually have lower resale value than general-purpose facilities. Good resale value can be critical, allowing a company flexibility to relocate when conditions change.

Company services should be listed. Capacities of parking lots, cafeterias, medical emergency facilities, male and female restrooms—in the right proportions—must be supplied. Adequate fire and police protection must be defined. Rail sidings, road access, and ship-docking facilities should be specified in the detailed facility-factor analysis. Access to the Internet and various telecom services is no longer considered an extra advantage; it is a necessity in almost all cases.

External appearance and internal appearance are factors. An increasing number of companies are using the factory as a *showroom*. Some service industries use elegant offices to impress their clients. Others use simplicity to emphasize frugality and utilitarian policies. Some consider appearance to be a frill. Others take appearance seriously and illuminate their building at night. Japanese management stresses cleanliness as a requirement for maintaining employees' pride in their company. When Sanyo acquired dilapidated facilities, they painted the walls and polished the floors. Morale was lifted. Production's output quality improved. Costs decreased.

10.6 Rent, Buy, or Build—Cost Determinants

The costs of land, construction, rental rates, and existing structures are all numbers that can be obtained and compared with a suitable model. That model must be able to reflect the net present value of different payment schedules over time. An appropriate discount rate must be determined in conjunction with the financial officers of the firm.

Cost–benefit analysis is a model analogous to a ledger with dollar costs listed on one side and dollar equivalents for the benefits on the other side. The lists are summed. Net benefit is the difference. With intangibles such as community support, workers' loyalty and skill, union–management relations, and flexibility to relocate, it is difficult to get dollar figures. Thus, *estimates of qualities* need to be made and used *in a scoring model.*

Location, structure, and site come as a package of tangible and intangible conditions that have costs—some of which are difficult to determine but which are worth listing:

1. The opportunity cost of not relocating.
2. The costs of location and relocation studies. The collection of relevant data is facilitated through the cooperation of regional chambers of commerce.
3. The cost of moving may have to include temporary production stoppage costs. Inventory buildups may be able to help manufacturers offset the effects of stopping production. Inventory cannot help stoppage of services.
4. The cost of land—often an investment. Renting, buying, or building has different tax consequences, which can play a part in reaching decisions.
5. The costs of changing lead times for incoming materials and outgoing products as a result of different locations.
6. Power and water costs differ markedly according to location.
7. Value-added taxes (VAT) are used in many European countries. VAT is proportional to the value of manufactured goods.
8. Insurance rules and costs are location sensitive.
9. Labor scarcities can develop that carry intangible costs.
10. Union-management cooperation is an intangible cost factor.
11. The intangible cost of community discord (or the benefit of community harmony) can be significant.
12. Legal fees and other costs of specialists and consultants are location sensitive, especially for small- and medium-sized organizations.
13. Workmen's compensation payments and unemployment insurance costs differ by location.
14. The costs of waste disposal, pollution and smoke control, noise abatement, and other nuisance-prevention regulations differ by locations.
15. Compliance with environmental protection rules differs by location (especially different countries).

16. The costs of damage caused by natural phenomena are affected by location, which determines the probabilities of hurricanes and earthquakes, as well as floods and lightning. Insurance damage rates have risen markedly and do not cover production stoppages.
17. Costs of reducing disaster probabilities—such as using raised construction to reduce flood damage risk.
18. Normal weather conditions produce costs associated with location: heating, air conditioning, snow removal, frozen pipes, and more absentee days because of the common cold.
19. Extreme weather conditions, such as cold and heat, cause facilities to deteriorate faster than normal weather conditions. Extreme wind conditions, such as tornadoes and hurricanes, require special building protections. Even humidity must be taken into account.

How can the many tangible and intangible costs be brought together in a unified way? Scoring models provide a satisfying means for organizing and combining estimates and hard numbers. This topic is covered in the next section.

10.7 Facility Selection Using Scoring Models

The scoring model of facility location provides a relative weight to each factor that affects the location decision. The objective is to choose the location considering all relevant factors. We describe the scoring model using hypothetical data given in Table 10.1. A six-step process is described below.

Step 1: All factors that affect the location are identified. For this example, these are listed in the column "Factor Name" in Table 10.1.

Step 2: Each factor is assigned a weight that gives the relative ranking of various factors. For example, labor productivity has a weight of 0.15 (or 15%). Nearness to the sources of raw materials is the most important factor (25%). The total of all weights should add up to 1.0 (100%).

Step 3: Alternative locations that have been under discussion are then listed. For this problem the following potential international sites are identified: Chile, Mexico, Honduras, and Brazil.

Step 4: Each alternative is evaluated and is given a score on a ten point scale (it could be a 100-point scale) for each alternative. The score for each location and for each factor is given in Table 10.2. For example, for labor productivity, Chile scored 8 points, Mexico scored 7 points, Honduras scored 3 points, and Brazil scored 6 points. These points are given by the managers based on their judgments. A team of experts may be involved in giving these scores.

Step 5: The total score for each location is calculated by multiplying the weight of each factor by the points it earned (weight × score) and then

Table 10.1 Location Factors and Weights

Factor Name	*Weight*
Labor productivity	0.15
Nearness to markets	0.18
Nearness to sources of raw material	0.25
Infrastructure facilities	0.12
Transportation facilities	0.08
Power availability	0.08
Political climate	0.03
Labor unions	0.02
Labor cost	0.04
Material cost	0.05
Total	1

Table 10.2 Evaluation of Various Location Sites

Scoring Model					
		Location Alternatives			
		Chile	*Mexico*	*Honduras*	*Brazil*
Factor Name	*Weight*	*Score Out of 10*			
Labor productivity	0.15	8	7	3	6
Nearness to markets	0.18	4	6	9	7
Nearness to sources of raw material	0.25	3	5	2	8
Infrastructure facilities	0.12	7	3	4	4
Transportation facilities	0.08	6	6	7	9
Power availability	0.08	5	8	6	7
Political climate	0.03	9	9	8	8
Labor unions	0.02	3	4	3	3
Labor cost	0.04	6	5	5	5
Material cost	0.05	7	2	1	2
Total (weighted sums)	1.00	5.31	5.51	4.64	6.52

adding this number for all factors. For example, the calculations for Chile are given below.

$$\begin{aligned}\text{Score for Chile} = 5.31 = {} & (0.15 \times 8) + (0.18 \times 4) + (0.25 \times 3) + (0.12 \times 7) \\ & + (0.08 \times 6) + (0.08 \times 5) + (0.03 \times 9) \\ & + (0.02 \times 3) + (0.04 \times 6) + (0.05 \times 7).\end{aligned}$$

Step 6: The total score for each location is calculated and then the location with the highest score is selected. For this problem, the scores are: Chile (5.31), Mexico (5.51), Honduras (4.64), and Brazil (6.52). Therefore, Brazil is the most attractive location based on these factors.

The scoring model is useful for a wide range of applications in addition to the location problem. It is appropriate for product-, process- and service-design decisions, equipment selection, warehouse-location, etc. The scoring model's multiplication method of evaluating alternatives by means of weighting factors is reasonable to handle multidimensional problems. The scoring model permits the simultaneous evaluation of tangible and intangible costs. The method allows intangibles to be dealt with in a quantitative fashion with as many factors considered as seem necessary. However, the most practical idea is to concentrate on the important factors.

There can be multiple decision makers in making location decisions. Each manager on the site-selection team will provide individual ratings (numbers) for all of the factors. If averages are used to combine the preference scores and the weights, the effect would be to decrease the differentials between locations. Averaging the scores of multiple decision makers reduces variability. This is because individual extremes tend to cancel each other. If averaging is used, the results can be compared with the results for the individuals. The managers can identify problems and areas to be further studied by comparing their preference scores and weights.

Scoring models organize a lot of information that is relevant for location decisions. Managers can study what is known and what is not; what is agreed upon, and what is not; what is important and what is not; whether consensus exists about what is important; what appears to require additional research, etc. Ultimately, the decision has to be made about accepting or rejecting the solution indicated by the scoring model.

Pooling and polling opinions of many people in the company about location decisions increases involvement and builds pride in the company. It promotes interfunctional communication about facility decisions. If the company intends to relocate employees who are willing to make the move—at least to certain places, this approach provides positive motivation.

Information about the choices should be made available throughout the company. If feasible, cost factors should be shared. When the factor lists are made up, broad participation leads to better idea generation. Notions appear that otherwise might be overlooked and the process moves faster. The decision to relocate is better shared than sprung as a surprise.

10.8 Location Decisions Using the Transportation Model

Transportation costs are a primary concern for a new start-up company or division. This also applies to an existing company that intends to relocate. Finally, it should be common practice to reevaluate the current location of an ongoing business so that the impact of changing conditions and new opportunities are not overlooked. When shipping costs are critical for the location decision, the transportation model (TM) can determine minimum cost or maximum profit solutions that specify optimal shipping patterns between many locations.

Transportation costs include the combined costs of moving raw materials to the plant and of transporting finished goods from the plant to one or more warehouses. It is easier to explain the TM with the following numerical example than with abstract math equations. A doll manufacturer has decided to build a factory in the center of the United States. More specifically, Missouri and Ohio are identified as the potential states. Several sites in the two regions have been identified. Two cities have been chosen as candidates. These are St Louis, Missouri, and Columbus, Ohio. Real-estate costs are about equal in both. The problem is to select one of the two cities. The decision will be based on the shipping (transportation) costs.

10.8.1 Shipping (Transportation or Distribution) Costs

The average cost of shipping (also known as the cost of distribution or cost of transportation) the components that the company uses to the Columbus, Ohio, location is \$6 per production unit. Shipping costs average only \$3 per unit to St Louis, Missouri. In TM terminology, shippers (suppliers, in this case) are called sources or origins. Those receiving shipments (producers, in this case) are called destinations.

The average cost of shipping from the Columbus, Ohio, location to the market—distributor's warehouse is \$2 per unit. The average cost of shipping from St Louis, Missouri, to the market—distributor's warehouse is \$4 per unit. The same terminology applies. The shipper is the producer (source or origin) and the receivers are the distributors or customers (destinations). The configuration of origins and destinations are shown in Figure 10.1.

Total transportation costs to and from the Columbus, Ohio, plant are \$6 + \$2 = \$8 per unit; for St Louis, Missouri, they are \$3 + \$4 = \$7. Other things being equal, the company should choose St Louis, Missouri. However, the real world is not as simple as this.

The problem becomes more complex when there are a number of origins competing for shipments to a number of destinations. We will illustrate the complexity of the problem and its solution using the example of Rukna Auto Parts Manufacturing Company.

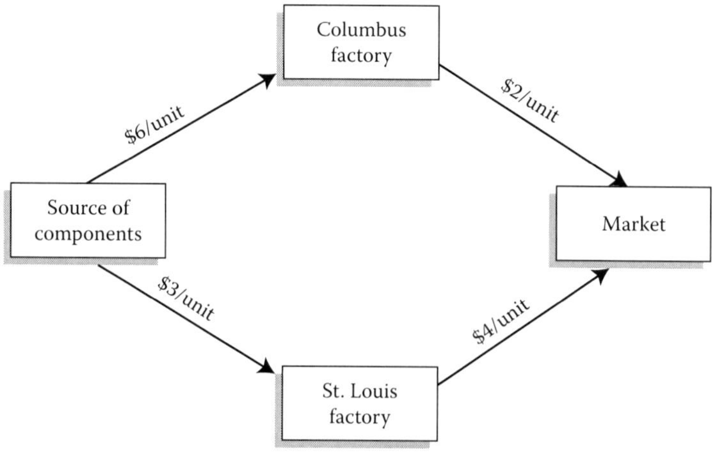

Figure 10.1 Configuration with one supplier, two factories, and one market.

10.8.2 Rukna Auto Parts Manufacturing Company

Rukna Auto Parts Manufacturing Company has three plants located in Miami (20,000), Tempe (40,000), and Columbus (30,000). The numbers in parentheses show the manufacturing capacity of each plant. Rukna supplies the auto parts to four distributors MKG, Inc. (25,000), ASN, Inc. (13,500), GMZ, Inc. (16,800), and Akla Inc. (34,700). The numbers in parentheses show the demand at each distributor. Therefore, the total supply (90,000) and demand (90,000) match. In a more general and complex problem the supply and demand may not match.

Rukna is experiencing an increasing demand for its products. It is expected that the demand will increase by about 20% within next 2 years. The growth is expected to be equally (can be different in a more general case) distributed among the four distributors. Therefore, the demand at the four distributors will be MKG, Inc. (30,000), ASN, Inc. (16,200), GMZ, Inc. (20,160), and Akla, Inc. (41,640) for a total of 108,000. Management of Rukna is considering setting up a new plant. Two alternative locations under consideration are Dallas and Chicago each with a capacity of 18,000 units. The management of Rukna needs to select one of the two proposed locations. The objective is to minimize the total cost of transportation.

Rukna first needs to determine the costs of transportation per unit from each of the plants, including the new ones, to each distributor. Table 10.3 shows the cost of transportation per unit from each plant to each distributor. For example, it costs $3.00 per unit to ship one unit from Miami to ASN, Inc. This table also shows the capacity of each plant and the demand of each distributor.

Table 10.3 Distribution Cost per Unit for Rukna Auto Parts Manufacturing

	MKG, Inc. (Cost in \$)	*ASN, Inc. (Cost in \$)*	*GMZ, Inc. (Cost in \$)*	*Akla, Inc. (Cost in \$)*	*Capacity*
Miami	1.00	3.00	3.50	1.50	20,000
Tempe	5.00	1.75	2.25	4.00	40,000
Columbus	2.50	2.50	1.00	3.00	30,000
Dallas (proposed)	1.50	2.50	3.75	1.75	18,000
Chicago (proposed)	3.20	1.75	2.0	4.35	18,000
Demand with 20% increase	30,000	16,200	20,160	41,640	

The problem has to be solved in two parts—one for each new site. The two problems are defined as follows.

Problem 1: Find the optimal distribution strategy and the corresponding cost for shipping auto parts from Miami, Tempe, Columbus, and *Chicago* to the four distributors.

Problem 2: Find the optimal distribution strategy and the corresponding cost for shipping auto parts from Miami, Tempe, Columbus, and *Dallas* to the four distributors.

The TM of linear programming, discussed in Appendix A can be used to solve this problem. The TM model finds the optimal distribution strategy that specifies the number of units to be shipped from each plant to each distributor so as to minimize the total cost of transportation.

Table 10.4 shows the optimal solution to problem 1 where Chicago is chosen as the new plant's site. This table shows the number of units to be shipped from each plant to each distributor. The total transportation cost is \$254,830 which is obtained by first multiplying the total number of units shipped and the cost of transportation per unit for each combination of the plant and distributor. These costs are then added together to get the total cost.

For example, the transportation cost between Miami and MKG, Inc. is \$2160 = 2160 (units shipped) × \$1.00 (unit transportation cost). Similarly the transportation cost between Tempe and Akla, Inc. is = \$95,200 = 23,800 (units shipped) × \$4.00 (unit transportation cost). The unit transportation costs are given in Table 10.3. The method to solve TM problems is described in Appendix A.

Table 10.4 Optimal Distribution Strategy—Chicago Plant

	MKG, Inc.	*ASN, Inc*	*GMZ, Inc.*	*Akla, Inc.*	*Total Shipped*
Miami	2160	—	—	17,840	20,000
Tempe	—	16,200	—	23,800	40,000
Columbus	9840	—	20,160	—	30,000
Chicago	18,000	—	—	—	18,000
Total	30,000	16,200	20,160	41,640	108,000

Table 10.5 Optimal Distribution Strategy—Dallas Plant

	MKG, Inc.	*ASN, Inc*	*GMZ, Inc.*	*Akla, Inc.*	*Total Shipped*
Miami	20,000	—	—	—	20,000
Tempe	—	16,200	—	23,800	40,000
Columbus	9840	—	20,160	—	30,000
Dallas	160	—	—	17,840	18,000
Total	30,000	16,200	20,160	41,640	108,000

If Dallas is chosen as the new site, the optimal quantities to be shipped from each plant to each distributor are given in Table 10.5. The total cost of distribution in this case $219,770. See Appendix A for solving the TM problem.

Since adding Dallas to the current set of existing plants gives a smaller total distribution cost as compared to Chicago, Dallas is the chosen site.

10.9 Location Decisions Using Breakeven Models

Breakeven models (discussed in more detail in Appendix A) can be used for making intelligent location and capacity decisions. To use breakeven analysis, it is important to review variable (direct) costs, fixed (indirect) costs, and revenues.

10.9.1 Variable (Direct) Costs

Variable costs per unit are the costs of input resources that tend to be fully chargeable and directly attributable to the product. They are also called direct costs because they are paid out on a per-unit of output basis. Direct costs for labor and materials are typical. Total variable costs, $TVC = C^*Q$, are the result of multiplying the variable cost per unit C by the number of units or volume (Q) produced.

10.9.2 Fixed (Indirect) Costs

Fixed costs have to be paid, whether one unit is made or thousands. For this reason, administrative and supervisory costs are usually treated as fixed charges. Together with purchasing and sales, they are all bundled together as *overhead costs*. Salaries and bonuses paid to the CEO and to every other manager are overheads because they cannot be assigned to specific products and services.

10.9.3 Revenue

Total revenue, $TR = P^*Q$, is the volume Q multiplied by the price per unit P. When goods and services are sold in the marketplace, they generate revenue. This breakeven model treats revenue as a linear function similar to variable costs. Each unit sold generates the same amount of revenue equal to the price (P). An assumption of breakeven analysis is that all units made and stored are delivered and sold.

The breakeven analysis question is: How many units of throughput need to be sold in order to recover costs and breakeven? The results change over time as factors vary. Fixed costs often grow, prices change, and variable costs per unit decrease with proper P/OM attention.

We illustrate the use of breakeven analysis for plant location decisions with the help of the following example.

10.9.4 Example—Musuk Spices Company

Musuk Spices Company (MSC), Delhi, India, manufactures high-quality spices primarily for the domestic market. However, MSC sees a big potential in the export market and therefore wishes to set up a manufacturing plant exclusively for its export division. MSC is considering two locations for establishing the new plant. The two locations are Bhopal and Agra in India. MSC has determined that the fixed costs per year will be \$450,000 and \$300,000 per year for Bhopal and Agra, respectively. The variable costs per pound are expected to be \$10/lb. for Bhopal and \$14/lb. for Agra, respectively. The selling price is expected to be \$30/lb. Musuk wants to pick up one of the two locations.

Using the concept of breakeven analysis, described in Appendix A, the breakeven points for the two locations are calculated below using the following equation.

$$\text{Breakeven point (BEP)} = \frac{\text{Fixed cost}}{\text{Selling price} - \text{Variable cost}},$$

$$\text{BEP (Bhopal)} = \frac{450{,}000}{30 - 10} = 22{,}500\,\text{lbs},$$

$$\text{BEP (Agra)} = \frac{300{,}000}{30 - 14} = 18{,}750\,\text{lbs}.$$

It means that Bhopal will generate profits only if the demand volume is more than 22,500 lbs. whereas Agra will start generating profits if the volume is more than 18,750. This analysis, however, does not tell us which plant is more desirable. To select one of the plants, we need to compare the profits from the two plants for various quantities and find the point of indifference. Suppose Q lbs. of spices are produced and sold by each plant. Then the revenues from the two plants will be:

$$\text{Revenue (Bhopal)} = Q(30 - 10) - 450{,}000 = 20Q - 450{,}000,$$
$$\text{Revenue (Agra)} = Q(30 - 14) - 300{,}000 = 16Q - 300{,}000.$$

By equating the two revenues we can find the value of Q at which the two revenues are equal. We get the following value of Q:

$$Q = \frac{450{,}000 - 300{,}000}{20 - 16} = \frac{150{,}000}{4} = 37{,}500.$$

This means that the plant at Bhopal is more desirable if the expected volume of sales is more than 37,500. If the expected sales are less than 37,500 then Agra is more desirable. Either one of them can be chosen if the sales are exactly 37,500.

This is in spite of the fact that Bhopal's BEP is greater than that of Agra. There is a catch which is that if sales are not yet certain and could be more than 37,500, but also could be less, then, we need a good forecast and a decision model to determine what is best to do. Also, if demand is volatile, sometimes up one year and down the next, then, we might be risk averse and pick Agra.

10.10 Facilities Layout

Layout is the physical arrangement of facilities within a manufacturing plant or a service facility. The layout of a plant specifies where various machines and equipment will be placed. Layout affects the productivity and costs of transportation (materials handling) within the plant.

It is hard to believe that until the 1980s, workers in American and European auto-assembly plants had to bend down to install the new tires on the wheels, and then tighten the tire bolts. The auto-assembly conveyors ran flat out near the ground. The layout made the job not much different from changing a flat tire on the road. Visitors to Japanese auto plants saw that the conveyor lifted the car up at the point of tire attachment. This carried the car above the worker and allowed putting the tire on at shoulder level. Workers could stand tall and attach the tires to the wheels without bending. Thereafter, conveyors that lifted the car were installed in world-class auto-assembly plants. This simple anecdote illustrates how good layout interacts with good job design. It also shows that layout of the workplace is a 3D situation lending itself to creative solutions.

It is evident how job design in the flow shop is partnered with layout. Productivity increases and quality improves when the layout and job are properly designed in the job shop. Travel time and distance traveled are reduced. Paths are kept clear. Queues at workstations are reduced. Production scheduling, shop-floor layout, and job design are connected in a systems sense. The benefits of improved quality derived from better layout must be continuously monitored because process changes are constantly occurring.

Intelligent service layouts are critically important. Disney has spent large sums of money on the configuration of line protocol and signage for its attractions. The idea is to make the wait as pleasing as possible. As the line loops around people get a chance to do "people watching." It turns out to be a form of entertainment enjoyed by many. In that way, good line design can reduce frustration. Mirrored walls in front of elevators allow people to watch themselves. That is also a frustration-reduction sport embraced by a goodly number of the elevator-waiting public.

Hospital corridors, steps, and elevators have to be designed with the best ride available for patients being transferred from (say) X-ray rooms to their wards. Robotic orderlies follow paths for delivery of required medicines. Optimal hospital layout interacts with technological invention. Smart restaurant layout pays off with fast deliveries to the tables and satisfied employees as well. Good school layout means that students do not have to walk miles between classes so class schedules and school layout interact in this case. The design of highways for cars is at a tipping point with respect to driverless autos. What is optimal for drivers may be different for drone autos.

10.10.1 Opportunity Costs for Layout Improvement

Proper workplace layout design can provide product and process quality improvements (QIs), throughput, and cost benefit productivity improvements (PIs), and health benefits (HBs) for employees. All of these result in higher profits. Evaluation of all jobs and workplaces along the supply chain will lead to layout improvements.

The paradigm used for such evaluations is like a balance scale or a seesaw. All of the costs of doing the job with the new layout X are put on one side of the ledger. All of the costs of not using layout X are put on the other side of the ledger. The lowest cost–highest benefit side (based on QI, PI, and HB) wins the contest. This describes the opportunity cost trade-off model succinctly.

Improving layout and job design involves opportunity costs trade-off analysis. Opportunity costs are the costs of doing less than the best. By doing something that is not the best, the cost to be paid is the difference in net benefits. There are opportunity costs for not having the best possible layout design.

The equation for this trade-off model states that layout design improvement should be made if:

CPI < OC (QI + PI + HB), where CPI = Cost of layout plan improvement

OC = Opportunity costs incurred for not having used the best possible layout with respect to QI, PI, and HB.

OC(QI) = Opportunity costs for quality improvements

These OC(QI) could have been obtained if layout design improvement had been made. If not made, they are the opportunity costs incurred by not rectifying quality deficiencies.

For example, improved conveyor layout would yield a better product, which is then translated into:

- Larger market share
- Greater revenues
- Fewer warranty claims
- Less service calls at company expense
- Higher prices that could be charged
- Smaller company discounts on ticket prices
- Better worker attitudes and morale
- Less dealer discontent
- More effective advertising campaigns

OC(PI) = Opportunity costs for improved productivity

These OC(PI) could have been obtained if layout design improvement had been made. They also are the opportunity costs incurred by not rectifying factors that deteriorate productivity. Not having to bend down and work in a crouched position would enable workers to work faster as well as better. The cost of installing four tires to each car could be reduced. It is possible that one or more workers could be freed up to do other jobs in the time now allotted for the repetitive job at the production rate. Job observations and time studies can be used to get good estimates for the opportunity costs for productivity. Unlike OC(QI), the costs OC(PI) and OC(HB), discussed next, are often measurable.

OC(HB) = Opportunity costs for health benefit savings

OC(HB) might have been obtained if layout design improvement had been made. For example, auto conveyors raise the cars so that tire assembly can be done without bending down. Costs associated with bending and crouching include back problems for workers that result in medical claims, higher health insurance, absenteeism, and lost time on the job. Computer operators experience a problem with their hands and wrists that is called carpal tunnel syndrome. Redesign of the computer keyboard, pads for the wrists to rest on, and other redesigns of the wrist support system have improved the situation. However, the strain of such repetitive work has led to the conclusion that keying jobs on the computer are best redesigned to include a regular rest period every hour. Job design, work schedules, and job layout must be treated together as an integrated system to deal with this issue.

10.10.2 Layout Types

There are at least six types of layouts that will be found in plants and offices. These are:

1. *Job shop process layouts.* Similar types of equipment or jobs are grouped together. The lathes are in one place and the presses in another. For services, filing is in one room and copy machines are in another. Inspectors are in one place; designers are in another. Job shop process layouts should facilitate processing many different types of work in relatively small lots. Mobility of equipment enables P/OM to set up intermittent flow shops when that configuration is attainable. Space must be allocated for work completed at one station and waiting for access to another station. Equipment mobility and layout flexibility allow this configuration to be rearranged to suit the various order mixtures that can occur.
2. *Product-oriented layout.* This layout is typical of the flow shop. Equipment and transport systems are arranged to make the product as efficiently as possible. The layout is designed to prohibit disruptions to the flow. The product-oriented layout is most often associated with assembly lines.
3. *Cellular layout.* This layout is used with a team of people and machines working together to produce a dedicated family of parts, as in group technology. The layout is engineered to facilitate efficient transfer of work between stations in the cell. Setup and transfer are programmed by the system for fast changeovers that enable small runs of a limited number of parts or items.
4. *Group technology layout.* GT layout is used to efficiently produce families of parts without emphasis on computer programming controls as in cellular layouts. Layout is based on the advantage of having similarity of design features.
5. *Combinations of product and process* orientation are very common. Some of the products in the job shop achieve demand volumes that allow them to be run for long periods of time as intermittent flow shops. Modular product-designed parts often have the high volumes that allow cellular manufacture or group technology layouts with, in some cases, serialized flow shop advantages. At the same time, other jobs in the shop remain at low volumes that are only suited to the process layout of job shops. The combination produces a mixed-layout orientation, which is also called a hybrid layout.
6. *Fixed-position (or static) layout.* In this layout the product does not move. The machines, materials and people are brought to the product. Examples include, shipbuilding and a skyscraper like Burj Khalifa in Dubai.

The complaint department of a large organization with a number of different products has the job shop process layout. Within that group, there are certain subgroups that handle high frequency requests with flow shop dedication. Sixty percent of the complaints can be treated in the highly repetitive fashion. The remaining 40% require special treatment for customer satisfaction. The analogy can be extended to families of complaint types and the use of group technology process layout.

Volvos are manufactured in Sweden on fixed platforms. The work does not move. Workers move to the platform and around it. Similarly, shipbuilders and homebuilders move around the work. Modular housing exemplifies hybrid layout where the parts are made in a factory and then brought to the site where they are assembled, similar to Volvos.

With fixed platforms, workers carry or drive their tools to where they are needed. Commercial airplanes are moved along a production line, albeit slowly. Commercial air-line layout is a different form of hybrid, one that combines fixed and moving positions. The *fixed-position* layout is necessary for power plants, refineries, and locomotives. It is much more controversial in the Volvo case which shall be explained below.

Volvo uses the fixed platform because Swedish workers find it more interesting to participate in building the whole car than doing a repetitive step along an assembly line. Worker motivation is improved because they all meet together around the platform. This is in contrast to assembly line work where each station is individual and relatively isolated. In Sweden, worker motivation is important because unemployment benefits are a large percentage of earned wages and readily available. Bored workers are likely to prefer the government benefits to the company wages.

The United Auto Workers (UAW) arranged an interesting experiment in which a group of GM workers were given the opportunity to work with Volvo for six months. Most of these workers said they preferred to work on the GM flow shop assembly layout in spite of, and because of, its repetitive nature. The Volvo layout required knowing and doing too many things. A few of the GM workers preferred the Volvo system and they remained with Volvo. The rest (and it was most) returned to Detroit.

Group technology (GT) for families of parts is a layout that many companies, such as Caterpillar, John Deere, and Cummins Engine, have found rewarding. GT cells are usually part of hybrid layouts. The group technology concept uses a product layout with the capability of producing an entire family of parts. Think about how clothing manufacturers size their clothes; paint manufacturers fill quart and gallons cans; shoemakers size their shoes.

Special equipment is required that has rapid changeover capabilities to make parts that are identical except for size and/or other dimensional characteristics. Crankshafts, motors, and pumps are typical parts for GT. The production cells use flexible machine-tool organization. Group technology layout is usually a self-contained part of a process layout job shop.

10.10.3 Layout Models

The plant layout problem can be tackled with different kinds of models. There is good engineering knowledge for designing flow shop processes. Usually, it is expensive to do the job well, and worth it. There is less solid knowledge about achieving excellence for job process layouts and less monetary justification. Here are four points to bear in mind:

1. Job shops and the batch production environment are subject to major changes in the product mix. What is optimal for one set of orders may be downright poor for the jobs in the shop one month later. Therefore, layouts that are likely to be good for the expected range of order types are better than those that are excellent for one type and not good for others.
2. The degree to which a few order types dominate the job shop will modify the statement in point 1 and allow some product layouts to be mixed in with process layouts. It will be found that the per unit costs of operating product layouts will be significantly less than the costs associated with process layouts. The comparison is even more extreme if GT cells can be set up.
3. Flexibility is desirable, but on the other hand, it is expensive and disruptive to keep moving equipment around the plant. A balance has to be found between these two objectives. If the character of the batch work changes a great deal over time, it is best to go for the most general form of process layout. If layout is to be changed from time to time, it is essential to set up the layout system with this capability in mind. Modular office layouts that are well planned have remarkable flexibility for making quick changes without paralyzing the workforce.
4. Use quantitative models with caution. Creative thinking and common sense pay off in achieving good layouts. Precise measurements count when trying to squeeze a machine into a tight spot. More questionable is the use of elaborate mathematical models to minimize total distance traveled or handling costs.

10.10.4 Layout Criteria

What criteria determine a satisfactory layout? Seven measures of layout effectiveness are presented below:

Capacity—throughput rate. Goal: Maximize total output volumes and throughput rates.

Balance—for perfect balance, the throughput rates of consecutive operations are 100% aligned and synchronized. Applicable for flow shops and projects; more difficult to evaluate for job shops—although balancing work rates for job shops is always desirable. Goal: Perfect balance.

Amount of investment and operating costs. Goal: Minimum expenditures.

Flexibility to change layouts. Goal: Maximize ease of change.

Amount of work in process (WIP). Goal: Minimize units of inventory.

Distance that parts travel; saving an inch traveled thousands of times per day sums to sizeable savings. Goal: Minimize total travel time and distance.
Storage for WIP and how much handling equipment is required to move parts from one place in the facility to another. Goal: Minimize space used for storage and minimize moving equipment.

10.10.5 Floor Plan Models

All seven of these criteria can be evaluated crudely yet with reasonably correct perceptions by using floor plan models. The floor plan drawings are graphic methods of trial and error that are used by interior decorators. They use paper cutouts that represent the furniture.

Plant layout models can be 2D or 3D. Often 2D floor plans, with cutouts made from templates, are used to represent the various pieces of equipment. When conveyors are employed, overhead space requirements may be important, and 3D "dollhouse" models are preferred. Otherwise, it is flat-planning which signifies that detailed studies have been completed and are being evaluated.

These techniques are useful for an incremental approach to a satisfactory layout. They do not bring any quantitative power to bear. Although an optimal layout based on the numbers may be far from satisfactory, the quantitative information might assist creativity.

10.11 Load–Distance Models

A relatively simple quantitative approach examines alternative layout plans in terms of the frequency with which certain paths are used. Usually, the highest frequency paths are assigned the shortest plant floor path distances to travel. The objective is to minimize the total unit distances traveled.

The physical space is divided into areas that conform to floor layouts including different work areas, stairs, elevators, rest rooms, etc. The floor space locations are designated as A, B, C, D, and E in Figure 10.2.

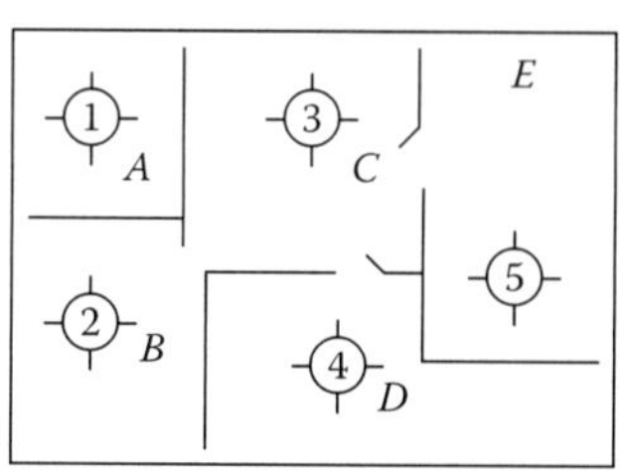

Figure 10.2 Layout (with areas A, B, C, D, and E, and work centers 1–5 assigned to the areas).

The average distance that must be traveled between areas A, B, C, D, and E are shown in Table 10.6. For example, distance between location A and D is 32 feet; between location B and C is 16 feet etc. Table 10.6 is called a distance matrix. This matrix is symmetric so the distance from A to D (32 feet) is the same as the distance from D to A. However, symmetry does not always hold because of one-way passages, escalators, conveyor belts, and gravity feed delivery systems.

Layout planning involves locating work centers at specific locations on the plant floor. Work centers are equipped to do specific kinds of jobs—like the press shop, copy room, hamburger grill, and darkroom. Manufacturing a product requires materials to move between the work centers and, therefore, between the physical locations. The materials are at different stages of work in process. Further, they represent various orders and job types. For this problem, there are five work centers designated as 1, 2, 3, 4, and 5.

The problem is to determine which work center is to be assigned to which location so that the total distance traveled is minimized. One possible assignment is: A(1), B(2), C(3), D(4), and E(5) where A, B, C, D, and E are the locations and the numbers in parentheses are the work centers assigned to them. We will call it Layout-1. This assignment is shown in Figure 10.2. In this discussion, we have used letters of the alphabet A, B, and C, etc. to represent physical locations and numbers 1, 2, and 3, etc. to designate work centers. However, these are not universal conventions. We could have used the letters of the alphabet to represent work centers and the numbers to represent the physical locations.

What is needed next is a measure of the amount of materials or the number of trips between the various work centers that different jobs entail over a period of time say monthly, quarterly, etc. It may be difficult, if not impossible, to include all jobs that are processed in a manufacturing department. Therefore, the layout is designed around jobs that are most important either because they are the most frequent, involve the largest number of units, are the most difficult and costly to transport, or are the most profitable. Often, no more than 20% of the jobs consume 80% of the transport resources. For the jobs that are most important, the objective

Table 10.6 Distance Matrix—Distances in Feet between Locations

		To				
		A	B	C	D	E
From	A	0	10	20	32	40
	B	10	0	16	18	20
	C	20	16	0	12	15
	D	32	18	12	0	10
	E	40	20	15	10	0

Table 10.7 Load Matrix—Number of Units being Moved between Work Centers

		To				
		1	2	3	4	5
From	1	0	100	60	80	20
	2	40	0	50	10	90
	3	80	90	0	60	30
	4	120	10	40	0	70
	5	110	5	5	30	0

is to minimize the total distance traveled. Matrices appropriate to these special jobs should be created and studied.

Table 10.7 presents the information relating to the number of units that move between work centers based on an analysis of the jobs selected to design the layout. For example, 100 units move from work center 1 to work center 2 and 60 units move from work center 1 to work center 3 and so on. This matrix is not symmetrical. For example, 80 units move from work center 3 to work center 1. See Section 10.13 (Finding the Load Matrix) to see how the Load Matrix is obtained. The word "Load" is a generic term. It may represent the number of units moved, weight or volume of the product that is moved, or the number of trips that are made between departments.

We can combine the load and distance matrices into a load-distance matrix given in Table 10.8. The load–distance matrix multiplies the distance and load for

Table 10.8 Load–Distance Matrix for the Layout-1

	A (1)	B (2)	C (3)	D (4)	E (5)
A (1)	0 $= 0 \times 0$	1000 $= 100 \times 10$	1200 $= 60 \times 20$	2560 $= 80 \times 32$	800 $= 20 \times 40$
B (2)	400 $= 40 \times 10$	0 $= 0 \times 0$	800 $= 50 \times 16$	180 $= 10 \times 18$	1800 $= 90 \times 20$
C (3)	1600 $= 80 \times 20$	1440 $= 90 \times 16$	0 $= 0 \times 0$	720 $= 60 \times 12$	450 $= 30 \times 15$
D (4)	3840 $= 120 \times 32$	180 $= 10 \times 18$	480 $= 40 \times 12$	0 $= 0 \times 0$	700 $= 70 \times 10$
E (5)	4400 $= 110 \times 40$	100 $= 5 \times 20$	75 $= 5 \times 15$	300 $= 30 \times 10$	0 $= 0 \times 0$

Table 10.9 Load–Distance Matrix for the Layout-2

	A (1)	B (4)	C (5)	D (3)	E (2)
A (1)	0	800	400	1920	4000
B (4)	1200	0	1120	720	200
C (5)	2200	480	0	60	75
D (3)	2560	1080	360	0	900
E (2)	1600	200	1350	500	0

each combination of the work center and the location area. For example, 100 units move from work center 1 (located at A) to work center 2 (located at B) for a distance of 10 feet (distance between location A and B). Look for the cell combination A(1) (the row number) and B(2) (the column heading) in Table 10.8. The total distance traveled by all items that move from A(1) to B(2) is, therefore, 1000 unit-feet. Similarly, total distances traveled from A(1) to C(3), A(1) to D(4) and A(1) to E(5) are 1200, 2560 and 800 unit-feet, respectively. In this way the distances traveled between each combination of work centers can be calculated. The total number of unit-feet traveled in this layout is 23,025.

Is this is the best layout? We cannot say yes or no until some alternative layouts are evaluated. There could be another layout which is better than the current layout. For this problem there are 5! (factorial) ways ($5 \times 4 \times 3 \times 2 \times 1 = 120$) to assign five work centers to five locations. Consider, for example the following assignments: A(1), B(4), C(5), D(3), and E(2). We will call it Layout-2. Calculations show that an improvement results when these new assignments are used. These new assignments are the basis of Table 10.9. The total number of unit-feet traveled in this layout is 21,725. Unit-feet traveled has been decreased by 1300. That is an improvement of almost 6%. It is likely that additional decreases can be achieved by studying the matrices and trying to eliminate incorrect assignments of heavy-unit volumes to large distance movements. The improved version is based on a reasonable heuristic discussed in the next section.

This analysis has tested only two configurations out of 120 possibilities. There remain 118 other possible assignments of work centers to locations. It may not be feasible to test all different layouts. Instead, heuristics rules are used to develop a reasonably good (not necessarily the optimal) layout. The heuristic rules are discussed in the next section.

10.12 Heuristics to Improve Layout

Two heuristics (rules of thumb for improving the system's performance) are useful when searching for policies that might help to improve plant or office layout.

The heuristic methods are logical, sensible, and clever rules for finding good solutions to complex problems.

1. Assign work centers with large unit flow rates between them to locations as close as possible.
2. Assign work centers with small unit flow rates between them to locations as distant as possible.

In Table 10.7, look at those pairs of work centers that have the largest work-flow rates between them. These are in descending order: Work center 4 to work center 1 has a flow of 120 units. Work center 5 to work center 1 has a flow of 110 units. Work center 1 to work center 2 has a flow of 100 units. Assign these work centers to locations that are as close as possible. Note (in Table 10.6) that the distance between A and B is 10 feet and between C and D is 12 feet.

Using the data in Table 10.6, the biggest distances are between locations A and D (32), and between A and E (40). This means that Layout-1, A (1), B (2), C (3), D (4), and E (5), assigned 120 units to the second biggest distance (viz., 32 between A and D) in the matrix of Table 10.7. Further, Layout-1 assigned work center 1 to location A and work center 5 to location E. It means that 110 units were assigned to move the largest distance between locations—viz., 40 between A and E.

To follow the heuristic "assign the largest number of units to the shortest distance" would mean assigning work center 1 to A and work center 4 to B. That was done in Layout-2, which reassigned work center 4 to location B. Because work center 1 was already assigned to A, the next best distance from A is 20 at C. Therefore, work center 5 was assigned to C in Layout-2.

The second part of the heuristic calls for assigning the centers with the smallest work-flow rates between them to the locations that are as distant as possible. The two smallest flow rates between centers (in the matrix of Table 10.7) are as follows: Work center 5 to work center 2 has a flow of 5 units. Work center 5 to work center 3 has a flow of 5 units. From Table 10.6, the shortest distances from C to D is 12 and from C to E is 15. Therefore, assign work center 3 to location D and work center 2 to E. This is Layout-2 (the revised assignment) shown in Table 10.9.

The heuristic did its job, but it is not clear whether further improvement is still possible. Trial and error with the improved matrix can be used to test further shifts.

The literature on layout has suggested the use of various mathematical techniques and the use of computerized algorithms. See, for example, Buffa et al. (1964) and Tompkins et al. (1996).

10.13 Finding the Load Matrix

The load matrix gives the number of units of all jobs that are moving from one work center to another. Consider the problem given in Table 10.10. There are five

Table 10.10 Load and Processing Sequence for a Five-Job Problem

		Processing Sequence—Operation Number							
Jobs	*Load*	1	2	3	4	5	6	7	8
Job 1	100	A	B	C	A	C	D	F	E
Job 2	200	C	A	B	D	B	E	F	D
Job 3	300	C	B	A	B	C	D	E	F
Job 4	50	B	A	C	D	F	E	F	C
Job 5	150	A	B	D	C	E	F	D	C

jobs, Job 1 through Job 5. The number of units to be produced of each job is given in the column "Load". The load can represent the number of units or number of trips. There are six work centers that are labeled as A, B, C, D, E, and F. The sequences in which these work centers are required to produce these jobs are also given in Table 10.10. For example, 200 units of Job 2 are to be produced and these units move through the following sequence of the work centers: C-A-B-D-B-E-F-D. Instead of the number of units, this could have been the number of trips of Job 2 between various work centers. The number of trips essentially depends on the number of units produced and the number of units included in one trip. Material handling equipment can transfer one or more units at a time which determines the number of trips.

The load matrix for this problem is given in Table 10.11. Consider, for example work centers A and B. The load matrix shows that 750 units are being moved from work center A to work center B. Let us see how to get the number 750.

Table 10.11 Load Matrix

		To					
		A	B	C	D	E	F
From	A	0	750	150	0	0	0
	B	350	0	400	350	200	0
	C	300	300	0	450	150	0
	D	0	200	300	300	0	150
	E	0	0	0	0	0	700
	F	0	0	50	350	150	0

The information given in Table 10.10 will be used. The movement from A to B occurs for the following jobs.

- Job 1 (100 units), after completing its first operation in work center A, goes to work center B.
- Job 2 (200 units), after its second operation has been done in work center A, goes to work center B.
- The third operation of Job 3 (300 units) is processed in work center A and then this job goes to work center B.
- The first operation of Job 5 (150 units) is processed in work center A and then this job goes to work center B.

Job 4 never goes from work center department A to B; although it goes from B to A but that is a different order. Therefore, the total number of units that go from work center A to work center B is 750 (= 100 + 200 + 300 + 150). In the same way, the number of units that go from work center A to work center C is 150 and so on.

Summary

The subject of facility management comprises choosing the location, determining the structure and the site, selecting equipment, and preparing the plant or office layout. There are numerous qualitative factors to take into account. Then, the quantitative modeling can begin. At the heart of facility decisions is recognition that each facility is part of a supply chain network. The big picture must be kept in mind to optimize the supply chain and not the facility alone.

The TM is used to minimize shipping costs or maximize profits where production location and market differentials exist. Scoring models are developed, which can be applied for site or equipment selection. The workplace layout and job design interact strongly. Floor models that minimize distance traveled are explained.

After quantitative modeling, it is necessary to return to the qualitative factors to evaluate the relevance and quality of the solutions offered by the models. The systems approach is explained because teamwork among all participants in the supply chain is desired in making location, site, and equipment decisions.

Review Questions

1. With four work centers and four locations there are how many possible layouts?

2. Finding the best locations for police and fire stations is a pressing urban problem. Discuss the nature of this problem and what variables are likely to be important. How does this compare to the facility layout problem?
3. Why do services usually profit by being close to the customer?
4. Is it true that plant selection should, at least in part, be based on estimates of labor costs for different areas of the country? Explain.
5. Gasoline station location often is based on traffic density studies. What would be the criteria for a good location in terms of traffic density and patterns?
6. With the scoring model, what happens when a weight of zero is chosen for one of the factors?
7. One of the best known, most used, and most abused scoring models is the credit scoring model. The credit score is used by lenders to impute the likelihood that a customer will default. Discuss the strengths and weaknesses of credit scoring as it is presently used.
8. Company X has been renting a facility in the Columbus, Ohio area. This is a great general location because a 600 mile circle drawn around the city contains a large percentage of X's customers. The building in which X rents has been sold and the new owners intend to refurbish it and raise the rents significantly. The president of X has hired you as a consultant to help in relocation. She points out that zoning is of great importance. Prepare a list of questions that you must ask so that you can approach this problem in a systematic way.
9. Some of the following industries tend to form high-density clusters in specific geographic areas. Identify those industries that have strong clusters and try to find the rational explanation for these specific clusters. In the same sense, identify those industries that do not have strong clusters and try to find the rational explanation for the lack of clustering.

a. Financial services
b. U.S. automobiles
c. Stockyards
d. Steel
e. Textiles-bathing suits
f. Semiconductors
g. Aerospace
h. Garlic processing
i. Motion pictures
j. Publishing
k. Tobacco products
l. Petroleum
m. Credit card processing
n. Non-U.S. automobiles
o. Theme parks and resorts
p. Advertising agencies
q. Soy products
r. Pharmaceuticals

Problems

1. The cost of improving plant layout is \$100,000 to be depreciated over a 5-year period. The estimated annual improvements in profit are \$6000 from better

quality, $4000 from higher productivity, and $11,000 from health benefits. Is the layout improvement recommended?

2. The cost of improving office layout is $27,000 to be treated as a one-time expense. The office manager does not have any way of estimating improved profits resulting from quality changes but says that the new layout will save time and miles of walking for the 40 office employees. She estimates that the new layout will cut the amount of walking for the average employee by 100 miles per year. How much will the layout improvement be worth?
3. A feasible solution for a 3-plant, 3-market problem is given below.

	Markets			*Supply/Day*
Plants	MA	MB	MD	
P1	40	20		60
P2		40	30	70
P3			60	60
Demand/day	40	60	90	

The costs of shipping per unit are given in the following table.

	Markets		
Plants	MA	MB	MD
P1	$23	$32	$25
P2	$16	$25	$19
P3	$30	$38	$42

What is the total cost associated with this assignment?

4. The American Company has two factories, A and B, located in Wilmington, Delaware, and San Francisco, California. Each has a production capacity of 550 units per week. American's markets are centered in Los Angeles, Chicago, and New York City. The demands of these markets are for 150, 350, and 400 units, respectively, in the coming week. A matrix of shipping distances is prepared. Determine the shipping schedule that minimizes total shipping distance. Estimates of shipping distances (in miles) are shown in the following matrix, along with supply and demand.

Matrix of Shipping Distances in Miles

	Los Angeles	*Chicago*	*New York City*	*Supply*
Wilmington	3000	1000	300	550
San Francisco	400	2000	3000	550
Demand	150	350	400	

5. Using the load-distance model, determine a good arrangement for work centers at locations for the numbers given in the following matrices.

   Distance Matrix: Distances (feet) between locations A, B, C, D, and E from row i to column j.

i/j	A	B	C	D	E
A	0	10	20	32	40
B	10	0	16	18	20
C	20	16	0	12	15
D	32	18	12	0	10
E	40	20	15	10	0

   Number of units flowing between work centers 1, 2, 3, 4, and 5 from row i to column j:

i/j	1	2	3	4	5
1	x	50	60	80	20
2	40	x	50	10	90
3	80	90	x	60	30
4	50	10	40	x	70
5	60	5	5	30	x

   If the cost of a unit moving one foot is one dollar ($1 per unit per foot traveled), what is the total daily work-flow cost of the solution obtained?

6. Consider the data given in the table below. This table gives various factors to be considered in a location decision. Four alternatives under consideration are also listed in the table. The table also includes the score for each factor for each location. Calculate the total score for each location. What is the best location?

Scoring Model					
		Location Alternatives			
		Chile	*Mexico*	*Honduras*	*Brazil*
Factor Name	*Weight*	*Score Out of 10*			
Labor productivity	0.06	8	7	3	6
Nearness to markets	0.08	4	6	9	7
Nearness to sources of raw material	0.15	3	5	2	8
Infrastructure facilities	0.19	7	3	4	4
Transportation facilities	0.09	6	6	7	9
Power availability	0.07	5	8	6	7
Political climate	0.04	9	9	8	8
Labor unions	0.03	3	4	3	3
Labor cost	0.04	6	5	5	5
Material cost	0.25	7	2	1	2
Total	1.00				

7. There are five jobs, Job 1 through Job 5. The number of units to be produced of each job is given in the column "Load" in the table below. There are six processing departments (work centers) that are labeled as A, B, C, D, E, and F. The sequences in which these processing departments are required to produce these jobs are also given in the table below. For example, 200 units of Job 2 are to be produced and the sequence of the processing departments is C–A–B–D–B–E–F–D.

Load and Processing Sequence for a Five-job Problem

		Processing Sequence—Operation Number							
Jobs	*Load*	1	2	3	4	5	6	7	8
Job 1	100	A	B	C	A	C	D	F	E
Job 2	200	C	A	B	D	B	E	F	D
Job 3	300	C	B	A	B	C	D	E	F
Job 4	50	B	A	C	D	F	E	F	C
Job 5	150	A	B	D	C	E	F	D	C

Each one of these processing departments A, B, C, D, E, and F requires 400 square feet. The plant in which these departments will be located has six rooms labeled as Room 1, Room 2, Room 3, Rooms 4, Room 5, and Rooms 6. The size of each of these rooms is 20 feet × 20 feet (400 square feet).

The figure below shows the floor plan of the building with the six rooms. The problem is to find which processing department should be located in which room. Assume that the movement of the material takes place only in horizontal and vertical directions from the center of the rooms. For example see the two arrows showing movements between Room 1 and Room 2 and Room 3 and Room 6; the distance between Room 1 and Room 2 will be 20 feet (center of Room 1 to center of Room 2).

Floor Plan

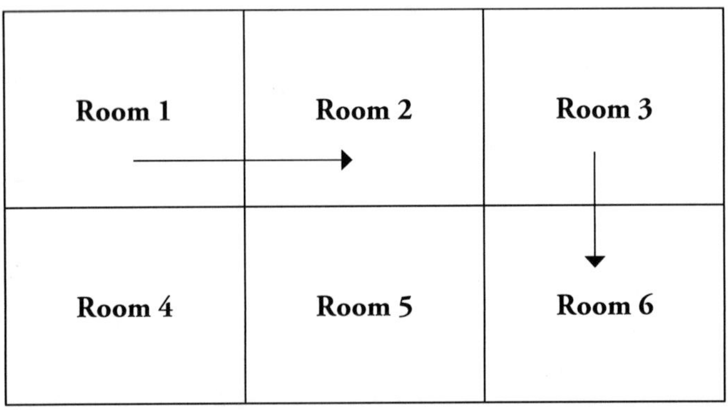

Find the total unit-feet traveled for the following three room assignments. Which one is the best?

a. Assignment 1

Room	Room 1	Room 2	Room 3	Room 4	Room 5	Room 6
Department	A	B	C	D	E	F

b. Assignment 2

Room	Room 1	Room 2	Room 3	Room 4	Room 5	Room 6
Department	A	C	B	D	E	F

c. Assignment 3

Room	Room 1	Room 2	Room 3	Room 4	Room 5	Room 6
Department	A	C	D	B	E	F

8. The Columbus Technology Corporation (CTC) plans to add an extension to its current manufacturing plant. CTC will locate four departments, each of size 30 feet × 30 feet in the following configuration.

Department 1 30 feet × 30 feet	Department 2 30 feet × 30 feet
Department 3 30 feet × 30 feet	Department 4 30 feet × 30 feet

Sequences of processing assembled products through these departments are given in the following table.

Sequence of Processing Assemblies Through Departments		
Assembly SKU	*Sequence of Departments*	*Units Produced per Year*
Q655	1-3-2-4	1800
Z432	2-4-3-1	600
T691	1-2-3-4	2200
K518	2-1-3	850

a. Complete the Load Matrix.
b. Complete the Distance Matrix.
c. What is the total unit-feet traveled per year?
d. What is the total unit-feet traveled per year if departments 1 and 3 are interchanged?

Readings and References

Benjaafar, S., S. Heragu, and S. Irani, Next Generation Factory Layouts: Research Challenges and Recent Progress, *Interfaces*, Nov–Dec, 2002, pp. 58–78.

Bozer, Y. A., and R. D. Meller, A Reexamination of the Distance-Based Layout Problem, *IIE Transactions*, 29(7), July 1997, pp. 549–560.

Buffa, E., S., G. C. Armour, and T. E. Vollmann, Allocating Facilities with CRAFT, *Harvard Business Review*, 42, March–April 1964, pp. 136–158.

DeForest, M. E., Thinking of a Plant in Mexico?, *The Academy of Management Executive*, 8(1), 1994, pp. 33–40.

Denton, B., Decision Analysis, Location Models, and Scheduling Problems, *Interfaces*, 30(3), May–June, 2005, pp. 262–263.

Fulton, M., New Factors in Plant Location, *Harvard Businesss Review*, 4(17), May–Jun 1971, pp. 166–168.

Houshyar, A., and Bob W., Comparison of Solution Procedures to the Facility Location Problem, *Computers & Industrial Engineering*, 32(1), Jan 1997, pp. 77–87.

Lee, R. S., and J. M. Moore, CORELAP-Computerized Relationship Layout Planning, *Industrial Engineering*, 18, March 1967, pp. 195–200.

Meller, R. D., and K.-Y. Gau, The Facility Layout Problem: Recent and Emerging Trends and Perspectives, *Journal of Manufacturing Systems*, 15(5), 1996, pp. 351–366.

Mentzer, J. T., Seven Keys to Facility Location, *Supply Chain Management Review*, 12(5), May 2008, p. 25.

Muther, R., K. McPherson, Four Approaches Computerized Layout Planning, *Industrial Engineering*, 2, 1 970, pp. 39–42.

Partovi, F. Y., An Analytic Model for Locating Facilities Strategically, *Omega*, 34(1), Jan 2006, p. 41.

Price, W. L., and M. Turcotte, Locating a Blood Bank, *Interfaces*, Sep–Oct 1986, pp. 17–26.

Roodberger, K. J., I. F. A. Vis, A Model for Warehouse Layout, *IIE Transactions*, 38(10), Oct 2006, pp. 799–811.

Snyder, L. V., Facillity Location Under Uncertainty, *IIE Transactions*, 38(7), July 2006, p. 547.

Swamidass, P. M., A Comparison of the Plant Location Strategies of Foreign and Domestic Manufacturers in the U.S, *Journal of International Business Studies*, 21(2), Second Quarter, 1990, pp. 301–317.

Tompkins, J. A., J. A. white, Y. A. Bozer, E. H. Frazelle, J.M.A. Tanchoco, and J. Trevino, Facilities Planning, Wiley, New York, 2nd edition, 1996.

Zeng, A. Z., M. Mahan, and N. Fleut, Design an Efficient Warehouse Layout to Facilitate the Order-Filling Process, *Production and Inventory Management Journal*, 43(0), 3rd–4th Quarter 2002, pp. 83–88.

Chapter 11

Innovation by P/OM for New Product Development (NPD) and Sustainability

Reader's Choice—If you always do what you always did, you will always get what you always got.—Albert Einstein

Atasu, A., Daniel, V., Guide Jr., R., and Van Wassenhove, L., So What If Remanufacturing Cannibalizes My New Product Sales? *California Management Review*, Winter 2010, 52(2). If the remanufactured item is a perfect substitute, then the customer is willing to pay the same as if it were new. Since company costs are lower, profit is improved. Remanufacturing can be a legitimate manifestation of innovation and new product development (NPD).

Bonabeau, E., Bodick, N., and Armstrong, R.W., A More Rational Approach to New Product Development, *Harvard Business Review*, March 2008. Type I errors occur when "good" NPD projects are terminated by mistake. Type II errors occur when managers ignore evidence that NPD projects are likely to fail. There are various reasons why managers "turn a blind eye" to impending NPD failure. This article advocates minimizing the cost of both errors as early as possible.

Christensen, C.M. and Overdorf, M., Meeting the Challenge of Disruptive Change, *Harvard Business Review,* March–April 2000. Sustaining technologies are innovations that make a product or service perform better in ways that customers in the mainstream market already value; disruptive innovations create an entirely new market through the introduction of a new kind of product or service. The disruption effect of innovations is established by this article.

Laurie, D.L., Doz, Y.L., and Sheer, C.P. Creating New Growth Platforms, *Harvard Business Review,* May 2006. Successful innovation is viewed as a succession of alterations of the original invention. Hence, the idea of creating "new growth platforms (NGPs) by identifying and populating families of strategic opportunity" is discussed.

Toffel, M., The Growing Strategic Importance of End-of-Life Product Management, *California Management Review,* Spring 2003, 45(3). EOL laws have proliferated over the past decade, imposing a host of new requirements on manufacturers. Innovations, in this case, may deal with recycling features that will never be of concern to customers. In addition to being green, they often can improve profits.

Unruh, G. and Ettenson, R., Winning in the Green Frenzy, *Harvard Business Review,* November 2010. As motivation to be green in both product and process increases, there is pressure to accept green standards. The authors warn: do not let competitors define what "sustainable" means, modify existing standards, or create the best standards to use. Do not imitate—innovate.

Zipkin, P., The Limits of Mass Customization, *Sloan Management Review,* Spring 2001. Mass customization requires "three main elements" which are elicitation (can customers define what they want?), process flexibility, and delivery logistics. The market may not be aware of the benefits let alone be willing to pay for them.

Once upon a time, or should we say "In the old days" production was manufacturing and operations were what surgeons did to remove tonsils. There were also logistical, mathematical, and secret operations. All that has changed as service operations became a fundamental part of P/OM. Operations has become the proper way to describe activities intended to make a product or deliver a service whether for profit or not. Also, boundaries of responsibility and authority were expanded to permit the crossing of functional lines and traditional disciplines to better resolve problems. All operations that are relevant to the solution of a problem must be included in the definition of that problem and in the methods used to solve that problem. Consequently, the connection between processes and products led to the inevitable understanding that new products must fall within the joint domains of P/OM and marketing. NPD is at the core of the competitive evolution of the business enterprise. That core must be sound so that the dynamics of evolution produce processes that are beneficial and sustainable in an entirely global society.

This chapter provides an understanding of the role of innovation in new product development (NPD). This chapter also explains why NPD is a critical function of production and operations management (P/OM). Striving to innovate is the appropriate state of mind (means to an end) needed to achieve successful NPD. Also, in this chapter the relationship of innovation to *sustainability* will be developed.

From Wikipedia: "Sustainability requires the reconciliation of environmental, social equity and economic demands—also referred to as the "three pillars" of sustainability or (the 3 Es)."

After reading this chapter, you should be able to:

- Make it clear why adaptability is essential to innovation.
- Detail how innovation applies to humanitarian operations and crisis management.
- Discuss why sustainability has become an important goal for all organizations.
- Explain why innovation is essential for successful sustainability.
- Discuss the three pillars on which sustainability stands.
- Describe the application of patent protection for innovations.
- Make evident why continuous innovation is essential for success.
- Explicate how coordinated teamwork is critical for continuous innovation.
- Describe how product innovation failures can be avoided.
- Explain the importance of the lifetime value (LTV) of a loyal customer.
- Calculate LTVs and explain what NPV has to do with LTV.
- Describe the structure and importance of the product planning platform.
- Illuminate the character of product modularity.
- Explain mass customization and product life cycles.
- Describe the dynamics of the brand switching matrix and show calculations for equilibrium market shares.

- Explain blue versus red ocean strategies.
- Detail the nature of closed-loop supply chains and explain why they are increasingly more important.
- Discuss the roles of imitators and innovators.

11.1 Introduction to NPD and Innovation

P/OM uses innovation to design the *processes* which are either entirely new or adaptive (making incremental or major changes from old systems to new ones). This process design aspect of P/OM is based on strong linkages to product design. The goal culmination is the responsibility of both marketing and P/OM working closely together—the appropriate idiom is—hand in glove.

For proper coordination of functions we need to keep in mind that marketing and P/OM use different lingos. It is understandable that managers of P/OM and marketing have hurdles that must be overcome to cooperate fully in every aspect of the transitions from original designs to new ones. Strong interactions are necessary to deal competently with all NPD projects.

Decisions made by both marketing and P/OM must be shared in a *transparent environment* to achieve organizational success. Forecasting is often done separately by both marketing and P/OM. *This is unacceptable.* Forecasting responsibilities must be shared. Differences in P/OM and marketing expectations must be explained leading to acceptance of a common vision. Other aspects of coordination, including the critical importance of teamwork, will be addressed later on in this chapter.

11.2 Organizations Must Be Adaptable

Success as defined by organizations dedicated to making a profit for their shareholders is measurably different from success defined by not-for-profit organizations. For example, the U.S. Centers for Disease Control and Prevention (CDC) succeed when disease is eliminated and good health is pervasive. For example, benefits (not profits) describe goals of the Red Cross, OXFAM, and The Salvation Army. On the other hand, profits define the goals of airlines, banks, computer manufacturers, hotels, restaurants, supermarkets, and cable companies.

At this stage in the text, every student should be able to define financial success as it applies to profit-making organizations. Defining successful innovations for non-profit organizations is more difficult. Often it means that costs have been reduced without any detrimental effect on services provided. Less tangible measures of what constitutes improved *benefits* need evaluations of services offered, and these can be viewed differently by various constituencies (e.g., different demographic segments). In spite of this problem, there is general agreement that innovations can provide a means to success for both profit and not-for-profit organizations.

In fact, both types of organizations have a lot to learn from each other. Organizational specifics always predominate in defining innovations. However, the state of mind that encourages creative change by means of innovation is similar across the boards. At the root of successful innovation are competence in the field, confidence in management's skills, and an ability to adapt to changing conditions. Adaptability is strongly related to comfort with alteration of the status quo. There is much relevant literature devoted to the attributes (such as flexibility) required for adaptability. McKeown's book on *Adaptability* (McKeown, 2012), lists 17 rules. Rule 2 is: All failure is failure to adapt. A cause of failure to adapt (cited by McKeown) is inability to recognize the *need for adaptation*. At this point, it should be noted that adaptability and sustainability are very closely related.

11.2.1 Innovation Is Necessary for Humanitarian Operations and Crisis Management (HO&CM)

P/OM competence in achieving desirable innovations for "society" is increasingly called upon. This is clearly the case in the area of humanitarian operations & crisis management. P/OM is the master of processes needed for acquisition, storage, and transportation of critical materials and personnel—from and to—the disaster zone. P/OM alone is capable of designing, implementing, and monitoring supply chain operations. P/OM is solely responsible for repair and maintenance.

The necessities of survival (physical and mental) must be addressed when disaster strikes. While each disaster is different, there are similar patterns of what occurs, and what must be done. P/OM has the knowledge, experience, and connections to innovate appropriate solutions. Crisis managers are systems thinkers looking at the big picture of what must be accomplished with the resources at hand. The *generic* methodology (of supply and demand) has been applied by P/OM over many years, but the HO&CM application is relatively new. No pat answers exist.

Many unique situations demand new approaches to ideation (i.e., idea generation). These include group "brain storming." It will be useful to look at the Wikipedia entry for Brainstorming. This term was explained and popularized by Alex Osborn in his 1953 book, *Applied Imagination* (Osborn, 1993). In HO&CM, solutions must be applicable to situations with specifics that have never been encountered before—even if the generic circumstances are entirely familiar. One flood is not the same as another to the crisis managers on the ground reacting to real people in experiencing tangible disaster conditions. Particularly important is the need to innovate with speed while avoiding the penalties of "making waste with haste."

The tradeoff between speed and accuracy requires new methods of making appropriate innovative decisions. Crowdsourcing (i.e., outsourcing for ideas from volunteers drawn from a broad demographic of the general public) can provide competitive advantage (CIO, 2013). Other possible approaches include: Genius Bars (as in Apple stores), Geek Squads (as in Best Buy stores), Knowledge Management Methods using data mining and intelligent repositories of relevant data, may be other methods for innovation. It may be useful to use Google to search for these terms. That will provide further clarification.

A large number of new product projects fail. The percent is significant even if elusive. A fair average of various reports might be about 70%. That is a lot of money lost and energy expended without a return. The opportunity costs (i.e., lost opportunities to be successful) are staggering. A great deal can be done to improve this record. As a beginning, all innovative organizations should learn how to be successful. In this regard, it is important to maintain a history of success and failures. Memory must be retained to promote learning. Who are the successful innovators? What methods did they use? Conversely, who and what caused failures? A successful innovator in one area often fails miserably in another domain. Data analytics can be fruitfully applied to an organization's innovation record. The feedbacks from forecasts to the design team are critical.

11.2.2 Innovation Is Essential for Sustainability

Managing world systems (controlling changes on a global scale) is a huge problem that is now called "achievement of sustainability." This refers to many facets of global dynamics, one of the most important of which is *protecting the environment* from critical disruptions (let us call it the first of three pillars). The climate has never been steady; it has always been in flux. Mark Twain said "climate is what we expect, weather is what we get." That idea still holds true. Weather is more fickle than climate.

The critical issue is whether the changes that are now being experienced are caused by people (e.g., increasing the carbon dioxide levels in the atmosphere). If the causes are natural (such as activity levels of the sun, or planetary orbits) then our efforts must be to adapt to (the extreme) situations that are likely to occur. In other words, when actions are taken by governments to control variables that have nothing to do with impending nature-caused catastrophes—such actions mislead people to believe they have controlled the situation. In fact, they have taken their eye off-the-ball. They are addressing the wrong situation. A basic P/OM rule is that one must adapt to inevitables. See Figure 11.1 where, along the top branch, planning for the *inevitable* is one of five options that must be addressed. Planning for the *improbable* takes a lot of courage and often pays off well. It takes a Walt Disney to strive for the *impossible*.

Sustainability is a very difficult concept because it involves defining "satisfactory conditions that should be maintained" and defining changes that are "not acceptable." Although this subject does involve philosophical issues, P/OM is not inclined to philosophy. Instead, as the process-master it requires standards and methods to measure deviations from acceptable levels. For example, P/OM knows what controls regulate CO_2 emissions and it can determine how much it will cost to meet a new standard, and if that is feasible. Thereby, the environment (pillar 1) and the economic factors (pillar 2) become interdependent.

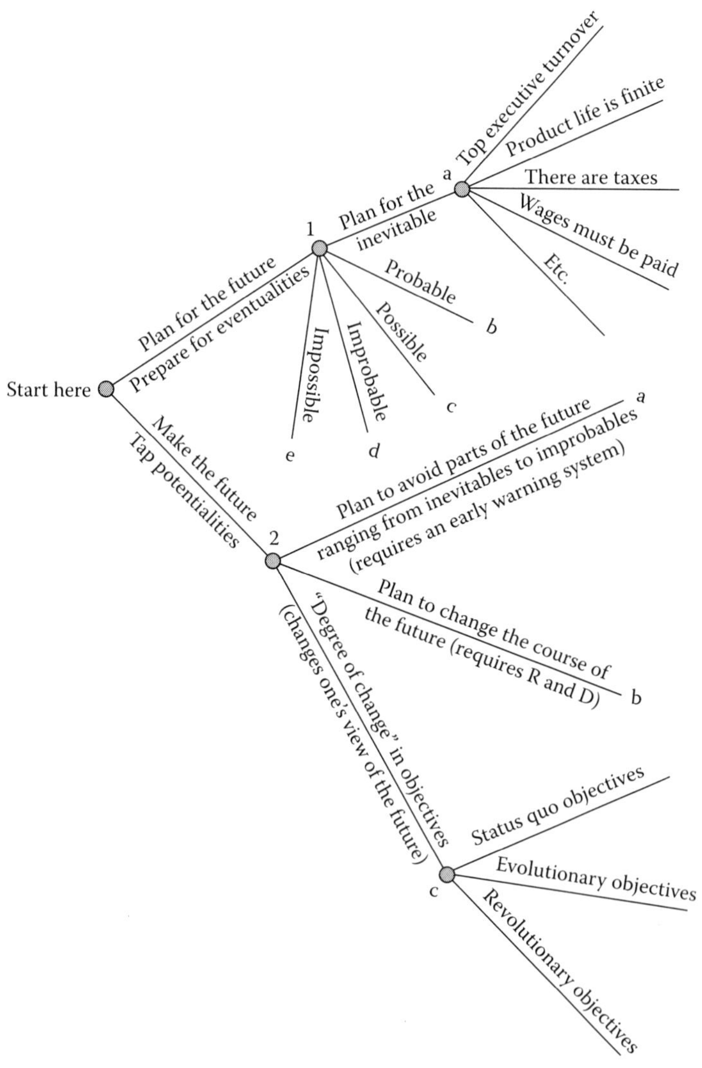

Figure 11.1 Planning tree of management options.

Pillar 2 of sustainability refers to economic forces. It is a contention of almost all economists that real business cycles in economic systems are normal and outside the control of governmental intervention. However, even the most passive interventionists believe that some steps can be taken to alleviate the pain of severe recessions. Proactive governors take steps to *shield their economies* from economic down turns. Shielding (protecting) from long-lasting damage is a strong form of sustainability. The jury has not yet passed judgment on what can and cannot be done to safeguard the economic well-being of developed and developing countries. P/OM's potential contribution is significant in both helping to innovate new products that increase revenue generation and to innovate process designs that are world-class competitively.

There are many theories about short, medium and long-term cycles. In the long-term category, there is the Kondratiev Wave. This 45–60-year cycle is important because it reflects economic fundamentals of innovation (see Wikipedia for Nikolai Kondratiev). When there is a major new invention (e.g., steam engines, internal combustion engines, light bulbs, and computers) investors rush in to gain a return on redesign for betterment and process improvements. Over five or six decades, return on investment (ROI) in aging technology diminishes in a marked way. Improvements become increasingly incremental. It is logical that investors search for alternatives. When a promising replacement innovation appears, such as LEDs (light-emitting diode) instead of tungsten filament, investors' attention switches from the old marginal technology to high-return, high-risk investments in new technology. P/OM's role in guiding these major economic transitions is evident, whether the cycle is named Kondratiev or something else.

There are many short-term cycles (e.g., the Kitchin inventory cycle of 3–5 years) that have been proposed. They have less effect on sustainability. Long-term cycles tracking changeovers in technological dominance are far more significant. P/OM can provide understanding of what is going on and guidance going forward. When applying this long-cycle concept to all economies, it is imperative to recognize that change is the only constant. Those who are chosen to manage the dynamics of long-term cycles (no matter how many years are involved) need to recognize the sustainability challenge. While particulars of any specific technological "evolution" are crucial, awareness of generic properties of long-term cycles can improve decisions about when to use innovation.

The third pillar of sustainability is even more ambiguous than the prior two pillars. It is termed "the social dimension." What can P/OM do to address societal needs? From p. 174 of an article entitled *Social Sustainability: One Company's Story,* by Jesse Dillard and David Layzell, in the referenced text (Dillard, 2009), "We describe how one company, Intel Corporation, frames and responds to its perceived social responsibilities. … Sustainability originates in manufacturing, has an operational flavor within the company, and is implemented and monitored through input/output ratios." The authors conclude that Intel uses the term *corporate responsibility* to address various aspects of social sustainability.

This difficult concept of societal necessities interacts with the other two pillars of sustainability. P/OM plays a crucial role in creating well-paid, steady jobs that help: raise the living standards of the world; maintain high levels of social health; contain environmental damage created by every human activity. Faced with destructive forces in nature, human malevolence, and economic cycle theories, innovation by P/OMS, in coordination with others on the team, is an essential ingredient of sustainability.

The best approach may be in terms of the goal of an ever improving *quality of life* (QOL). If QOL is not improving, at least it must not be degraded. That is a tried and true principle upon which to base sustainability. It translates into "preserving the environment, staying "green," and allowing only responsible developments. From both a corporate and governmental point of view, P/OM champions sustainability as a means of being and remaining successful. There are enough examples of organizations that could not sustain their early successes to serve warning on those who cannot innovate in the face of dire threats. Since technology development is moving faster and faster, the cycles of innovation and investment are required to speed up as well. P/OM project management must be adjusted accordingly.

Section 11.2 has addressed the fact that *every* organization, whether striving for future success or working to protect present achievements must adapt and master transitions. The speed of change has accelerated because of extensive innovations in communication methods and transportation technologies. The rate of change is accelerating for new products and their underlying processes. Transition management and innovation may be so similar in their definitions, and in their basic characteristics, that they become indistinguishable, one from the other.

11.3 Competition for New Ideas, Resources, and Customers

All organizations must innovate to stay relevant and to continue to succeed. That is because everything is changing in their environments (e.g., technologies, tastes, taxes, temperatures, and training). The number of noteworthy competitors for any market (goods and services including humanitarian operations) has increased. The world is no longer divided into territories (countries and even regions) with protected boundaries.

The costs of transporting materials have become so insignificant relative to other costs of doing business that location is often of secondary importance. For example, if the cost of moving products and resources great distances from low-cost producers to markets has less impact on total costs than the low production costs, the latter trumps in decision making. In the long run, however, the costs of transporting materials to production sites and the costs of transporting finished goods to markets is sheer waste. Better systems can be found that cut out transport costs.

11.3.1 Patent Protection of Innovations

New technology cannot be guaranteed to be protected even by strong patents. In large measure, the twentieth-century concepts of the U.S. Patent Office do not apply to technological developments of the twenty-first century. There is increasing internationalization of the patent system with national offices in Japan, China, Korea, Europe (EU), Germany, etc. Almost every country has its own patent laws and laws of many countries differ. There is a Patent Cooperation Treaty with many cases of disagreement in the courts. A new patent law went into effect (Spring 2013). It is called *America Invents Act* and it marks a fundamental change. The old first-to-invent rule has been replaced with the international standard, first-to-file. It is too soon to interpret the effect of this change.

11.3.2 Continuous Innovation (Step 1)

Before computer systems became powerful, and widely available, technological diffusion took time. Now, copying can be accomplished in days. What was once considered to be too complex to imitate is no longer valid. The economic benefits of patent protection are under-whelming (at the present time). *Continuous innovation* is the only sure way to stay ahead of the pack.

The first step is to cultivate an organizational attitude that fosters creating new ideas to replace old ideas. An organizational culture that embraces disruption (Christensen, 2002; Immelt, 2009) must cut across functional domains (Jassawalla, 2000). A culture that accepts failure as a learning opportunity (Thomke, 2012) is primed for success. It takes a great deal of anxiety or foresight, and flexibility for traditional organizations to make such changes.

The reason that anxiety is listed as a possible necessity for changing from a status quo firm to one that innovates continuously is that organizations are constantly challenged with extinction from outside competitors. The potential damages are not trivial. Management must take actions from inside the organization to counter the attacks from outside. Otherwise, deterioration of market conditions is inevitable. Competitive efforts to seduce loyal customers are inescapable.

Consider on-going battles between smart phone manufacturers. Apple, Blackberry, Microsoft, Motorola, Nokia, and Samsung (in alphabetical order) are vying with each other to grab market share by outdating their opponents' products. Challenges are frequently based on competitive intelligence (knowledge about weaknesses in competitors' products, often gained from focus group market research; sometimes gained by pressure, bribes, and spying). Anxiety may be a driving force when it gradually becomes evident *for those willing to see* that change is required for survival. The toy business is an excellent example of the need for continuous innovation. Mattel's Barbie Boutique is a tribute to strategic thinking about continuous updating of the battle plan aimed at coping with fierce competition. The similarities to military situations in combat are striking. There is more than enough anxiety for all participants to acknowledge.

As another illustration, the rising cost of kerosene fuel *forced airlines to innovate with their revenue base.* They decided to charge customers for a host of services including baggage, food, drinks, preferred seating, as well as the traditional ticket price. At the same time, having only one type of aircraft (or a few), cuts maintenance, inventory, and training costs. Quick turn-around time for an aircraft is equivalent to increasing capacity utilization (at which Southwest Airlines is masterful). Delta's purchasing department (a P/OM function) used hedging to buy fuel futures for a number of years in advance to lock in lower fuel costs. That was widely acknowledged as very innovative. Such P/OM innovations permit more services to be provided with no additional fees.

Obtaining and sustaining success requires understanding that all organizations are on a treadmill that moves quickly and abruptly changes speed. Only those that are fit and prepared can remain in the game. The best preparation for survival is based on successful *adaptation* which is a major form of innovation. Airlines adapted by changing long accepted rules about satisfying customers with free meals, blankets, etc. It turns out that customers will pay for amenities and they do not have to be satisfied to be customers because there is no other airline that can get them to where they are going. That strategy required reducing competitive choice by merging airlines having shared routes. This seems to run counter to governmental opposition to blatant monopolies.

From the discussions above, it is evident that *innovative* NPD applies to all forms of organizations both domestically and globally in highly differentiated ways. Though alternate goals are held by various organizations, it is common that they all want to succeed and that innovations provide crucial paths to success. As such, P/OM is both creator and controller in the achievement of innovations that are designed to enhance success for-profit and not-for-profit organizations.

11.3.3 Continuous Innovation (Step 2)

New product innovation failures must be avoided. Step 2 is to rigorously and continuously test innovations. JC Penney undertook a major innovation of retailing starting in November of 2011. Without testing the new concepts, JC Penney launched into the altered strategies which reduced sales by 4.3 billion and stock price by more than 50%. The CEO who championed the new plan (a combination of market pricing and operations management retail methods) was fired in April of 2013. Testing the new plan might have allowed it to be modified or rejected as unworkable. As any retailer will allow, many operational issues relevant to success or failure were introduced without testing the impact they would create. Always test innovations before activating them.

The fashion industry is always perpetrating change. Part of the success story is the continual testing of new concepts at fashion shows around the globe. Quick service restaurants have new product introductions on a regular basis to maintain the freshness of offerings but also to capture interest in new menu items for reasons of health or to "ride the wave" of interest in new foods. The toy industry must make commitments to new products long in advance of the Christmas season which accounts for a large percent of annual sales. Knowing the "lay of the land" requires a market research and forecasting capability that not only tests new concepts but provides directions of trends for the future. P/OM acknowledges that market research is the closest friend it has in the firm.

11.3.4 Coordinated Teamwork Is Essential

The NPD project team must move toward the goal of product release like a rugby team and not like a relay race. All members of the team work together *in synchronization* (Takeuchi, 1986). They do not wait for one sub-group to complete its work before engaging in their own related aspects of product development. Innovative NPD managers can find lots of ways to connect and link parallel activities to support synergies throughout the entire project. Effective leaders overcome functional differentiation; they cut across boundaries and follow non-linear paths (Jassawalla and Sashittal, 2000).

Narrow definitions of NPD project teams can be avoided by recognizing the breadth of viewpoints and knowhow that will facilitate breakthrough innovations. For example, suppliers of components and vendors of materials may have ideas that accelerate innovation beyond traditional boundaries. Therefore, good suppliers must be on the team (Handfield, 1999).

Teams will work quickly to uncover latent problems and to gain the advantages of everyone (from all disciplines) knowing about *the total system* as early as possible. Costs of discovery about product and project weaknesses are relatively low at the starting point of projects. They become increasingly greater after project directions have been set. That is because the costs of reversals are quite high in both time and money. Figure 11.2 illustrates the typical advantage of innovations that facilitate early-resource utilization in P/OM projects. Examples of benefits of understanding requirements for coordination of the total system that provide faster completions and lower costs include: new product development (NPD), design of processes for manufacture (DFM), design for assembly (DFA), and design for rapid project completion (DFRPC).

For example, it would have been relatively inexpensive and far less traumatic to fix the Tacoma Narrows Bridge which spanned Puget Sound in the state of Washington, USA, instead of having to replace it. It was opened in July 1940 and collapsed four months later because of aeroelastic flutter. It was known as *Galloping Gertie* because of extreme vertical motions of the roadway in the wind.

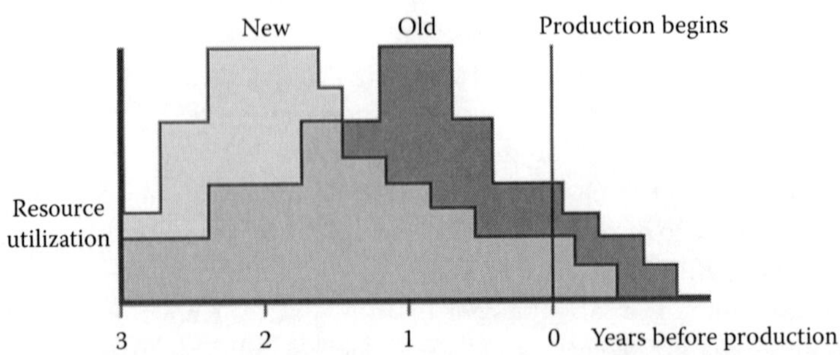

Figure 11.2 More resources are allocated initially to avoid error-correction costs.

The Bronx-Whitestone Bridge had similar problems which were fixed with eight stay cables (1940) and trusses (1943). The Tacoma Narrows Bridge problem could have been avoided with proper P/OM-type systems analysis. The important question to ask is "why was P/OM-engineering type systems analysis missing?" The fact that such an obvious question is still lacking in so many recent structural tragedies provides insight about managing innovations in the future.

11.4 Product Innovation Failures Can Be Avoided

Let us further examine the record of unsuccessful *innovations*. There are numerous examples of failed innovations. Yet, increasingly, new products are the only means to bring about positive change in a company's status—especially when there are competitive attacks. Innovation may be a necessity to survive but innovation failures are a risk that can be avoided with due diligence. Management knows that change is risky but good managers know that there is growing demand for them to take such risks. Smart risk-taking is a critical component of innovation ability.

The oldest (fallacious) version of sound management bureaucracy is that one avoids failure by taking no risks. The best way to avoid risks (again fallacious) is by making no changes. So, when faced by doom, the consummate bureaucrat looks away. When faced by change from outside, that is not a prescription for success. It is not a way to avoid failure. Successful strategies must embrace well thought out change. Random or careless changes will not succeed. How changes are designed and carried out will be crucial for success. JC Penney's might have been a successful innovation if testing (including market research of prices and retail operations) had been properly employed.

Before we look at other examples of failures, let us amplify what is meant by testing. Proper testing includes design of experiments followed by careful analysis. Smart information collection is vital. Ultimately, asking the right questions is essential. The building blocks of success require coordination of market research and P/OM planning using *in-depth market analysis of relevant forecasts—to assist with P/OM evaluation of inventories, facilities, and staff needed.* P/OM must be able to synchronize all processes (e.g., retailing methods are production strategies) with marketing strategies. Such prudence might have avoided most of the situations described below, as well as the JC Penney example given above.

An often cited example of a product change that had a potentially catastrophic negative effect was made by the (otherwise) brilliant strategic planning company, Coca Cola. For an explanation of how this could happen, it has been suggested to consult the word "hubris." Another way of explaining this situation is "a knee-jerk reaction."

Here is the story: The Coca Cola Company launched an innovation called "new Coke." As a guess, newly hired managers convinced anxious older management that innovation was essential to remain competitive. There was truth in that statement. Coke's arch competitor, Pepsi-Cola, was engaged in full-scale combat with an innovative product line that was attacking Coca Cola with alarming success. Under such circumstances, it is generally agreed that an innovation attack requires an innovation counter-attack. To ignore competitive attacks is a big mistake, but it is crucial to test with care before rolling out an entirely new product.

11.4.1 Failure Might Turn Out to Be an Unexpected Opportunity

Coke provided its competitive response by reformulating traditional Coke with "new Coke." This was announced by the Coca Cola Company on April 23, 1985. It was soon apparent that "new Coke" was a disaster. Formerly loyal customers hated it. The Coca-Cola Company then returned to the original formula. They called it "Coke Classic." It is said that huge marketing expenditures were required to recover from this fiasco. We have only common sense to back this up. However, there is no doubt that New Coke's failure was due to flawed innovation. Coke paid an enormous penalty for the new product concept error that had been made. Why had the entire new concept not been rigorously tested? No Coke explanation is available for what happened.

Coke handled this matter brilliantly. It took this fall; then recovered by renouncing the mistake and grew an ever-larger market share as a result. Although we do not know the real story behind the screen, we do know that Coke recovered and improved its market situation. There is every reason to believe that in so doing, the campaign for Coke Classic was tested, modified, and directed by rigorous analysis. A strong recovery, as in this case, required carefully orchestrated innovation.

Recovery from failure *often engenders stronger bonds between customers and company*. We do not know of any documented case where failure was used as a competitive strategy from which to recover with an even stronger market franchise, but it may be a feasible strategy if done correctly. Our own personal experiences tell us that this is the case.

11.4.2 Failure Caused by Inadvertent Process Change

Many processes are improved by innovations that consumers never see. Instead, they reap the rewards in terms of quality enhancements. Also, processes are changed to increase consistency of output. The result is a product innovation through process improvement. Sometimes, however, a process change is made without testing the consequences. Again, there can be bad results.

As an example, again for the beverage industry, the Perrier story follows. A headline in the *New York Times* dated February 10, 1990, said "Perrier Recalls Its Water in U.S. After Benzene is Found in Bottles." The article goes on to say "The impurity was discovered in North Carolina by county officials who so prized the purity of Perrier that they used it as a standard in tests of other water supplies."

Then, traces of benzene were found in the product in Holland and Denmark. Further analysis indicated that the product may have been contaminated for a long as six months. The French Ministry of Health certified that the spring which was the source of this chic bottled mineral water was not contaminated.

Where did the benzene come from? Eventually, Perrier found that the problem was caused by charcoal filters used in the bottling process. This situation reflects a serious P/OM failure in its responsibility for control of quality. It is not believed that the charcoal filters were intended as an innovation. They were defective. The process had been changed inadvertently. Perrier had to follow a recovery plan, and it took a great amount of time for them to succeed. P/OM has the greatest responsibility to maintain process consistency. Every maintenance operation could be the source of possible deviations from process specifications. Aircraft accidents have been traced to incorrect maintenance procedures (e.g., neglecting to put O-rings into jet engines). While we advocate learning from mistakes, in aircraft maintenance, the consequences of errors can be dire.

We have given a number of examples of changes to products and processes that failed for different reasons with serious repercussions. New product perceptions among customers must be carefully scrutinized using the best available market research capabilities. New Coke would not have survived such tests. Similarly, processes must not be allowed to change without in-depth analysis of the results. Proper tests of its charcoal filters would have saved the Perrier Company from huge penalties. Proper P/OM would have so advised general management that testing for process consistency costs very little in comparison to the costs of process inconsistencies.

11.4.3 Unintended Consequences of Product and Process Changes

Good innovations in production methods can provide beneficial (albeit hidden) rewards. For example, better cell-phone connections from towers and satellites, healthier foods with sanitary harvesting, cleaner manufacturing, and fewer auto recalls (caused by product design innovations interacting with vendor problems and process glitches). Although consumers generally do not get to see the producer's facility, and they may not be able to discern the product changes, eventually they experience both problems and advantages linked to changes. The unintended consequences of defective charcoal filters for Perrier provide an excellent example. Eventually, *intangibles* can have a major effect on customer loyalty. The effect of *tangible* (directly visible) innovations may be more immediate and dramatic, but not necessarily as long-term and powerful.

It is clear that product and process innovations can introduce changes that may be beneficial or harmful. What is not so evident is that they can be both. Good change is evident when enough customers want to buy the new product(s) at a price that ensures financial success. In familiar terms, total revenues are substantially greater than total costs (way **above breakeven**). The route to such success is never simple because (as in physics) every force (new products force change) gives rise to one or more counter forces (competitive newer products). In other words, successful innovations give rise to competition that can produce severe financial distress for the innovators. The harmful aspect of other beneficial innovations arises when the original innovator is not ready to counter the competitive onslaught.

Further complicating the picture, as more venture capital backs up new product introductions, markets are destabilized. As noted previously, Christensen first called attention to the powerful effect of "disruptive innovation" (Christensen, 2002). Production managers must take notice since development of new processes is costly and time-consuming, whereas consumers may shift their product loyalties overnight. In other words, a stable supply system can be disrupted by an unstable demand system. Further, such occurrences of severe destabilization are accelerating as powerful competitors make their stand in the global market. P/OM is both the challenger and the challenged. From the point of view of P/OM when challenged by innovations coming from outside, managers of operations must be alerted and warned to take nothing for granted.

Changes of many kinds are occurring faster and faster. There are more competitors offering an increased variety of choices. A new competitor has a *greenfield* advantage (i.e., unencumbered by investments in earlier technologies). Older firms have put money into machines and programs that are called legacy technology. The term "legacy" means old and superseded but difficult to replace because of the hierarchy of commitments and unamortized capital expenditures. Legacy problems define the burden of the *brownfield* competitor who cannot start from "scratch."

Innovation is reflected in the rate of change of new technologies replacing old *legacy technology*. Legacy inheritances (from an earlier management era) are constraints on innovation. Even if funds can be found to replace old software or hardware (superseded by improved versions) it is often difficult because many employees are still committed to using what was once "the best." They are comfortable with the old technology. Cost disruptions for retooling and relearning are only part of the trauma of switching from the familiar. The *greenfield competitor* enters the fray without the burden of previous commitments that foster hostile attitudes to change.

Sometimes the *greenfield* technology results in a product that is so superior that customers will purchase it even if it means throwing out old, but functioning, products. Customers in developing countries may not even have the old products to replace. For example, many places that never had hard-wired telephones—skip over that old technology and go directly to mobile cell phones. Product superiority may also relate to longer product life and fewer failures from age and fatigue.

Innovation to counter age and fatigue failure is exemplified by new LED lights which last much longer than the old tungsten bulbs. Though LEDs are more expensive than tungsten bulbs, they save money in power and maintenance costs. They also pollute less than compact fluorescent lamps (CFLs). The CFLs represent a transition step away from tungsten but cannot compete with the LED technology as a green alternative.

From Hartford Courant (courant.com—dated March 3, 2013): "The Energy Independence and Security Act of 2007 requires most screw-in light bulbs to use at least 27% less energy by 2014, notes Consumer Reports. CFLs, LEDs and some halogen bulbs, a type of incandescent, meet that requirement. Standard incandescents do not and are being phased out. As of January 1, 2012, 100-watt bulbs were no longer being made or imported but could be sold until supplies run out. ... California began its phase-out in January 2011. Europe and Australia in 2009." Similar statements can be made about 75-, 60-, and 40-W bulbs. The entire old lighting industry (product and process) has been innovated out of existence in a matter of a few years. The rapidity of this changeover is astounding, but it is driven by simultaneous cost and QOL improvements.

Capital expenditures for process equipment have been avoided by P/OM (in conjunction with finance) by going overseas to outsource production functions. From a P/OM point of view, it is important to recognize that lower wages are not the only driver for outsourcing. In fact, a major explanation of outsourcing abroad is the avoidance of *evanescent relevant investments*. In other words, innovation can outdate technology so quickly that satisfactory ROI is not achieved. Instability in demand creates uncertainty for P/OM investments. Today's winner may be forgotten tomorrow. The uncertain dynamics of innovation teach caution. Successful managers have learned the hard way—they cannot *rest on their laurels**.

11.5 Continuous Project Management Is a Successful Innovation

To begin with, products must be developed starting with a concept and its *goals*. Note that the concepts are best derived from a menu of former and existing products—which, if done properly, is called "*the conceptual platform*." Always augmenting that carefully constructed pre-existent menu of interrelated concepts is one of the ways in which project management is a continuing activity of production, marketing, and other members of the team.

Next, design alternatives are proposed to achieve the *goals*. These designs must be reviewed as quickly as possible, because speculation (about design alternatives) gets leaked and published. It can be useful to competitors and harmful to the innovator. Decisions have to be made rapidly to select designs that are best suited to the goals. Then, prototypes are created for consumer testing. If the test results align with the goals, the new products are *launched* into the market. Thus, the project, which started with product conception, now enters its pre-launch stage. This is when consumer tests (such as evaluation by focus groups) confirm that the prototypes are

* Laurels date back to antiquity as an award for quality. "Laurels" are wreaths of laurel—sweet bay leaves—placed on the victor's head in ancient Greece. Olympic medals are imprinted with a sprig of laurel. The *Nobel Laureate* award is a great, yet controversial, honor.

on target to achieve the objectives. Before the new product is launched, test results must confirm that the design is optimal (i.e., tweaking will not make it better).

Speed is essential because innovators are vulnerable to information leaks. There are a great many avenues for scientific and technical intelligence to be procured by potential competitors. Speculation about the details of innovations by such companies as Apple, Google, IBM, Intel, Qualcomm, and Samsung abound. Such "intel" (which has been known to involve actual spying) comes from many sources including in-house workers but more often from those who work for vendors that are thought to be supplying materials, parts, and components for the final product.

Testing (for example, with focus groups) puts potential versions of the product in the hands of consumers who are not obliged to keep secrets. An industry of journalists tracking developments has appeared which feeds on stock market frenzy of traders who want to know about innovations that can disrupt existing markets. There are also a plethora of patent contests which leads to such news items as "Jury awards Apple $1.049 billion in Samsung patent dispute ruling." That was August of 2012. In March of 2013 the amount was reduced to $598 million. We can expect many more legal maneuvers. See Section 11.3.1 on Patent Protection of Innovations.

The final project stage consists of ramping up the process output to match the scale of expected demand. If this is not done properly, the innovation can fail to achieve market success. Problems caused by supply not being in *sync* with demand occur when customers, expecting a new model, stop buying the prior model (e.g., a new version of Office software, next year's auto model, or a new iPhone release). Ideally, prior version inventories are reduced to match the expected demand at a lower price level. The project stages that are described above are depicted in Figure 11.3.

After launching, further testing (and forecasting) takes place to ascertain if changes are required. Note testing (which is based on samples of consumer responses) occurs repeatedly. It must be done continuously because market conditions are altered

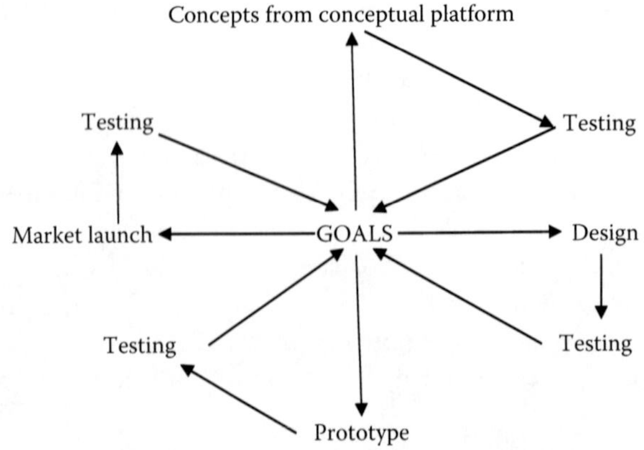

Figure 11.3 Continuous on-going project management cycle.

by competitive actions; there is much economic volatility; there are constant shifts in commodity prices, etc. Forecasting abilities must be honed to a fine art. Company adaptive responses can include product price changes, package and delivery method changes, and responsive modifications in production output schedules.

The entire repetitive system of the continuous project management cycle is replicated over and over. The conceptual platform is revisited with each completed cycle. Pre-planned functionality that has not yet been activated by design decisions is likely to be called into action. This on-going NPD methodology is the new performance standard for competent project managers. The old beginning and end of a project (with disassembly of the project group) is now passé. The next iteration—of concepts (derived from the pre-planned concept platform), designs, and prototypes—is always created to achieve upgraded **goals** by means of innovation.

11.5.1 New Service Development Applications

The same terminology applies to services. The project procedure described fits the development of goods and services (including non-profit goals). Innovative IT development is applicable as well. As an example, a gym starts with a limited array of equipment. It plans to gradually expand the variety of machines to apply to a multitude of aerobic exercises. Testing competitive appeal relative to other facilities in the neighborhood is sensible. Layout designs will be improved by means of focus group testing. As profits are reinvested, pre-planned acquisitions of new workout equipment will alter floor layouts and flow patterns of members which can lead to new designs.

The nature of the product (for example, laundry detergents vs. electric cars vs. coffee and tea shops) will alter the specific character of the NPD plan. Working out exact details for every cycle of the project requires competent project management with expertise in the product category. Myriad elements must be done right. When mistakes are made, failures occur that range from—badly chosen "open" hours, unappealing or unreliable equipment—to insufficient cleaning operations, bad staff scheduling, and overall, insufficient demand. For most goods, late deliveries and defective products can lead to merchandise returned and contract cancellations. In goods and service situations, quality defects can lead to injuries—resulting in legal actions with serious financial consequences. For all goods and services, product callbacks are signals of management incompetence. They will damage customer loyalty unless they are handled in a very caring way. Forgiveness is a wonderful trait of humanity, but it only appears when sincere apologies are accompanied by competent redress of an unforgiveable error.

11.5.2 Lifetime Value of a Loyal Customer

P/OM and marketing know that customer disappointment caused by faulty product can diminish or even extinguish company (i.e., brand) loyalty. They also understand how diminished levels of loyalty will affect their respective operations. For example, lower volumes of demand change the economics of process design. Price decreases might increase demand but the lower margins will affect the quality of materials (and/or skills) that can be used. A spiral of decreasing loyalties can ensue with grave long-term consequences for the *sustainability* of the system. Liability lawsuits for damage, injury, and death are amplifications of lifetime value (LTV) losses. They can be potential killers of a company without sufficient reserves. Such penalties are not included in the normal derivation of the LTV of a loyal customer.

Innovations may help to reinvigorate consumer demand. The analysis of LTV of loyal customers is valuable for managing sustainability. Loyalty will be defined as repeated reuse of the goods and/or services. It does not mean that a loyal customer purchases one company brand exclusively. How to calculate LTV will be spelled out in sufficient detail to enable students of this chapter to perform such analyses. How LTV and innovation interrelate will be explained.

LTV of an angry disenfranchised customer can amount to a substantial revenue loss. For example, the average loyal pizza customer for a well-known chain (brand) has been shown to have a LTV revenue of about $8000 (Heskett, 1994). The spreadsheet in Figure 11.4 depicts a 20-year lifetime expenditure of an average pizza buying family summing to $10,800 (i.e., $540 per year times 20 years). $540 was derived from twenty-seven $20 purchases per year. The total of column B ($10,800) is the sum of 20 payments of $540.

The $8,000.51 calculation of revenue, at the bottom of column D, is based on the sum of the net present values (NPVs) of each yearly payment with a prevailing interest rate of 3.045%. NPV compares the value of a dollar in hand now with the value of that dollar in the future (say a year from now). If a dollar will be paid every year for 20 years, NPV will calculate the sum of those values which will be less than $20.00 if the interest rates are greater than zero. The larger the interest rates, the smaller the sum will be—because delayed payments represent money which cannot be invested until the payment is received. Thus, the NPV of $1.00 paid at the end of one year (rather than at the beginning) is $0.97 with a 3% interest rate. All payments in Figure 11.4 are at the end of year with the interest rate of 3.045%. Figure 11.4 is an Excel spreadsheet with Column C based on f_x = NPV(0.03045,A1) = $524.04 for cell C2, f_x = NPV(0.03045,A1,A2) = 1032.6 for cell C3, etc. For Column D, $D_n = C_n - C_{n-1}$.

The discounted amounts become progressively smaller as payment is delayed another year into the future. For example, payment value at the end of the 20th year ($296.38) is 55% (actually 54.885%) of the fixed yearly payments of $540. Note that the value of $540 paid at the end of the first year is $524.04. Column C lists the *cumulative* discounted value of the payments in column D—year by year—which explains why the cumulative value in the 20th year equals the total of column D.

To catch the gist of what NPV accomplishes, let us pose the question—how much is an IOU (from the pronunciation of I owe you) worth that is dated January 1, 2015

	A	B	C	D
1	Year	Amount of annual expenditure	Cumulative Discounted expenditure	Discounted expenditure
2	1	$540	$524.04	$524.04
3	2	$540	1032.6	508.56
4	3	$540	1526.13	493.53
5	4	$540	2005.08	478.95
6	5	$540	2469.87	464.79
7	6	$540	2920.93	451.06
8	7	$540	3358.65	437.73
9	8	$540	3783.45	424.79
10	9	$540	4195.69	412.24
11	10	$540	4595.75	400.06
12	11	$540	4983.99	388.24
13	12	$540	5360.75	376.77
14	13	$540	5726.38	365.63
15	14	$540	6081.21	354.83
16	15	$540	6425.55	344.34
17	16	$540	6759.72	334.17
18	17	$540	7084.01	324.29
19	18	$540	7398.72	314.71
20	19	$540	7704.13	305.41
21	20	$540	8000.51	296.38
22	Total	$10,800		8000.51

Figure 11.4 Excel spreadsheet for revenue of a famous pizza company.

which states "I promise to pay bearer $1.00 on January 1, 2035"? That question can not be answered until an estimate is given for interest rates, year by year. To be consistent with our example and to simplify the situation, assume that interest will be constant at 3.045% over the next 20 years. Then, the IOU will be worth 55 cents. To derive this result, divide the 20th-year payment of $296.38 in the year-20 row cell of column D by the amount it would have been if the interest rate was zero, (i.e., $540)—which is the bottom cell in column B. The quotient is $0.55.

NPV calculations, such as the ones just used, are built into Excel spreadsheets. Click the "insert function" icon (f_x) in Excel. It will be found on the Formula Bar which appears above the row of column letters. The Insert Function box will pop up. Where it says "search for a function" write in NPV. Then, click "Go." Select the function NPV and click on "ok." The Function Argument Box is shown with windows for Rate and Value 1, Value 2, etc. Type the interest rate in the Rate window. In the Value 1 window, designate the column of values to which the NPV calculation will be applied (For example, B1:B20 in Figure 11.4). There is no need to enter numbers for Value 2. The NPV result is shown in the Total Row, Column D, in Figure 11.4.

Suppose the pizza chain delivers a pie with a roaming roach to a loyal customer who, as a result, will switch allegiance to another pizza chain. Assume this happens nationwide once per day. Although this is a very small number for a large pizza company, the resulting annual loss for this defect is almost $3,000,000. Other complaints will add to losses. For late deliveries or a damaged box, most customers would not be alienated, but some might be, and there are other issues such as the rudeness of employees taking orders. Market research testing is essential to find out what are the major triggers for alienation (i.e., customers switching from their accustomed brand).

Critical issues that apply to Section 11.5 include management fully endorsing the need for detail. It is always essential that projects proceed with careful checking. On the *dashboard* of relevant measures are dials to monitor sales during the early growth rollout. Are there enough new customers coming on board? Do these customers remain loyal after buying the brand? Do these customers provide total revenues that well exceed breakeven? To answer these questions, constant market research is essential to track consumer behavior. Plans for adjustment of strategies should be prepared beforehand. Costs must be monitored as well. How else can the managers be assured they are operating within budget? Monitoring costs permits correction of over-spending. Project design should provide up-to-the-minute information from accounting. In addition to cost control, the quality of production output is critical. In this regard, it is essential that the process deliver products that conform to what was promised in a consistent fashion.

A typical project question is: are planned deadlines being met? Without competent **coordination** of all elements, failure is likely to occur. Will new product upgrades be available as promised—on time? Are competitors' product characteristics being tracked? Payment (accounts receivable) timing must be perfect to provide sufficient cash flow. Generally, customers must pay on time so that reimbursement to vendors conforms to contracts and staff payroll can be funded as needed.

Functional integration is crucial to achieve the necessary *synchronization*. Production, finance, marketing, HR, accounting, and other organizational systems must be properly orchestrated and synchronized. With 100 artists playing their respective instruments, Beethoven's Sixth Symphony will sound like a cacophony of pots and pans, unless everyone follows the same score led by only one conductor. This orchestra analogy can be associated with successful leadership and coaching of team sports. Both apply directly to business success (Takeuchi, 1986).

11.5.3 Innovations to Restore Loyalty

How can innovations stymie (prevent or hinder) such damages and mitigate the effects of the lost loyalties? If there are not outright complaints, how can the company identify the problem? One answer is to ask for evaluations after specific customer experiences, such as hotel stays or meals in a restaurant. Making it clear that an honest reply can lead to improvements in service appears to be an effective inducement. Increasingly companies are doing this—and rewarding compliant customers with a free appetizer.

When a "severe" complaint arises, significant remediation must occur. A sincere effort to make amends often leads to significant re-bonding of customer and company. Strengthened bonds of loyalty have been noted when the customer appreciates the redress of grievances. The innovation required is finding the remedy that will satisfy alienated customers at a reasonable cost. For the large pizza chain previously discussed, remedies should cost less than $8,000.51. Otherwise, the remedy will cost more than the cure is worth. Also, a small percent of customers will never be appeased no matter what forms of compensation are offered. The Coke example (Section 11.4.1.) provides another case of recovery from failure and the benefits of redressing grievances.

Frequent flyer programs are an airline innovation that has been a great success. Airline loyalty programs were considered a remarkable innovation when in 1972 United Airlines gave plaques to their top-flight customers. In 1979, Texas Airlines created the first real frequent-flyer program (by present standards) using mileage tracking on which to base rewards. Today, credit card use for many kinds of purchases has augmented mileage achievements. This additional innovation has become another source of revenue for airlines—sorely tested by huge increases in fuel costs.

The fact that getting new customers is far more costly than holding on to existing customers has led to the adoption of retainment programs in numerous industries. Retail stores, restaurants, hotels, trains, and airlines have added loyalty incentives with discounted gift cards, upgrading of services, special deals. To retain customers who are intrigued with the "latest and the greatest" yearly upgrades of car designs and cell phones are commonplace. Innovations yet to come will increase the length of the loyal lifetime period—by offering rewards for staying year after year. Properly done, the cost of switching brands can play an important role. Not only must customers learn the intricacies of a new device when switching from one technology to another; they must also change the "cloud" that contains their programs, music, books, etc. This is another new class of innovations, namely, innovations of infrastructures (often called ecosystems) that are designed to make it costly for non-business consumers to switch suppliers.

How do consumer products such as toothpaste and shampoo deal with the holding vs. switching problem? As before, the cost of holding is significantly less than the cost of getting a customer loyal to another brand to switch to our brand. Coupons are bestowed on loyal customers to keep them from wandering. Strong brands are continually innovating claiming their new formula can make teeth even whiter and brighter. Marketing contributes greatly to reinforce loyal customers' beliefs that they have always selected the best brand. *Cognitive dissonance theory* developed by social psychologists (Festinger, 1957) explains how people can be strongly motivated to justify their purchase decisions—as being the best choice available—even if an impartial observer would be able to point out serious shortcomings with that product. Physical and psychological innovations are both in play when P/OM and marketing work to restore and/or amplify customer loyalties.

11.6 New Growth Platforms for Innovation

As we proceed, it must be reiterated that when NPD is continuous—projects do not start with an idea and then stop with a prototype. Projects are ongoing. Teams are seldom halted and disbanded—but they may well be redirected. Project groups are committed to creating streams of innovation and change. Specifically, it is necessary to be able to design a succession of upgraded configurations of the original invention. This is described as creating NGPs by Laurie, Doz and Sheer (Laurie, 2006).

The timing between successive launches of new members of the family will depend on a variety of factors related to evolution of demand in the marketplace. In particular, when a competitor directly attacks a recent innovation with improvements that disrupt the growth pattern of the original product, a counter-move is required. The success of the counter-attack will depend on the existence of a strong "platform." Lacking a good platform, the original innovator may be obsoleted by competitors. The product of the original innovator will become outmoded in design, style, and/or construction within a very short time frame. The platform provides conceptual and physical resources for continuity. Though continuity is fast becoming a standard for success, it is hardly a new discovery.

11.6.1 Innovative W. L. Gore & Associates, Inc.

W. L. Gore & Associates has been using continuous project systems with NGPs for many years without using that specific terminology. This company was founded in 1958 by Wilbert L. (Bill) Gore (who had been a research chemist with DuPont) and his wife Genevieve (Vieve) Gore. The company sees itself as "A uniquely inventive, technology-driven enterprise focused on discovery and product innovation." Gore's vision is clearly stated: "a commitment to innovation shapes everything we do." The company remains privately held and is very successful with annual sales of $3.2 billion derived from a constant stream of product innovations. Perhaps best known for Gore-Tex® brand fabrics, the company continuously invents electronic, medical, and industrial products. Gore has been granted more than 2000 patents worldwide. It is regularly listed in the top 25% of "The 100 Best Companies to Work For." Everyone participates in the innovation process which is conceptually equivalent to the product platform planning (PPP) process we describe in the next section.

11.6.2 Continuous Innovation Using Product Platform Planning (Step 3)

Our discussion now turns to explaining why NPD platforms (Step 3) are so important and how PPP is accomplished. PPP starts as a concept rather than a method for developing new products. It replaces planning for one product (goods and services) at a time. The one-at-a-time concept is now passé. The speed with which product development occurs has accelerated at such a great rate that planning a product in isolation from its further updates and revisions is untenable. In this regard, it is important to read a 15-page article, *Planning for Product Platforms* (Robertson and Ulrich, 1998).

Platform planning can be based on logical evolution (e.g., increasing screen size or decreasing thickness). It can also benefit from swapping out less good components for better ones (e.g., upgrading memory size or using more pixels). Entirely new functions can be added (e.g., GPS-Global Positioning System, or rear view camera on a car). New colors are often added to the original palette. Additional sizes make sense for a variety of products including those which are forms of clothing. New forms of service can be extended as product innovations. At-home healthcare monitors can permit doctors and nurses to evaluate a patient's condition without requiring a personal visit.

11.6.3 Continuous Innovation Using Product Modularity

Modularity is a production strategy that has been utilized in special ways for many years (Starr, 1965). It is based on the concept of swappable components any one of which can be plugged into the mother unit (and it does not have to be a computer). Each swappable component has its own capabilities for doing the job. For example, these components can represent different interior color designs for a car. Various motors produce different amounts of suction for a vacuum cleaner. Modular components have a standard interface which permits any one of them to be plugged into a space reserved for one or another of the swappable components.

Electric plugs within any one country are interchangeable in relatively standardized sockets but different systems prevail in various countries making it necessary to use adapters. Electric bulbs are highly modular and fairly well standardized but there are many exceptions. The Edison screw base has been made to dimensions prescribed by international standards since 1909. LED (light-emitting diode) lamps and CFL (compact fluorescent lamp) are made with Edison bases precluding major problems replacing incandescent lamps. Most Edison screws have a right-hand thread that tightens when the bulb is turned clockwise. However, to deter theft, some installations are based on left-hand threads (e.g., some hotels and the New York City subway system).

Additional examples include various size motors (for kitchen waste disposal units) that fit into the same container under the sink. In general, interchangeable motors (of varying HP) and power supplies of different wattages can be designed to provide modular-based variety. Many different models of cars, refrigerators, computers, vacuum cleaners, washer/dryers, etc., are built on the same production line. They are differentiated by the particular modules selected for assembly. Modularity in certain industries (such as construction and computers) has improved delivery dates, increased variety of choice, enhanced product quality, and lowered the cost of goods sold.

Modularity is not limited to physical products. There is process modularity which allows several models of automobiles (e.g., two-door and four-door) to be assembled on the same line with minimal changeover disruption. New PPP can take many forms including service modularity. For example, in a restaurant, the waiter takes the order and delivers the bill while the food is delivered by other members of the serving team. In the supply chain, vendors can deliver one unit just-in-time, or several units to a kanban system that will alert the vendor when the next order must be shipped to prevent out-of-stocks occurring. These are samples of a multitude of service innovations using the power of modularity (Baldwin, 2000).

Opportunities for innovation are amplified greatly by the concepts of modularity and platform planning—where the former is one important aspect of the latter. Marketing has major input regarding the best ways to differentiate product, at any time, and over time, by drawing on the platform potentials. However, platform permutations and component rearrangements must be coordinated with production. For example, new colors added to traditional palettes are seldom a simple process change, for example, going from white to grey to black may allow gradual modifications whereas the reverse (going from dark to light) is likely to need many more instances where the process must be stopped for complete take-down and clean-up. Modularity has been around for a long time and it has succeeded greatly but disappointed as well (Starr, 2010).

11.6.4 Platform Differentiation for Different Demographics

Every platform has special elements that address the requirements of unique demographic segments. Certain groups of consumers prefer vegetarian foods for a variety of reasons. Satisfying these customers may constitute one aspect of platform differentiation. Operating with the same food processing plant and much of the same equipment, how can the needs of those who do not want vegan dietary menus be satisfied? Production managers will state unequivocally that this cannot be accomplished unless the total system has been carefully analyzed and designed. In all instances, various sets of needs that differ by demographic segmentation can be satisfied only by a well-diversified product platform and equivalent production system.

The chocolate ice cream demographic segment differs from the vanilla, and there is an overlap to include those who are equally happy with both flavors. Some like Starbucks coffee and others say they prefer the Dunkin Donuts brew. Within a Starbucks store there are 100s of different drinks that can be ordered. In addition, coffee drinkers and tea drinkers represent quite different demographics. Allergies create thousands of segments. Those who cannot tolerate even a whiff of peanuts must have peanut-free food processing lines. There are many types of margaritas and hundreds of recipes for BBQ. It is surprising that such great discrepancies exist about which tastes best.

People exhibit strong opinions about what they like. Theories about why consumers agree or disagree about what is best abound but they explain "what is" and not "what will be." Those in the fashion field profit from style control over time and the great differences existing in choice of clothes by different age demographics. Supermarket aisles display an abundance of detergents in colorful containers. The powders and liquids inside look and function (more or less) alike but the package options and the marketing campaigns seem to make the difference. Coupons are known to be effective for some, but other (types of) customers prefer to pay more because they equate price and quality. Across a very broad spectrum, demographics play a big role. Well-designed product platforms enable companies to attract *and supply* customers across a great variety of boundaries.

Marketing managers decide which demographic segments are likely to become customers. They also determine whether innovations to the product line will chase away customers who were formally loyal. Once this complicated analysis is completed, decisions can be made about how the product line should evolve over time. Platform planning is the method used by smart manufacturers to produce profitable volumes of differentiated products configured to satisfy customers (belonging to unique demographic segments). If this sounds like personalized customized product (which is known as bespoke—in the U.K.) it differs in only one regard. The custom tailor cuts and fits one person at a time. The product platform, on the other hand, satisfies the needs of segments of customers as well as being able to update product design over time. P/OM participates in platform planning by developing processes that can increase variety and/or update competitive offerings. The latter is achieved by substitution of innovative new modules for outdated ones.

11.6.5 Mass Customization

Higher levels of successful innovation have evolved since the concept of product platforms have been accepted and integrated as a part of strategic planning. Technology has created equipment that is computer-programmed to take on a variety of on-line capabilities. Computer-assisted flexibility permits production that is called mass customization. It is an exceptional ability to differentiate product *as it is being made* instead of having all items of a given production run being as close to identical as statistical variation permits (i.e., mass production). As an alternative explanation of mass customization, it is when the efficiency of the production line is not affected by changing one or more components before each finished production item rolls off the line.

Extension of the product platform reveals the ultimate goal of new products tailored to the needs and desires of each customer. Mass customization is the polar opposite of mass production (a stream of identical items produced in minimum time and cost per unit). That defines mass customization as a stream of customized items, each unique within a set of design parameter constraints, produced in minimum time and cost per unit. The idea of mass customization is elusive because there are so few examples of this kind of process. A few good ones follow.

It is not difficult to imagine the mass production process to produce coins such as pennies, nickels, dimes, or quarters. The disk is blanked out of the appropriate metal sheet. The blanking dies corrugate (called reeding) the edge of the dime and quarter. This process also forms two rims (one circles the outside of the obverse face and the other circles the reverse [tail's] side). It is a systems factor to note that the rim aids blind people in distinguishing between different coins. If that fact is lost, a new manager hire might aspire to save money by removing the rimming.

The coin is imprinted on both sides. Rough edges are polished (flash trimmed). Millions of dimes can be run at one time, but to make quarters, the dime-making process must be stopped. The materials and dies have to be changed. Before restarting the process flow, the set-up system includes: stopping, taking-down, cleaning-up, setting-up, and restarting the process flow; that is the essence of mass production.

Mass customization, on the other hand, allows the orders to be filled (for bank demands) without having excessive production that requires lots of storage until coins are distributed for use. Upon demand for a specific coin—within seconds—the production process accesses the appropriate material, uses the proper dies, trimmers, and transports the product from raw material to finished items. This process can switch back-and-forth with no clean-up, take-down, start-up, and no loss of time.

The mass customization process is as economic to run (per unit made) as the mass production process. The advantage is that "you run what you need." There is no inventory with its idle value-added being put into storage. However, as Paul Zipkin points out, the investment in mass customization equipment (i.e., able to shift without set-up costs between pennies, nickels, dimes, and quarters) is much greater (Zipkin, 2001). Therefore, straight-forward business analysis must be used to determine when mass customization is a good deal. For the US Treasury Department, elicitation, process flexibility, and delivery logistics are all likely to be supportive of a "go for it" decision. See Readers' Choice at the start of this chapter for a definition of elicitation.

PPP permits mass customization of product by demographic segments (e.g., Dell, HP, Ford, GM, and Toyota can configure the product that you want). On the other hand, The Campbell Soup Company, using mass production, cannot deliver a can of soup that meets the specifications of a customer who wants no more than 50 mg of sodium. Regular chicken noodle soup from many brands has 890 mg of sodium which is exceedingly high for people required to have a low sodium diet. The market is growing for low-sodium products of many kinds.

Campbell's low sodium chicken noodle soup has 90 mg of sodium. With enough demand at the 90-sodium level, Campbell will use large runs that approximate mass production of its three flavors of low-sodium soup. There are also 25%-lower-sodium soups. As the lower-sodium soup market grows, production will find a way to increase the variety and volume of this specialty output. At the present time, mass production alone will be economically feasible.

Starbucks charges a lot for producing coffee drinks that fit each individual's request. Starbucks coffee-making process is not exactly mass customization but it approximates that capability. Their baristas are able to switch between several hundred varieties of drinks on-the-spot and within its 3-min legacy time for service. Starbucks has developed a remarkable PPP and that is one of the reasons people flock to its stores. Competitors offer fewer varieties at lower costs but the trick of Starbucks' success lies in the broad scope of Starbucks' customization as compared to that of its competitors.

"Mass customization has been the 'next big thing' in product strategy ... yet for years, mass customization has disappointed." This statement written by Forrester Research (the consulting firm) is historically accurate for a variety of reasons that align with points cited in the Zipkin reference. Now, Forrester explains that mass customization has "hit an inflection point." Perhaps, it is more recognizable to say "tipping point." The reason is that new technologies permit the fast setup changes required to "make to order" from menus of available possibilities. With such technology, customers are learning to elicit what they would like (the power of customer capability for customizing). Distributors are adapting their systems to be able to deliver small amounts quickly (e.g., Amazon's Prime delivery). The transformations taking place are important examples of innovations with respect to PPP leading to mass customization of new products. The U.S. Postal Service has a great opportunity to innovate at this time to become a new force for rapid delivery of small packages to individual's homes.

11.6.6 Product Life Cycles: Rollovers

Let us refer to the product life cycle figure that is shown in Chapter 3 with the title: Figure 3.10: *Evolution of Life Cycle Stages.* This figure is composed of Stage 1—rapidly accelerating growth, Stage 2—decelerating growth (after the inflection point), Stage 3—stabilization (with maturity), and Stage 4—deteriorating share (requiring replacement or line extension). Stage 4 replacements are called "single or solo-product rollovers." If the old product is maintained along with a newer version, that is called a "dual-product rollover." Strategic decisions about product rollovers have been discussed in the literature, for example, Analysis of Product Rollover Strategies in the Presence of Strategic Customers (Cakanyildirim et al., 2011). In *Successful Strategies for Product Rollovers* (Billington et al., 1998), the authors point out that short product life cycles increase the need for frequent product rollovers.

Discussion of life cycle stages requires patience because there are so many factors to consider for each stage. Coordination is essential to make sure that the rollover strategy works precisely the way it was planned. In a solo-product rollover, all inventory of the old product should be gone at the time of transition to the new product. Delays in production of the new product can cripple the solo-product rollover by exhausting supplies of the old product before new product replacements can begin to satisfy demand.

Preparation for the solo-product rollover involves having the replacement product ready to go to market. A good analogy is how runners pass their relay-race baton. It is critical to pass the baton without delay. In effect, the old product has done its job. The new product takes over. Proper timing is vital. The new product is a replacement—not a line extension. As a swap it does not have the burden of going through Stages 1 and 2 of Figure 3.10. The new product is designed to enjoy the mature stability of Stage 3—even though the interval of stability has been decreasing markedly in the past decade (Billington et al., 1998). There are many failures. The legacy market share of the old product may not be transferable to the new product. That legacy is not automatically renewable. The joint plans of marketing and production must be innovative to enjoy success.

In the dual-product rollover situation, it is necessary to achieve the right balance of price for the trade-off in attributes, so that new product demand does not cannibalize demand for the old product. Communication between marketing and production is crucial—but will not suffice without a carefully thought out plan. The plan must consider all contingencies. Expectations should be detailed beforehand and results monitored after the new product has been introduced. The perils of inept market research can lead to strategic marketing blunders. Production failures can lead to customer dissatisfaction and switching to the products of competitors. The effects of cannibalization can be disastrous. The business model analysis for dual-product rollovers must be based on anticipating the real behaviors of customer.

When done properly, line extensions have an opposite effect from cannibalization, that is, the total share of A (S_A) and A' ($S_{A'}$) is larger than that of the original share of A alone. A is also better able to fend off competition. These two points are the underlying purposes of the dual-product rollover. For example, Jell-O® was originally (1899) available in lemon, orange, strawberry, and raspberry. Lime was added in 1930. There are now about 30 flavors of gelatin and another 30 of different flavors of pudding. Many flavors are also available as sugar free or low calories mixtures. Changes occur regularly; new flavors are introduced and old ones are retired or refreshed. Consequently, the estimate of 160 different Jell-O® products is variable. What is not variable is the fact that this huge variety sums to a very large and stable share of the gelatin and pudding market for Kraft Foods. The estimates for Jell-O's® share of market vary between 80 and 90%.

11.7 The Dynamics of Brand Share

Section 11.7 develops mathematical models to describe how brand share changes under various conditions such as line extensions. We begin with the classical

Table 11.1 Number of Customers Switching and Repeating in Two Consecutive Purchase Periods for Brands A and B before Any Innovations are Introduced

Time 1	*Brand A at Time 2 = A(2)*	*Brand B at Time 2 = B(2)*	*Total*
Brand A(1)	36 repeaters: A(1) to A(2) = 60%	24 switchers: A(1) to B(2) = 40%	60
Brand B(1)	24 switchers: B(1) to A(2) = 60%	16 repeaters: B(1) to B(2) = 40%	40
Total	60	40	100

Markovian switching model as shown in Table 11.1. Let us *assume* a simple market of two brands with Brand A having 60% of the market and Brand B having 40%. Table 11.1 shows the repurchase history of 100 consumers over a period of two consecutive purchases. The market is stable since the number of consumers switching from Brand A to Brand B equals the number switching from B to A. Equal switching in-and-out for both brands is the necessary requirement for stability.

To achieve the stable market shares, Brand A requires a 60% repeat rate (36/60) which is a measure of brand loyalty. Brand B requires a 40% repeat rate (16/40). Brands with very large market shares (such as Jell-O® with 80–90%) require a huge amount of customer loyalty, that is, very high repeat rates to sustain that share. Brand A has 40% of its customers switching to B, that is, 40% = 100(24/60). Brand B has 100(24/40) = 60% of its customers switching to Brand A.

11.7.1 Product Renewal Innovation

Brand B wants to grow its market share. It has used process innovation to improve the quality of its product. Brand B has spiffed up its package and is advertising that it is *new and better*. This *kind of new* belongs to the category of innovation called "renewed." Market research results indicate that this renew strategy has increased switching from Brand A to Brand B from 40% of Brand A's share (Table 11.1) to 50% (Table 11.2). Since the number of consumers switching between brands is not equal, brand sizes will alter. In this case, Brand B will grow at the expense of Brand A until switching in-and-out is equal for each brand. Almost stable is the situation shown in Table 11.2 where Brand A has been reduced in size to 54 and B' has grown to 46. Here is the way in which the new shares are obtained: A(2) = 0.5(60) + 0.6(40) = 54; and B' = 0.5(60) + 0.4(40) = 46.

Calculating the next set of brand share values from the matrix in Table 11.2 will show Brand A with a share of 54.6 and Brand B' with a share of 45.4. This step represents the third purchase occasion. What might be the penalty cost of

Table 11.2 Number of Customers Switching and Repeating in Two Consecutive Purchase Periods for Brand A and Renewed Brand B′ (Apostrophe Indicates Restaged Brand)

Time 1	*Brand A at Time 2 = A(2)*	*Brand B′ at Time 2 = B′(2)*	*Total*
Brand A(1)	30 repeaters: A(1) to A(2) =50%	30 switchers: A(1) to B′(2) = 50%	60
Brand B′(1)	24 switchers: B′(1) to A(2) = 60%	16 repeaters: B′(1) to B(2) = 40%	40
Total	54	46	100

this redesign by Brand B to Brand A which has experienced a 5.4% decrease in its market share? Assume that the total market demand is 10 million units per month and each unit sells for $5. A 5.4% loss of share translates into a decrease in brand volume of 540,000 units per month. This constitutes lost revenue for Brand A of $2,700,000 per month. If Brand A can figure out how to recoup its losses, Brand A will push back. Problem No. 2 at the end of this chapter requests demonstration of this calculation as well as that of the fourth purchase occasion.

Table 11.3 shows the stable shares that are reached after five successive purchase period iterations. Another decimal place is generated every iteration because the solution shows that share values are infinitely repeating decimals. The number of customers switching in-and-out of both brands is nearly perfectly balanced by the fifth iteration.

There are many questionable assumptions such as the persistence of the switching and repeating rates. Brand A will fight back to regain the 5.46 share points that it has lost to Brand B′.

Table 11.3 Number of Customers Switching and Repeating after Five Consecutive Purchase Periods for Brand A and Renewed Brand B′ (Apostrophe Indicates Revamped or Restaged Brand)

Time 1	*Brand A at Time 2 = A(2)*	*Brand B′ at Time 2 = B′(2)*	*Total*
Brand A(1)	27.273 repeaters: A(1) to A(2) = 50%	27.273 switchers: A(1) to B′(2) = 50%	54.546
Brand B′(1)	27.2724 switchers: B′(1) to A(2) = 60%	18.1816 repeaters: B′(1) to B(2) = 40%	45.454
Total	54.5454	45.4546	100.0

Although imperfect, this simple model captures the relevant flows of loyal customers and those who want to try alternatives. It is a good model for simulating "what if we do this and they do that." Much more complex assumptions can be built in and tested. It is also easy to use Excel spreadsheets for analytic calculations. Analytic examples will be found in the problem section.

11.7.2 Brand Switching Matrix Equilibrium Analysis

This material is optional. It is intended for those students who want to understand the way in which equilibrium calculations are obtained. These calculations which use the data in Table 11.2 are straight-forward. There are three equations and only two variables, viz., share of A (S_A) and share of B' ($S_{B'}$). Shares are shown at time (1) and at time (2).

Here are the three basic equations:

$$S_A(2) = 0.5^*S_A(1) + 0.6^*S_{B'}(1) \tag{11.1}$$

$$S_{B'}(2) = 0.5^*S_A(1) + 0.4^*S_{B'}(1) \tag{11.2}$$

$$100 = S_A + S_{B'} \tag{11.3}$$

At equilibrium, $S_A(2) = S_A(1)$ and $S_{B'}(2) = S_{B'}(1)$. Using this condition, Equation 11.1 becomes:

$$0 = -0.5^*S_A + 0.6^*S_{B'} \tag{11.4}$$

Or:

$$0.5^*S_A = 0.6^*S_{B'} \tag{11.5}$$

So:

$$S_A = 6/5(S_{B'}) = 1.2\ S_{B'} \tag{11.6}$$

By incorporating this result into Equation (11.3), we obtain:

$$100 = 1.2\ S_{B'} + S_{B'} = 2.2\ S_{B'} \tag{11.7}$$

Thus:

$$S_{B'} = 100/2.2 = 45.4545455 \tag{11.8}$$

And:

$$S_A = 100 - S_{B'} = 120/2.2 = 54.5454545 \quad (11.9)$$

The resulting (rounded) shares of 54.5 and 45.5 will generate a push back from Brand A. In other words, the model describes what would occur if Brand B's move was uncontested. That is not likely to occur in the always evolving point-and-counterpoint of the competitive marketplace.

11.7.3 How Quickly Does Equilibrium Occur?

The length of a purchase cycle determines how long it takes *in real time* for the disequilibrium to stabilize. For illustration, compare the short purchase cycle of a bar of soap or a can of soda to the medium-length purchase cycle of hotel visits to the long purchase cycle of a laptop computer.

There is the accompanying question of how many purchase cycle iterations must occur before stabilization. For example, say that Brand A (an upscale hotel brand name) plans to introduce A′ (medium-priced hotels). A and A′ are identified with the company name, for example, Starwood Hotels include Sheraton, Westin, St. Regis, and W Hotels. A′ can be thought of as a line extension in a dual-product rollover. To visualize the dynamics of equilibrium convergence, assume that Brand A starts with 100% of its share of market. Table 11.4 presents the holding and switching rates.

Holding is 40% for both. Therefore, switching is 60% for both. After the first purchase cycle (i.e., the average time between hotel stays) A and A′ reverse market shares and then oscillate slowly to a 50:50 division of the marketplace. The shares for consecutive iterations are: (100, 0); (40, 60); (52, 48); (49.6, 50.4); (50.08, 49.92); (49.984, 50.016); (50.0032, 49.9968); (49.99936, 50.00064), etc.

By the seventh iteration, shares have stabilized at 50:50 for the hundredth decimal level. That is many years for cars and computers. On the other hand, it is less

Table 11.4 Number of Customers Switching and Repeating in Two Consecutive Purchase Periods for Brand A and Line Extension Brand A′ (Apostrophe Indicates Dual Line Extension)

Time 1	*Brand A at time 2 = A(2)*	*Brand A′ at time 2 = A′(2)*	*Total*
Brand A(1)	40 repeaters: A(1) to A(2) = 40%	60 switchers: A(1) to A′(2) = 60%	100
Brand A′(1)	0 switchers: A′(1) to A(2) = 60%	0 repeaters: A′(1) to A′(2) = 40%	0
Total	40	60	100

than a full year for soap and toothpaste. With rapid innovation, it is unlikely that stabilization will occur when the purchase cycle interval exceeds a year. Innovators need to understand appropriate time lines to be able to evaluate ROI in production and operating equipment as well as marketing expenses.

One additional point is that brand shares with line extensions such as $S_{(A+A')}$ are designed to take the share away from competitors resulting in a larger *total* share than each brand would have alone. It is also anticipated that the size of the market will increase thereby resulting in a bigger share of a larger pie. Thus, for Jell-O®, it is not a weakness that adding lime flavor takes some sales away from strawberry. This is because the lime flavor added some customers who would never buy strawberry. Both flavors have the same profit margins and the total share is larger than the shares of either flavor by itself.

This result may be contrasted with the iPad-mini which is viewed as a brand extension of the full-size iPad. There is evidence that the mini cannibalizes some sales of the full-size. Since the latter has a higher profit margin than the mini, there is a degree of trepidation. On the other hand, competitors have mini tablets. The question has been answered by the actions of the company. It is believed that total profit with the line extension is greater than without it.

11.7.4 Blue versus Red Ocean Strategies and Closed-Loop Supply Chains

Market place dynamics have been accelerated by global investors recognizing they can have the *greenfield* advantage with an area of technology that is developing but still under development. Old text books (and managers) were used to a different environment. It was one in which there was time to enjoy the stability of market equilibriums. However, as technology moves from incipient to mature quickly, the greenfield advantage disappears. What Chan and Mauborgne (of INSEAD) call traditional (red ocean) strategy comes into play. Red oceans are swarming with competitors trying to gain advantage by cutting costs and grabbing a few points of market share. That does define a huge portion of business strategy (Kim, 2005).

In contrast, these authors call for methods to discover blue oceans which are relatively untapped (uncontested) market spaces. They give examples including Australian wine (Yellow Tail), Callaway Golf, Cirque de Soleil, and Southwest Airlines. Finding new ways to do things can be profitable, at least for a while. Competitors turn blue oceans red quickly. Apple enjoyed iPhone blue ocean dominance for a short time only because Samsung was quick to imitate and improve.

An excellent example of P/OM providing blue ocean innovative capabilities lies in the field of closed-loop supply chains (CLSCs). The definition of CLSC requires understanding the systems breadth of the entire life cycle management of products and services (Guide Jr., 2009). The NPD platform must be viewed as spanning the creation of the family of products as well as the renewing and restaging of all members of the family. This includes all aspects of remanufacturing, and recycling parts and components (Atalay et al., 2010). The goal is to recover value at every stage of both forward and reverse supply chains. For example, design for economic disassembly has received attention from all auto manufacturers for various reasons including the great number of take-back laws that presently exist and new ones that are coming into force every year.

There is a lot of waste in the supply chain. Companies that excel in managing closed-loop supply chains have significantly better profit margins. There are plenty of examples but we will focus at this time on the automotive sector. BMW is one of the leaders in innovative recovery, reuse, and disposal of materials from retired vehicles. BMW has led in disassembly with recycling goals of better than 95% (by weight) of its cars.

Eventually close to 100% of all materials used in appliances, cans, cars, computers, TVs, etc. will be returned for recycling and/or disposal by the manufacturers. Chrysler, Ford, and GM actively participate in the United States Council for Automotive Research (USCAR) which directs research into end-of-life-of-vehicle (called ELV) by the Vehicle Recycling Partnership (VRP). Ford's Global Director, Sustainability and Vehicle Environmental Matters, and Ford's CFO explain how the company is engaged in many aspects of closed-loop life cycle planning (consistent with ONE Ford plan). These issues are related to the three pillars of sustainability (see Section 11.2.2). As *The Growing Strategic Importance of End-of-Life Product Management* states, we must predesign for efficient end-of-life reuse (Toffel, 2003).

11.8 Innovators and Imitators

The strategic success of new product planning using the planning platform to guide innovation is vital for overcoming the many hurdles that exist in the marketplace. Companies that are masters of innovation must be totally prepared to adjust for both marketing and production surprises. What turns a novice into a master of innovation? Novices are often taken by surprise because they are new to a field and lack experience. In today's marketplace conditions are changing so rapidly that it may be impossible for anyone to become a master without the means to keep abreast. That explains why both production and marketing rely on good market intelligence (viz., market research). Examples of failure to listen to the marketplace are legion as are cases where the company lacked versatility and flexibility. No one can understand the marketplace without listening. Then, upon hearing, the *versatile organization* responds with an appropriate plan (see adaptability in Section 11.2).

It is widely acknowledged that good new ideas are rarely exclusive. Important technological discoveries are seldom well-kept secrets. Every technological advance sets the stage for new inventions to appear. These are quickly assimilated by many specialists. The speed with which new technology is transferred around the globe reflects the growing number of individuals who are working on the same problem.

It also shows that knowledge has been disseminated which permits new technology to be absorbed and replicated with little effort. Everyone is on the same page. In large part, this is because computers store the relevant knowledge now, whereas years ago, it was the engineer and scientist who were the sole specialists capable of trying to keep up with new developments.

The playing field is not restricted to gadgets because new technology impacts every field. In services, healthcare is changing rapidly with IBM's Watson's enhanced diagnostic capabilities (www.ibm.com/innovation/us/watson/), smart phones providing medical apps (Chen, 2010), and surgical robots (also, couriers and drug dispensers). The travel industry has been transformed by new technology for information retrieval, reservations, and bookings. Innovations impacts on supply chains are altering export and import of *everything that we eat, wear, view, or hear.*

Smart cars are becoming homes away from home, and work is done under the rubrics of BYOD (bring your own device) because it has become impossible for IT departments to regulate the variety of telecommunication and computing devices that employees own and use productively.

11.8.1 Teaching and Learning Innovations

Traditional classrooms are not gone—but they are being replaced increasingly by smart, supportive on-line systems as which are getting ever better at doing the intended job—higher quality at a lower cost. Blended systems are being developed which use some Internet and some classroom time. Traditional educational systems are being replaced by an array of technologies that have many advantages including the ability to use asynchronous learning. Asynchronous means that the teacher and the student can be on different schedules and yet communicate with each other. Asynchronous systems allow each student to decide when, where and what he or she wants to study. The level of P/OM innovation in the product delivery system is astonishing.

Teaching and learning methods on the Internet were originally called "distance learning" because they had been used in places where education was hard to come by such as Australia's Outback. Overtime, it has been recognized that Internet on-line learning is not just a good alternative when nothing better can be done. It is, in fact, a superior method of education—under a great number of circumstances—and when properly utilized.

In all these examples, the applications of new technology are occurring in ways that could never have been imagined before the existence of that new technology—but the new technology was not conceived with those applications in mind. It is extremely important to dispel the belief that we can use market research to forecast what customers want technologists to develop. Customers are not able to ask for things that they cannot imagine. They could not opt for microwaves or for air-conditioners before they became technologically feasible. Some good R&D practitioners are innovators and can peer into the future. A source of innovative ideas has been the science fiction writing community. When P/OM and marketing managers look for break-through innovations, the likely sources will be from those who work in process and product design and in the R&D labs who can translate what complaints plague customers into technological developments that might alleviate their problems.

This discussion brings to mind the Gravity Research Foundation established in 1948 by Roger Babson (founder of Babson College). Babson sponsored the development of new products that would function when shielded from the effects of gravity. In other words, he supported the development of products that worked on principles that do not exist. No technology has been found that can remove the effects of gravity. The Foundation disappeared after Babson's death in 1967 but the story remains as an illustration of the fact that new technology can give rise to new products; whereas, hopes, wants, and wishes do not. In the same vein, customers wanted bigger and stronger fans until the new technology of air-conditioning was fully developed. Florida will soon become the third most populous state in the USA because air-conditioning is cheap. Before air-conditioning, Florida was thinly populated and would have remained so. The impact of technological innovation is far-reaching and unpredictable.

11.8.2 Competitors Are Working on the Same Areas of Innovation

Duplication is wasteful unless from it springs the best new product designs from the customers' points of view. From prior discussions about innovation it is evident that many competitors, and potential competitors, are likely to be working on new products that are quite similar to each other. Some are focused on *incremental* improvements of existing products in the market (known as the kaizen-orientation). Others are *radical innovations* that will *disrupt the market* if they succeed in gaining a foothold—leading to an eventual *tipping point***.** See *The Innovators Dilemma: When New Technologies Cause Great Firms to Fail*, and *The Innovator's Solution: Creating and Sustaining Successful Growth*, Christensen, 1997, 2003). See also, *The Tipping Point: How Little Things Can Make a Big Difference* (Gladwell, 2000).

It is hardly surprising that many potential competitors are working on the same technological problems. Myriad guesses are under discussion on the web. Being first with a workable and economical technological breakthrough used to provide a huge advantage. Many factors that allowed one firm to "get there first" constituted significant "barriers to entry" for subsequent competitors. This is no longer the case. The second to enter is an imitator. The economics of the imitator have some advantages if imitation is used to correct shortcomings of the first to produce a better product. The struggle between the first and the best can favor the best. The imitator may be able to price the new product significantly below the originator (first). Being first does not protect the originator's market from successful attack by imitators.

11.9 Innovators' Production Strategies

The organization that completes a new product entry first has some potential advantages from being first. However, it takes *smart planning, care*, and *cooperation* between business functions to consolidate that advantage. If advertisements fail to alert customers about the new product's availability, the "first to finish" advantage may not occur. Retailers who are not convinced that this new product has significant sales potential will not stock and display these new items. One should bear in mind, with retail there will always be a shelf space problem; it takes a lot of experience to know how to pressure retailers into allocating space to a new product. To avoid this problem, Apple created Apple stores as part of its total new product line innovation. Apple management understood that they could never get necessary product exposure using traditional retail avenues.

Amazon's Internet system does not face similar floor-space constraints. That is why Amazon stock is so highly valued. However, all mail-order companies recognize "space-as-a-resource" limitations in a mail-order catalog. The same applies to radio, TV, and pop-up ads used on the Internet. Whatever method is employed to educate potential consumers about new product attributes, it must reach and convince a broad enough market to provide first entry advantages.

On-time delivery is critical. When production output falters and does not feed the pipeline with new products as demanded by paying customers, it is difficult to repair sales lost by insufficient stock. In the same way, consistent quality is essential. Price eventually plays a major role in the determination of demand levels. At first, the market for high quality may seem indifferent to price, but once the wealthy market segments are saturated, price begins to count. In other words, P/OM should always be prepared to deal with problems related to price elasticity. The sales of many new products (especially of high quality) seem unaffected by elasticity constraints. P/OM is not under pressure to reduce process costs. Later on, when competition grows with lower price and equal and better quality, all operating costs begin to matter.

11.9.1 Lean Production Methods Are a Form of Process Innovation

It is wasteful to incur costs which do not add value to the product for the customer. Yet, many organizations have numerous costs that produce no tangible benefit. Programs to cut such waste are called "lean." They were developed by the Japanese manufacturers and especially Toyota. Elimination of waste (time and material) and smoothing work flows (balancing process rates) is at the core of lean. The Japanese goal of removing waste (mura) and unevenness (muda) has been highly successful in both manufacturing and service industries (such as healthcare). When operating costs begin to matter, lean is often brought into the system. It should always have been endorsed because it is an important aid in developing new products properly and faster (Adler et al., 1996).

Innovators face the challenge of moving too swiftly to be the first to market. *Haste makes waste* has stood the test of time. Imitators are watching and waiting for innovators to trip up. They are usually ready to take advantage of mistakes that "originators" often make. Although innovators know about potential pitfalls, driven by the need for speed they fail to protect themselves. Among the originators' mistakes are: bad timing, over-claiming, under-performing, and failure to up-grade fast enough when competitors' products enter the market (Billington et al., 1988).

Bad timing occurs when the new product is offered to consumers before they fully understand why they want the new product. Early growth of demand is too slow. Conversely, there is bad timing when demand overpowers supply and customers are irritated by unexpected delivery delays. Shipping product that must be recalled because adequate quality controls were not exercised is a serious downer. Numerous examples of consumers being upset by having to wait for delivery or by having to return recalled product can be cited. The production supply system must be synchronized to an intelligent marketing demand creation system. If these two functional systems are not coordinated, various kinds of serious failures are likely.

Over-claiming product attributes is difficult to avoid when advertising. The consumer gets quite used to this. Imitative competitors take advantage of over-claiming by incorporating advances based on "being second" and thereby discovering how to *leapfrog* the original product. The "leapfrogging effect" is where those who are lagging the innovators gain an advantage over first movers by imitating the originators with improvements that are based on consumer reactions to the original design. The originator's initial investment (legacy situation) prevents rapid responses to the leapfrogging imitator.

In other words, imitators can get ahead of the original innovator by providing in their imitation more of what was promised but was not sufficient in the "original" product. Innovators' investment risks can be reduced (but not entirely) if careful analysis is employed. Over-claiming and under-performing by the innovator are mistakes that are closely related. Any unmet expectations of customers are like manna from heaven for competitors who build their attraction by providing the missing qualities.

Summary

Successful NPD requires innovation development that results in products (goods and services) that please enough customers to exceed the breakeven point by a large enough margin to make the ROI acceptable to investors. Process innovation can play a major role in achieving competitive advantage. As an example, lean production methods are a form of process innovation.

Clever innovations are not enough. Unusual products can be deemed very innovative but find too few buyers. Because of the speed with which new products are accepted by the marketplace and then rejected in favor of more recent competitive entries, it is essential to innovate on a platform that has inherent plans for variations that provide a stream of successors. As the older versions mature, they are replaced with the newer conceptions. Innovations can fail which leads to learning opportunities. However, there are cases where alienation of loyal customers can occur. Calculating the LTV of a customer allows management to assess the penalty cost of losing loyal customers. This in turn permits the analyst to ponder, "How much can be spent to persuade a lost customer to return?"

The dynamics of competitive life cycles is captured by stochastic brand share matrixes. Sustainability is related to the ebb and flow of shares over time. If revenue and then cash flow drop below a sustainable threshold, the company cannot continue. These constraints are not limited to profit-making organizations. They apply to humanitarian operations and crisis management as well as not-for-profit and governmental organizations. Whatever the organizational objectives, questions of NPD, innovative solutions, and sustainability require understanding and utilizing the full-scope of the systems approach.

Review Questions

1. What is the importance of innovation to the firm?
2. What is P/OM's role in achieving innovation?
3. Why it is important for marketing and production (P/OM) to work together in an open and cooperating environment?
4. What are the three pillars of sustainability?
   a. How do the three pillars relate to innovation?
   b. How do the three pillars link with organizational success?
5. In the introductory material that starts Section 11.2, it is stated: "At the root of successful innovation are competence in the field, confidence in management's skills, and ability to adapt to changing conditions." Discuss these three points.
6. What does innovation has to do with humanitarian operations and crisis management?
7. Figure 11.1 depicts the planning tree of management options. Explain the P/OM rule that one must adapt to that which is inevitable.
8. Why should P/OM be aware of the Kondratiev Wave?
9. How are the Patent Cooperation Treaty and the America Invents Act related to P/OM decisions?
10. Explain continuous innovation and give some examples.
11. NPD and platform planning are of major importance in competitive markets. Explain why this is the case.

12. Why is it said that "The NPD project team must move toward the goal of product release like a rugby team and not like a relay race."
13. For successful NPD more resources should be allocated early on. Why is this so?
14. Explain what is meant by the unintended consequences of product and process changes, and give some examples.
15. What is "disruptive innovation?"
16. What is the *greenfield* advantage?
17. What is the *brownfield* disadvantage?
18. Testing appears four times in Figure 11.3. Are these all the same kinds of testing procedures? Is forecasting involved with any of these testing steps?
19. Contrast mass production and mass customization.

Problems

1. One of the greatest *innovations* of all time was the "cost of keeping comfortable" in hot climates. Air conditioning allowed regions like Florida to become highly productive and well populated. For such a large system, it is difficult to discuss true costs and benefits. However, there are some ways to try to get a reasonable estimate. The Gross State Product (GSP) of Florida in 2013 was $803.2 billion. Only California, Texas, and New York had larger GSPs. Assume that without air-conditioning, Florida and Michigan would be about the same, viz., $413.6 billion. The population of Michigan was 10.3 million which is just about half that of Florida (viz., 20.2 million). With these assumptions, what, if anything, can be said about the value of air conditioning?
2. The table below is similar to Table 11.2 in the text. It shows the number of customers switching and repeating in two consecutive purchase periods for Brand A and restaged Brand B. In this problem, we indicate that a brand is restaged by an apostrophe, for example, B′.

Time 1	*Brand A at Time 2 = A(2)*	*Brand B′ at Time 2 = B′(2)*	*Total*
Brand A(1)	30 repeaters: A(1) to A(2) =50%	30 switchers: A(1) to B′(2) = 50%	60
Brand B′(1)	24 switchers: B′(1) to A(2) = 60%	16 repeaters: B′(1) to B(2) = 40%	40
Total	54	46	100

Based on the fourth purchase occasion, what would be the penalty cost of this redesign by Brand B to Brand A? As in the text, assume that the total market demand is ten million units per month and each unit sells for $5.

3. Calculate the LTV of a customer similar to the average pizza buying family described in the text. In this case, the family is larger and buys more pizza on each occasion but less frequently. They buy \$60 worth of pizza once a month (on average). Use the interest rate of 4% for the NPV calculation in an Excel spreadsheet.
4. Continuing with the calculation of LTV. Change the example in the text of the angry disenfranchised customer by assuming that the lifetime horizon is 25 years instead of 20 years.
5. Alter the LTV example in the text of the angry disenfranchised customer by assuming that the prevailing interest rate over the lifetime horizon is 2.5% instead of 3.045%.
6. For the LTV example in the text, the expenditure of the average family for Pizza is \$15, with 32 purchases per year.
7. This problem on innovation is related to the quality of goods made in a job shop. After the setup is completed in the job shop, the first widgets made are likely to be defectives. The skillful employee adjusts the machinery. The order size calls for 250 units. The expected percent defective for start-up is 10%. What would the range of defectives be if the expected percent had been estimated to be between 8 and 12%?
8. For the example in 7, assuming the actual percent defective is 10%, how many units should be made on average? Calculate the exact number. You must ship 250 units. If there are extra units they are sent gratis.
9. A consultant suggests a process innovation. It would cost \$95 for each set up. It would enable all employees to adjust the equipment so that the expected percent defectives would be constant at 3%. If each defective costs \$5, under what circumstances is this process innovation justified?
10. A famous hotel has studied the LTV of business travelers in the Fall season. They have found it to be \$5000 over a horizon of 10 years. There are two main causes for alienating these customers which can be repaired by spending \$100,000 per year. How many alienated customers will have to be avoided (per year) to make this expenditure sensible? Does this result surprise? Explain what is meant by "Innovations to Restore Loyalty."
11. Brand C is a new brand name in the flavored coffee creamer market. We note that flavored coffee creamers have a relatively short purchase cycle (meaning they are bought often). Brand C will destabilize the coffee creamer market for a number of product purchase cycles. During that time, it will attract first time triers from both Brands A and B. To be successful, Brand C must develop loyal users who will repeat purchase regularly.

    The table below provides the switching and holding rates. Generate the changes in brand share that will occur for all three brands over time—and until equilibrium is achieved. How stable is the equilibrium state—in your opinion?

Number of Customers Switching and Repeating In Two Consecutive Purchase Periods for Brands A, B, and New Brand C

	Brand A @ t(2)	*Brand B @ t(2)*	*Brand C @ t(2)*	*Total*
Brand A @ t(1)	30%	20%	50%	60
Brand B @ t(1)	50%	20%	30%	40
Brand C @ t(1)	60%	30%	10%	00
				100

12. Many studies have been done regarding LTV. The LTV of a loyal customer is usually relatively easy to calculate. The average customer buys a certain amount of our product on a regular basis. For example, say it is one bar of soap every two weeks for $3.00. Our records indicate that loyal customers have a 10-year period of persistence.
    What is the LTV of this average customer?
13. Using the information in problem 12, if something offends the customer, the firm stands to lose less than the amount calculated because of the impact of NPV. Explain why this is so and develop a spreadsheet to show the LTV value for a range of interest rates from 1% to 5%.
14. What if the interest rate in the LTV—NPV calculation was set at zero percent?
15. In this problem, every row cell has a proportion equal to that (brand) column's share. Markets often start out with highly differentiated products. Over time, product design features begin to converge so that the product is viewed by consumers as a commodity. This effect which destroyed brand advantages for TV sets is well-underway for both personal computers and laptops. Tablets are not likely to be immune for very long.

Number of Customers Switching and Repeating In Two Consecutive Purchase Periods for Brands A, B, and C When a Commodity Market Exists

	Brand A @ t(2)	*Brand B @ t(2)*	*Brand C @ t(2)*	*Total*
Brand A @ t(1)	20.25 repeaters = 45%	10.35 switchers = 23%	14.4 switchers = 32%	45
Brand B @ t(1)	10.35 switchers = 45%	5.29 repeaters = 23%	7.36 switchers = 32%	23
Brand C @ t(1)	14.4 switchers = 45%	7.36 switchers = 23%	10.24 repeaters = 32%	32
	45	23	32	100

How will this market behave over time? What strategic lessons can be learned?

Readings and References

Adler, P.S., A. Mandelbaum, V. Nguyen, and E. Schwerer, Getting the Most out of your Product Development Process, *Harvard Business Review*, 74, March–April 1996, pp. 4–15.

Atasu, A., V. Daniel, R. Guide Jr., L.N. Van Wassenhove, So What If Remanufacturing Cannibalizes My New Product Sales? *California Management Review*, 52(2), Winter 2010.

Baldwin, C.Y., and K.B. Clark, *Design Rules: The Power of Modularity*, Cambridge, MA: MIT Press, 2000.

Billington, C., H.L. Lee, and C.S. Tang, Successful Strategies for Product Rollovers, *Sloan Management Review*, 39(3), Spring 1998, pp. 23–30.

Cakanyildirim, M., C. Liang, and S. Sethi, Analysis of Product Rollover Strategies in the Presence of Strategic Customers, October 1, 2011. Available at SSRN: http://ssrn.com/abstract = 1837773 or http://dx.doi.org/10.2139/ssrn.1837773

Chen, J., Y. Park, and G.J. Putzer, An Examination of the Components that Increase Acceptance of Smartphones among Healthcare Professionals, *Electronic Journal of Health Informatics*, 5(2), 2010, p. e16.

Christensen, C.M., The Innovators Dilemma: When New Technologies Cause Great Firms to Fail, *Harvard Business Review Press*, May 1997.

Christensen, C.M., and M. Overdorf, Meeting the Challenge of Disruptive Change, *Harvard Business Review*, 78, March–April 2000, pp. 1–10.

Christensen, C.M., and M.E. Raynor, *The Innovator's Solution: Creating and Sustaining Successful Growth*, Harvard Business School Press, September 2003.

CIO, Innovation from Outsiders, CIO, May 1, 2013, p. 13.

Dillard, J., V. Dujon, and M.C. King, eds., Understanding the Social Dimension of Sustainability, *Routledge Studies in Development and Society*, October 2009.

Festinger, L., *A Theory of Cognitive Dissonance*, Palo Alto, CA: Stanford University Press, 1957.

Gladwell, M., *The Tipping Point: How Little Things Can Make a Big Difference*, Boston, MA: Little Brown, 2000.

Guide, V.D.R. Jr. and L.N. Van Wassenhove, OR Forum—The Evolution of Closed-Loop Supply Chain Research, *Operations Research*, January–February 2009.

Handfield, R.B., G.L. Ragatz, K.J. Petersen, and R.M. Monczka, Involving Suppliers in New Product Development, *California Management Review*, 42(1), Fall 1999.

Heskett, J.L., T.O. Jones, G.W. Loveman, W.E. Sasser Jr., and L.A. Schlesinger, Putting the Service—Profit Chain to Work, *Harvard Business Review*, March–April 1994.

Immelt, J.R., V. Govindarajan, and C. Trimble, How GE is Disrupting Itself, *Harvard Business Review*, October 2009.

Jassawalla, A.R., and H.C. Sashittal, Strategies of Effective New Product Team Leaders, *California Management Review*, 42(2), Winter 2000.

Kim, W.C., and R. Mauborgne, Blue Ocean Strategy: How to Create Uncontested Market Space and Make the Competition Irrelevant, *Harvard Business Review Press*, February 2005.

Laurie, D.L., Y.L. Doz, and C.P. Sheer, Creating New Growth Platforms, *Harvard Business Review*, May 2006.

McKeown, M., *Adaptability: The Art of Winning in an Age of Uncertainty*, London, UK: Kogan Page, May 2012.

Osborn, A.F., *Applied Imagination: Principles and Procedures of Creative Problem-Solving*, 3rd Edition, Scituate, MA: Creative Education Foundation, February 1993.

Robertson, D., and K. Ulrich, Planning for Product Platforms, *MIT Sloan Management Review*, 39(4), Summer 1998, pp. 19–31.

Starr, M.K., Modular Production—A New Concept, *Harvard Business Review*, November–December 1965.

Starr, M.K., Modular Production—A 45-year-old concept, *International Journal of Operations & Production Management*, 2010.

Takeuchi, H., and I. Nonaka, The New Product Development Game, *Harvard Business Review*, January 1986.

Thomke, S., Enlightened Experimentation: The New Imperative for Innovation, *Harvard Business Review*, May 2012.

Toffel, M.W., The Growing Strategic Importance of End-of-Life Product Management, *California Management Review*, 45(3), Spring 2003.

Zipkin, P., The Limits of Mass Customization, *Sloan Management Review*, Spring 2001.

Appendix A: Quantitative Models

In this appendix, we discuss the following two quantitative models for managerial decision-making: breakeven analysis (BEA) and the transportation model (TM) of linear programming. We have made references to these two techniques in various chapters of this book.

A.1 Breakeven Models

Breakeven models are one of the most significant methods for making management decisions in all functional areas of management. In P/OM, breakeven models can be used to make the following decisions among others: location, capacity, investment in plant and equipment, investment in information technology, and outsourcing (make or buy).

These models are applicable when we deal with a situation in which there is a fixed cost which is independent of the output volume, a variable cost that changes per unit of output and a revenue amount per unit sold. A discussion of these costs is in order before we discuss the breakeven models.

Variable (Direct) Costs

Variable costs per unit are the costs of input resources that tend to be fully chargeable and directly attributable to the product. They are also called direct costs because they are paid out on a per-unit of output basis. Direct costs for labor and materials are typical. Variable costs can often be decreased by improving the way that jobs are done. Material costs can sometimes be reduced using value analysis to improve material usage or find better materials at lower costs. Changing order sizes can increase discounts, and grouping purchases can result in lower transportation costs for full carloads. Lower cost suppliers directly decrease the variable cost of inventory being held and stored. P/OM has many avenues to pursue as it strives to continuously improve quality while reducing variable costs.

Fixed (Indirect) Costs

Fixed costs have to be paid, whether one unit is made or thousands. For this reason, administrative and supervisory costs are usually treated as fixed charges. Together with purchasing and sales, they are bundled together as overhead costs. Salaries and bonuses paid to the CEO and to every other manager are overheads because they cannot be assigned to specific products and services. Additionally, insurance, facility costs, electric charges, maintenance of equipment are all fixed costs *if and only if* they are not related to how many units have been produced.

Revenue

When goods and services are sold in the marketplace, they generate revenue. This breakeven model treats revenue as a linear function similar to variable costs. Each unit sold generates the same amount of revenue equal to the price per unit. An assumption of BEA is that all units made and stored are delivered and sold.

The BEA question is: How many units of throughput need to be sold in order to recover costs and to breakeven? The results change over time as factors vary. Fixed costs often grow, prices change, and variable costs per unit decrease with proper P/OM attention.

A.1.1 Linear Breakeven Equations

We use the following symbols for BEA:

p = price per unit
vc = variable costs per unit of production
FC = fixed costs per period T
V = number of units made and sold in time period T
TVC = total variable costs = $(vc)V$
TC = total costs per period T, where

$$TC = FC + TVC = FC + (vc)V, \tag{A.1}$$

TR = total revenue per time period T, where

$$TR = (p)V, \tag{A.2}$$

TPR = total profit per period T, where

$$TPR = TR - TC = (p)V - FC - (vc)V = (p - vc)V - FC. \tag{A.3}$$

Breakeven point (BEP; also known as breakeven volume), V_{BEP}, is calculated by setting profit equal to zero, thus $TPR = 0$ at the BEP. To find V_{BEP}, set Equation A.3 equal to zero. Therefore, $(p - vc)V_{\text{BEP}} - FC = 0$ or $(p - vc)V_{\text{BEP}} = FC$. The value of V_{BEP} is then given by

$$V_{\text{BEP}} = \frac{FC}{p - vc}. \tag{A.4}$$

The difference between p and vc is also called margin. So, the BEP is also given as fixed cost/margin.

Example: Suppose the fixed cost of production is \$450,000 per year; and the variable cost per unit and the selling price per unit are \$ 10.00 and \$ 30.00 per unit. The BEP is calculated as

$$V_{\text{BEP}} = \frac{450,000}{30 - 10} = 22,500.$$

The implication of the number 22,500 is that if actual production is equal to 22,500, there is no profit and no loss. If actual production is less than 22,500, then there will be a loss; whereas a production volume above 22,500 will generate a profit. The following sections further explain the profit/loss analysis using graphical representation.

A.1.2 Breakeven Charts

From raw materials suppliers to producers, distributors, retailers, and delivery systems to ultimate consumers, supply chain capacity decisions must be made. Participants must have matched input–output and throughput capacities along the way. Breakeven charts can be constructed for each participant in the system.

Breakeven charts were developed in the 1930s by Walter Rautenstrauch (1949), Professor of Industrial Engineering at Columbia University. BEA has become an influential model of P/OM, providing important information for capacity decisions. BEA can be done either in mathematical terms (as discussed earlier) or in a graphical form as discussed in this section. Figure A.1 illustrates the basic chart for one product in the supply chain. It can be a good or a service that is being sold. It must relate to a specific period of time.

In Figure A.1, the ordinate (y-axis) is dimensioned in dollars of fixed cost for line 1, total costs for line 2, and dollars of revenue for line 3. The abscissa (x-axis) of the chart represents supply chain volume in units (as used in Section A.1.1) or the percent of capacity utilized for a given period of time. The x-axis terminates at 100% of capacity. This can be translated into whatever volume of units is equivalent at 100% for the given period of time.

Three Breakeven Lines

Fixed costs (FC) of line 1 are derived in terms of a specific period of time such as a week, month, or year, which is consistent with the amount of capacity expressed by the x-axis. Fixed charges do not change as a function of increased volume or

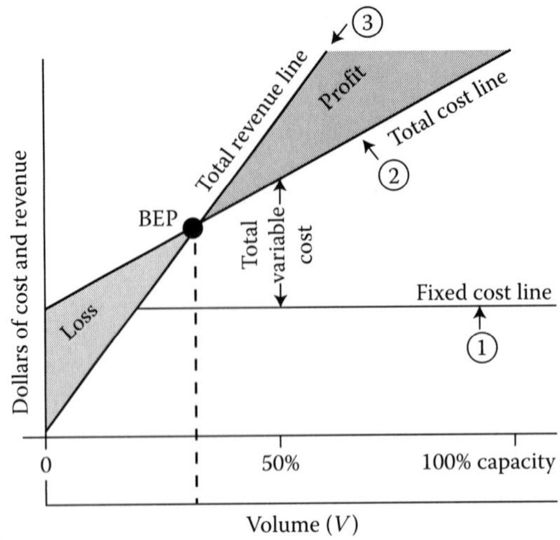

Figure A.1 Linear breakeven chart.

increased capacity utilization. Line 2 in Figure A.1 is the total cost line where total cost is the sum of fixed and total variable costs. Total cost increases linearly as throughput volume gets larger. Variable costs are added to fixed costs, which are usually assumed to be the same at all volumes, even at zero production level. Vertical distances in the triangular area lying between the fixed cost line and the total cost line measure the variable costs. Line 3 in Figure A.1 is the total revenue line. It is a linear function that increases with greater throughput volume that is sold.

Profit, Loss, and Breakeven

In Figure A.1, both profit and loss are shown. The lower-shaded area between the total cost line 2 and the total revenue line 3 represents loss to the company. Loss occurs to the left of the BEP. The upper-shaded area between the same lines represents profit to the company. Profit occurs to the right of the BEP. Because it is a linear system, maximum loss occurs at the leftmost side of the diagram. With growth of output volume, the loss decreases until it reaches zero at the BEP. Then profit starts and it increases linearly throughout the range of positive profits until maximum profit is achieved at 100% of capacity.

The definition of the BEP is that volume (or percent of capacity) at which there is no profit and no loss. Literally, it is the point at which the total costs of doing business exactly balance the revenues, leading to the name "breakeven."

The amount of loss is decreasing and cost recovery is improving as volume sold increases up to the BEP. At breakeven, cost recovery is completed. The BEP occurs at a specific sold volume of production (in units) or a given percent utilization of supply chain (e.g., plant) capacity (sold). The values of BEP are critical for the diagnosis of healthy supply chain systems. The concept is equally applicable to manufacturing and services and it always applies to a specific period of time. After calculating the variable cost, fixed cost, and revenue, the BEP can be calculated as given by Equation A.4 (or read off the graph in Figure A.1).

Analysis of the Linear Breakeven Model

Figure A.1 has been redrawn in Figure A.2 so that the *y*-axis becomes both profit and loss. This is accomplished by rotating the total cost line to be parallel with the *x*-axis. The slope of the line in Figure A.2 reflects the rate at which profit increases as additional units of capacity are sold. Above breakeven, it means making profit faster but below breakeven it means creating losses faster. Because companies expect to operate above breakeven, the larger slope is preferred.

During the years 2001 and 2002, there were many dot.com (Internet) firms that failed. These companies could not generate revenues to get them above their BEPs. They spent money on facilities and equipment. Their burn rate of investors' money was too fast. Customer acquisition occurs over a long planning horizon and it was not fast enough to compensate for the cash drain (outflow).

Consider the position of the BEP in Figure A.2. If it moves to the right because of changes in the costs and revenues, then the supply chain must operate at a higher level of capacity before it starts to make a profit. Conversely, by reducing the BEP, profit can be made at lower volumes. Companies want to operate above breakeven, so lower BEPs are preferred.

Figure A.3 shows two profit–loss lines marked A and B, where A and B are labels associated with specific (and unique) supply chain process alternatives. Alternative A has a lower BEP than alternative B. This makes A more desirable than B with

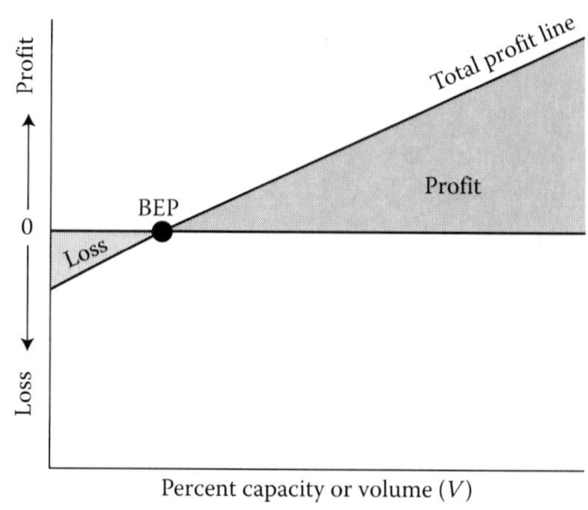

Figure A.2 The profit and loss breakeven chart.

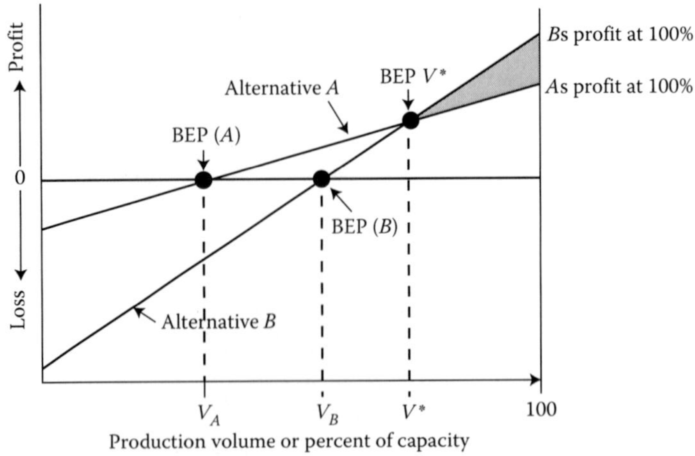

Figure A.3 Profit and loss as a function of throughput volume (or percent of supply chain capacity used).

respect to the position of the BEP. On the other hand, B generates more profit once the point V^* has been reached.

At the demand volume represented by point V^*, both alternatives yield equal profit. V^* is the point of equivalent profit for alternatives A and B. If greater demand than V^* can be generated, alternative B is preferred. If V^* cannot be achieved, then choose alternative A. Figure A.4 represents the four different situations that can arise in a 2×2 matrix. Two of the cells lead to clear-cut decisions. An alternative that has a lower BEP and higher profit-rate slope is the clear winner. An alternative that has a higher BEP and a lower profit-rate slope is unequivocally rejected.

The two other cells are ambiguous situations requiring a forecast of the expected demand volume. If the volume forecast is above the BEP of the two alternatives (see V^* in Figure A.3), then choose alternative B, which has the highest slope-profit rate. If the volume forecast is below BEP = V^* (in Figure A.3), then

		Breakeven point	
		High	Low
Profit rate (slope)	High	Use forecast	Accept alternative
	Low	Reject alternative	Use forecast

Figure A.4 Matrix showing four possible results with respect to BEP and profit rate (slope).

choose alternative A. Below BEP = V^*, alternative A returns the greatest profit and also, as the volume drops further, A produces the least loss because it has the smallest slope. Other forecasting conditions can occur. The forecast can span the volumes that fall above and below BEP = V^*. Then, the procedure is to determine the expected profit by combining the profit and loss function with the probability distribution of demand using a decision matrix. This decision matrix approach is not discussed in this book.

BEA provides important parameters for evaluating supply chain capacity decisions. The critical rate of return on capacity is based on the slope of the profit line. The equally critical BEP is determined by the ratio of fixed costs to the per-unit margin contribution which means that total margin contribution at breakeven completely recovers the fixed costs and no more.

A.1.3 Interfunctional Breakeven Capacity Planning

Every functional manager should be involved in breakeven capacity planning. There is something for every functional manager to be concerned about with the breakeven model. Without true interfunctional planning, all relevant information cannot be secured. Successful coordination of process activities is not likely to be achieved without the systems approach.

Accounting: Very careful assignment of fixed charges to each specific product is important because BEA is done for each product in the supply chain and not for a product mix. Incorrect assignment of overhead to the fixed costs will affect the breakeven value. It is generally believed that the basic proportions are evident and that roughly the right numbers are used. Activity-based costing may also help move some charges from the fixed to the variable category. This is possible when the cost per unit made or serviced can be identified by the accounting system and moved to the variable category.

There are some classification problems for variable costs. Differentiating between direct and indirect labor costs provides an example. Office work is indirect because such costs usually cannot be attributed to a particular unit of output on a cost per piece or cost per service-rendered basis. If a direct link can be made so that overhead costs can be assigned directly, then such costs should not be equally shared in the overhead pool with other products. It may also allow P/OM to assign appropriate variable costs to each product.

Marketing: Setting product and service prices to create the required demand volume while still generating a fair profit is marketing's responsibility. The way the demand volume changes as a function of the price set is described by the "price elasticity" of the product or service. Price elasticity is affected by customer quality expectations and the degree to which competitive products are substitutable. Advertising and promotion strategies are costs usually assigned to overheads. All such market drivers are crucial to the P/OM domain because they determine output volumes, thereby affecting supply chain capacity plans.

Production: How much supply chain capacity as well as what kind of supply chain capacity needs to be addressed? What work configurations are to be used? Where should the capacity be located? Price will play a part in the revenue line. Technology employed affects fixed and variable costs. Method of depreciation is another accounting issue that interacts with the finance-P/OM decision concerning technological investments to be made. The P/OM tie-in with marketing and finance is inescapable. The domains of cost concerns are

1. P/OM is accountable for variable cost per unit, a function of the technology employed. Throughput volume and technology used are strongly correlated. P/OM should participate in decisions about both. These responsibilities determine total variable cost.
2. P/OM and finance are mutually involved in major determinants of fixed cost. At the root of this relationship are the technology options and their effect on work configurations that can be used. Work configurations affect the variable costs and quality that can be achieved.
3. P/OM and marketing have to work out the price per unit that customers are to be charged and the demand volume that should result. These decisions lead to estimates of total revenue, total profit, and contribution margin.

A.2 Transportation Model of Linear Programming

In this section, we will study the transportation model (TM) of linear programming. We will use the example of Rukna Auto Parts Manufacturing Company described below. A spreadsheet will be included in the instructors' resources CD (IRCD).

Demand and Supply

Rukna Auto Parts Company has three plants located in Miami, FL (20,000), Tempe, AZ (40,000), and Columbus, OH (30,000). References to the states are omitted henceforth. The numbers in parentheses show the manufacturing capacity of each plant. Rukna supplies auto parts to four distributors MKG, Inc. (25,000), ASN, Inc. (13,500), GMZ, Inc. (16,800), and AKLA, Inc. (34,700). The numbers in parentheses show the demand of each distributor. Therefore, the totals of supply (90,000) and demand (90,000) match. In a more general and complex problem, the supply and demand may not match.

Transportation Cost

Table A.1 shows the cost of transportation per unit from each plant to each distributor. For example, it costs $3.00 per unit to ship one unit from Miami to ASN, Inc. This table also shows the capacity of each plant and the demand of each distributor.

Table A.1 Distribution Cost Per Unit for Rukna Auto Parts Company

		Distributors				
		MKG, Inc.	*ASN, Inc.*	*GMZ, Inc.*	*AKLA, Inc.*	*Capacity*
Plants	Miami	\$1.00	\$3.00	\$3.50	\$1.50	20,000
	Tempe	\$5.00	\$1.75	\$2.25	\$4.00	40,000
	Columbus	\$2.50	\$2.50	\$1.00	\$3.00	30,000
	Demand	25,000	13,500	16,800	34,700	

Problem

Rukna wants to develop a distribution strategy to minimize the cost of transportation (distribution). Specifically, Rukna wants to know the number of units to be shipped from each plant to each distributor.

Solution

The transportation method of linear programming requires a complete description of the problem that includes identification of the decision variables, objective function, and the constraints.

Decision Variables

A decision variable for this problem is the number of units to be shipped from a plant to a distributor. Since there are three plants and four distributors, there will be a total of 12 decision variables. We will designate these variable as X_{ij} ($i = 1, 2, 3$ and $j = 1, 2, 3, 4$), where i represents the plants and j represents the distributors. These decision variables are listed below:

Number of units shipped from Miami to MKG: X_{11}
Number of units shipped from Miami to ASN: X_{12}
Number of units shipped from Miami to GMZ: X_{13}
Number of units shipped from Miami to AKLA: X_{14}
Number of units shipped from Tempe to MKG: X_{21}
Number of units shipped from Tempe to ASN: X_{22}
Number of units shipped from Tempe to GMZ: X_{23}
Number of units shipped from Tempe to AKLA: X_{24}
Number of units shipped from Columbus to MKG: X_{31}
Number of units shipped from Columbus to ASN: X_{32}
Number of units shipped from Columbus to GMZ: X_{33}
Number of units shipped from Columbus to AKLA: X_{34}

Objective Function

The objective is to minimize the total cost (*TC*). It is obtained by using the following two steps:

Step 1: Find the cost of transportation for each combination of plant and distributor by multiplying the units shipped with the corresponding cost. For example, the cost of moving X_{11} units from Miami to MKG will be $1.00 \times X_{11}$, where 1.00 is the cost of transportation of one unit from Miami to MKG as given in Table A.1.

Step 2: Add all the costs obtained in step 1 to get the value of total cost. *TC* can be written as

$$\begin{aligned} TC = {} & 1.00X_{11} + 3.00X_{12} + 3.50X_{13} + 1.50X_{14} + 5.00X_{21} + 1.75X_{22} \\ & + 2.25X_{23} + 4.00X_{24} + 2.50X_{31} + 2.50X_{32} + 1.00X_{33} + 3.00X_{34} \end{aligned}$$

Constraints

While solving the problem, the following constraints should not be violated:

1. *Capacity constraints*: The total shipments from a plant cannot exceed the plant capacity.
2. *Demand constraints*: The total shipments received by a distributor should be equal to its demand.

Capacity constraints: Total shipments from Miami are: $X_{11} + X_{12} + X_{13} + X_{14}$. This total should be less than or equal to 20,000 (capacity of the Miami plant). In other words, the capacity constraints can be written as

$$X_{11} + X_{12} + X_{13} + X_{14} = 20{,}000 \text{ (Miami Plant)}$$

$$X_{21} + X_{22} + X_{23} + X_{24} = 40{,}000 \text{ (Tempe Plant)}$$

$$X_{31} + X_{32} + X_{33} + X_{34} = 30{,}000 \text{ (Columbus Plant)}$$

Note: If the total capacity of all three plants together is greater than the total demand, then the " = " sign in the above constraints will be replaced by " ≤." This means that capacity of one or more plants will not be fully utilized.

Demand constraints: Total shipments received at MKG, Inc. are: $X_{11} + X_{21} + X_{31}$. This total should be equal to 25,000 (demand at MKG, Inc.). In other words, the demand constraints can be written as

$$X_{11} + X_{21} + X_{31} = 25{,}000 \text{ (MKG, Inc.)}$$

$$X_{12} + X_{22} + X_{32} = 13{,}500 \text{ (ASN, Inc.)}$$

$$X_{13} + X_{23} + X_{33} = 16{,}800 \text{ (GMZ, Inc.)}$$

$$X_{14} + X_{24} + X_{34} = 34{,}700 \text{ (AKLA, Inc.)}$$

Note: If the total demand of all four distributors together is greater than total supply or capacity of all three plants, then the " = " sign in the above constraints will be replaced by " ≤." This means that demand of one or more distributors will not be fully met.

Non-negativity Restrictions

The values of the decision variables cannot be negative because we cannot ship less than zero unit. The non-negativity restrictions are written as

$$X_{ij} \geq 0 \text{ for every } i = 1, 2, 3 \text{ and } j = 1, 2, 3, 4.$$

The objective function, the constraints, and the non-negativity restrictions constitute the problem. The above process, where we identified the objective function, the constraints, and the non-negativity restrictions, is called formulation of the problem.

This problem can be solved using the function "solver" in Excel.

Solution to Rukna Auto Parts Company

The problem as formulated above is solved by using "solver." The solution to the problem is given in Figure A.5. We will first explain the solution and then describe the process to arrive at this solution.

Cells C4–F6 (highlighted cells) contain the values of the decision variables. For example, cell C6 gives the number of units shipped from Columbus to MKG (13,200 units). All costs are given in cells C13–F15. For example, cell C15 ($2.50) gives the shipping cost per unit from Columbus to MKG. The total transportation costs for each combination of the plant and distributor are given in cells C19–F21. The values in these cells (C19–F21) are obtained by multiplying the number of units shipped by the shipping cost per unit in the respective cells for each combination of the plant and distributor. For example, the value in cell C21 ($33,000) is obtained by multiplying the value in cells C6 ($13,200) and C15 ($2.50).

The value of the objective function is given in cell G22 ($203,525). This is the minimum value of cost of transportation for this problem.

The Process to Use Solver

Figure A.6 gives the setup of the Excel spreadsheet at the start of the process. The specific steps are listed below.

1. All decision variables are set to zero or left as blanks (shown as blanks in Figure A.6) in cells C4–F6.

		A	B	C	D	E	F	G	H	I
1			Rukna Auto Parts Company							
2				Distributors						
3				MKG	ASN	GMZ	AKLA	Total Shipped	Constraint Type	Capacity
4	Plants	Miami		11,800	0	0	8,200	20,000	=	20,000
5	Plants	Tempe		0	13,500	0	26,500	40,000	=	40,000
6	Plants	Columbus		13,200	0	16,800	0	30,000	=	30,000
7		Total Received		25,000	13,500	16,800	34,700	90,000		90,000
8		Constraint Type		=	=	=	=			
9		Demand		25,000	13,500	16,800	34,700	90,000		
10										
11				Unit Cost of Transportation						
12				MKG	ASN	GMZ	AKLA			
13	Plants	Miami		$1.00	$3.00	$3.50	$1.50			
14	Plants	Tempe		$5.00	$1.75	$2.25	$4.00			
15	Plants	Columbus		$2.50	$2.50	$1.00	$3.00			
16										
17				Total Transportation Cost						
18				MKG	ASN	GMZ	AKLA	Total Cost		
19	Plants	Miami		$11,800	$0	$0	$12,300	$24,100		
20	Plants	Tempe		$0	$23,625	$0	$106,000	$129,625		
21	Plants	Columbus		$33,000	$0	$16,800	$0	$49,800		
22		Total Cost		$44,800	$23,625	$16,800	$118,300	$203,525		

Figure A.5 Solution to Rukna Auto Parts Company.

2. Total units shipped from the three plants are given in cells G4, G5, and G6. These are calculated cells, where G4 = C4 + D4 + E4 + F4, G5 = C5 + D5 + E5 + F5, and G6 = C6 + D6 + E6 + F6. These three equations are entered for each of the three spreadsheet cells.
3. Capacity of each plant is entered in cells I4–I6. The constraint type is given in cells H4–H6. The constraints will be represented as G4 = I4, G5 = I5, and G6 = I6. These constraints are entered in the solver matrix.
4. Total units received at each distributor are given in cells C7, D7, E7, and F7. These are calculated cells, where C7 = C4 + C5 + C6, D7 = D4 + D5 + D6, E7 = E4 + E5 + E6, and F7 = F4 + F5 + F6. These four equations are entered for each of the four spreadsheet cells.
5. The constraints types are given in cells C8, D8, E8, and F8. The demand constraints are represented as C7 = C9, D7 = D9, E7 = E9, and F7 = F9. The demand constraints are entered in the solver matrix.
6. All costs are entered in cells C13–F15.
7. The total transportation cost for each combination of the plant and distributor are shown in cells C19–F21. These are calculated cells and are obtained by multiplying the number of units shipped by the shipping cost per unit for each combination of plant and distributor. To begin with, the values in all of these cells are zero because the units shipped are zero.

	A	B	C	D	E	F	G	H	I
1				**Rukna Auto Parts Company**					
2				Distributors					
3			MKG	ASN	GMZ	AKLA	Total Shipped	Constraint Type	Capacity
4	Plants	Miami					0	=	20,000
5		Tempe					0	=	40,000
6		Columbus					0	=	30,000
7		Total Received	0	0	0	0	0		90,000
8		Constraint Type	=	=	=	=			
9		Demand	25,000	13,500	16,800	34,700	90,000		
10									
11			**Unit Cost of Transportation**						
12			MKG	ASN	GMZ	AKLA			
13	Plants	Miami	$1.00	$3.00	$3.50	$1.50			
14		Tempe	$5.00	$1.75	$2.25	$4.00			
15		Columbus	$2.50	$2.50	$1.00	$3.00			
16									
17				**Total Transportation Cost**					
18			MKG	ASN	GMZ	AKLA	Total Cost		
19	Plants	Miami	$0	$0	$0	$0	$0		
20		Tempe	$0	$0	$0	$0	$0		
21		Columbus	$0	$0	$0	$0	$0		
22		Total Cost	$0	$0	$0	$0	$0		

Figure A.6 Rukna Auto Parts—solver setup.

8. The value of the objective function, total of cells C19–F21, is given in cell G22. The equation for total cost: =SUM($C$19:$F$21) must be in the spreadsheet.

After the spreadsheet is set up as described above, we are ready to use "solver."

Under the "data" tab in the Excel file, you will find the icon for "solver." You might have to install the solver add-in if you do not find the "solver" icon on your computer. (Information about installing is available by writing "install solver" in the help window of Excel. The video explains that File > Options > Add-Ins > Manage (Excel Add-ins) > Go will create a new Feature called Analysis under the Data Tab.)

Click on "solver." You will see the dialog box shown in Figure A.7. The dialog box will be empty when you first click "solver." You have to provide the required inputs. Figure A.7 shows the dialog box after all inputs have been provided.

The cell that gives the value of the objective function (G22) is entered in the box for "Set Objective" by clicking on that cell in the spreadsheet. The keyboard can also be used.

The objective is to minimize the total cost. Therefore, the button "Min" is chosen.

In the box "By changing Variable Cells," the addresses of the cells of the decision variables (C4–F6) are entered. (Click on C4; then drag and drop the adjacent cells into the solver matrix. The keyboard can also be used.)

In the box under "Subject to the Constraints," various constraints are added. (The symbols can be entered from the keyboard or using drag and drop procedures.)

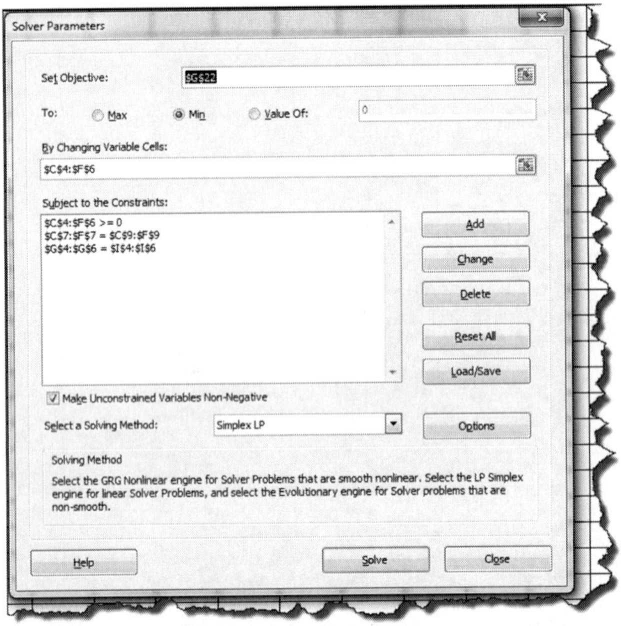

Figure A.7 Solver dialog box for Rukna Auto Parts Company.

In Figure A.7, the three constraints are as follows:

Constraint 1: $C$4:$F$6 ≥ 0 is the non-negativity restriction.

Constraint 2: $C$7:$F$7 = $C$9:$F$9 is the demand constraint.

Constraint 3: $G$4:$G$6 = $I$4:$I$6 is the capacity constraint.

The solving method chosen is Simplex LP. Leave the box checked for non-negative variables.

Click on "Solve" and you will get the solution represented in Figure A.5.

Note: The "$" sign is automatically added by solver when you click a particular cell that has to be added to the dialog box.

Reference

Rautenstrauch, W. and R. Villers, *The Economics of Industrial Management*. New York: Funk & Wagnalls Co., 1949.

Appendix B: The z-Table

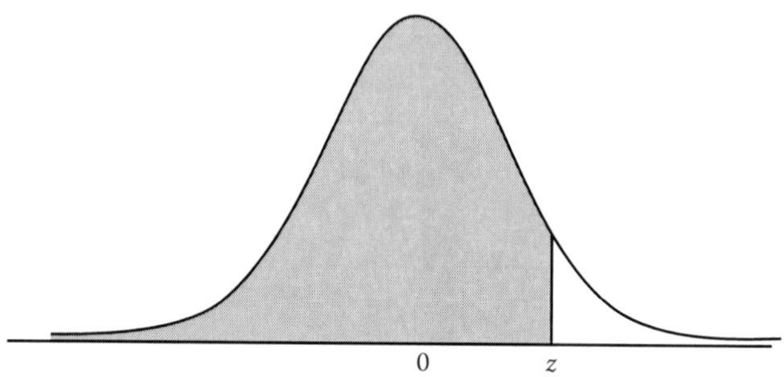

Normal distribution

The entries in the z-table give the area to the left of the z-value under the standard normal curve.

Reading the z-value: A z-value is read as a combination of a number in the first column and a number in the top row. For example, the number 1.2 in the first column combined with the number 0.06 in the top row gives value of z equal to 1.26. The corresponding table entry is 0.8962. The area of the tail of the distribution to the right of z would then be $1.0000 - 0.8962 = 0.1038$.

z	0	0.01	0.02	0.03	0.04	0.05	**0.06**	0.07	0.08	0.09
0	0.5000	0.5040	0.5080	0.5120	0.5160	0.5190	0.5239	0.5279	0.5319	0.5359
0.1	0.5398	0.5438	0.5478	0.5517	0.5557	0.5596	0.5636	0.5675	0.5714	0.5753
0.2	0.5793	0.5832	0.5871	0.5910	0.5948	0.5987	0.6026	0.6064	0.6103	0.6141
0.3	0.6179	0.6217	0.6255	0.6293	0.6331	0.6368	0.6406	0.6443	0.6480	0.6517
0.4	0.6554	0.6591	0.6628	0.6664	0.6700	0.6736	0.6772	0.6808	0.6844	0.6879
0.5	0.6915	0.6950	0.6985	0.7019	0.7054	0.7088	0.7123	0.7157	0.7190	0.7224
0.6	0.7257	0.7291	0.7324	0.7357	0.7389	0.7422	0.7454	0.7486	0.7157	0.7549
0.7	0.7580	0.7611	0.7642	0.7673	0.7704	0.7734	0.7764	0.7794	0.7823	0.7852
0.8	0.7881	0.7910	0.7939	0.7969	0.7995	0.8023	0.8051	0.8078	0.8106	0.8133
0.9	0.8159	0.8186	0.8212	0.8238	0.8264	0.8289	0.8315	0.8340	0.8365	0.8389
1	0.8413	0.8438	0.8461	0.8485	0.8508	0.8513	0.8554	0.8577	0.8529	0.8621
1.1	0.8643	0.8665	0.8686	0.8708	0.8729	0.8749	0.8770	0.8790	0.8810	0.8830
1.2	0.8849	0.8869	0.8888	0.8907	0.8925	0.8944	**0.8962**	0.8980	0.8997	0.9015
1.3	0.9032	0.9049	0.9066	0.9082	0.9099	0.9115	0.9131	0.9147	0.9162	0.9177
1.4	0.9192	0.9207	0.9222	0.9236	0.9215	0.9265	0.9279	0.9292	0.9306	0.9319
1.5	0.9332	0.9345	0.9357	0.9370	0.9382	0.9394	0.9406	0.9418	0.9492	0.9441
1.6	0.9452	0.9463	0.9474	0.9484	0.9495	0.9505	0.9515	0.9525	0.9535	0.9545
1.7	0.9554	0.9564	0.9573	0.9582	0.9591	0.9599	0.9608	0.9616	0.9625	0.9633

z	0	0.01	0.02	0.03	0.04	0.05	0.06	0.07	0.08	0.09
1.8	0.9641	0.9649	0.9656	0.9664	0.9671	0.9678	0.9686	0.9693	0.9699	0.9706
1.9	0.9713	0.9719	0.9726	0.9732	0.9738	0.9744	0.9750	0.9756	0.9761	0.9767
2	0.9772	0.9778	0.9783	0.9788	0.9793	0.9798	0.9803	0.9808	0.9812	0.9817
2.1	0.9821	0.9826	0.9830	0.9834	0.9838	0.9842	0.9846	0.9850	0.9854	0.9857
2.2	0.9861	0.9864	0.9868	0.9871	0.9875	0.9878	0.9881	0.9884	0.9887	0.9890
2.3	0.9893	0.9896	0.9898	0.9901	0.9904	0.9906	0.9909	0.9911	0.9913	0.9916
2.4	0.9918	0.9920	0.9922	0.9925	0.9927	0.9929	0.9931	0.9932	0.9934	0.9936
2.5	0.9938	0.9940	0.9941	0.9943	0.9945	0.9946	0.9948	0.9949	0.9951	0.9952
2.6	0.9953	0.9955	0.9956	0.9957	0.9959	0.9960	0.9961	0.9962	0.9963	0.9964
2.7	0.9965	0.9966	0.9967	0.9968	0.9969	0.9970	0.9971	0.9972	0.9973	0.9974
2.8	0.9974	0.9975	0.9976	0.9977	0.9977	0.9978	0.9979	0.9979	0.9980	0.9981
2.9	0.9981	0.9982	0.9982	0.9983	0.9984	0.9984	0.9985	0.9985	0.9986	0.9986
3	0.9987	0.9987	0.9987	0.9988	0.9988	0.9989	0.9989	0.9989	0.9990	0.9990
3.1	0.9990	0.9991	0.9991	0.9991	0.9992	0.9992	0.9992	0.9992	0.9993	0.9993
3.2	0.9993	0.9993	0.9994	0.9994	0.9994	0.9994	0.9994	0.9995	0.9995	0.9995
3.3	0.9995	0.9995	0.9995	0.9996	0.9996	0.9996	0.9996	0.9996	0.9996	0.9997
3.4	0.9997	0.9997	0.9997	0.9997	0.9997	0.9997	0.9997	0.9997	0.9997	0.9998

Index

F

G

H

I

J

K

L

O

P

Q

T

U

V

W

X

Z